The Ocean Carbon Cycle and Climate

NATO Science Series

A Series presenting the results of scientific meetings supported under the NATO Science Programme.

The Series is published by IOS Press, Amsterdam, and Kluwer Academic Publishers in conjunction with the NATO Scientific Affairs Division

Sub-Series

I. Life and Behavioural Sciences	IOS Press
II. Mathematics, Physics and Chemistry	Kluwer Academic Publishers
III. Computer and Systems Science	IOS Press
IV. Earth and Environmental Sciences	Kluwer Academic Publishers
V. Science and Technology Policy	IOS Press

The NATO Science Series continues the series of books published formerly as the NATO ASI Series.

The NATO Science Programme offers support for collaboration in civil science between scientists of countries of the Euro-Atlantic Partnership Council. The types of scientific meeting generally supported are "Advanced Study Institutes" and "Advanced Research Workshops", although other types of meeting are supported from time to time. The NATO Science Series collects together the results of these meetings. The meetings are co-organized bij scientists from NATO countries and scientists from NATO's Partner countries – countries of the CIS and Central and Eastern Europe.

Advanced Study Institutes are high-level tutorial courses offering in-depth study of latest advances in a field.
Advanced Research Workshops are expert meetings aimed at critical assessment of a field, and identification of directions for future action.

As a consequence of the restructuring of the NATO Science Programme in 1999, the NATO Science Series has been re-organised and there are currently five sub-series as noted above. Please consult the following web sites for information on previous volumes published in the Series, as well as details of earlier sub-series.

http://www.nato.int/science
http://www.wkap.nl
http://www.iospress.nl
http://www.wtv-books.de/nato-pco.htm

Series IV: Earth and Environmental Sciences – Vol. 40

The Ocean Carbon Cycle and Climate

edited by

Mick Follows
Department of Earth Atmosphere and Planetary Sciences,
Massachusetts Institute of Technology, Cambridge, MA, U.S.A.

and

Temel Oguz
Institute of Marine Sciences,
Middle East Technical University, Erdemli, Turkey

Kluwer Academic Publishers

Dordrecht / Boston / London

Published in cooperation with NATO Scientific Affairs Division

Proceedings of the NATO Advanced Study on
The Ocean Carbon Cycle and Climate
Ankara, Turkey
5–16 August 2002

A C.I.P. Catalogue record for this book is available from the Library of Congress.

ISBN 1-4020-2086-4 (PB)
ISBN 1-4020-2085-6 (HB)
ISBN 1-4020-2087-2 (e-book)

Published by Kluwer Academic Publishers,
P.O. Box 17, 3300 AA Dordrecht, The Netherlands.

Sold and distributed in North, Central and South America
by Kluwer Academic Publishers,
101 Philip Drive, Norwell, MA 02061, U.S.A.

In all other countries, sold and distributed
by Kluwer Academic Publishers,
P.O. Box 322, 3300 AH Dordrecht, The Netherlands.

Printed on acid-free paper

Printed in the Netherlands.

Contents

Preface

Our desire to understand the global carbon cycle and its link to the climate system represents a huge challenge. These overarching questions have driven a great deal of scientific endeavour in recent years: What are the basic oceanic mechanisms which control the oceanic carbon reservoirs and the partitioning of carbon between ocean and atmosphere? How do these mechanisms depend on the state of the climate system and how does the carbon cycle feed back on climate? What is the current rate at which fossil fuel carbon dioxide is absorbed by the oceans and how might this change in the future?

To begin to answer these questions we must first understand the distribution of carbon in the ocean, its partitioning between different ocean reservoirs (the "solubility" and "biological" pumps of carbon), the mechanisms controlling these reservoirs, and the relationship of the significant physical and biological processes to the physical environment. The recent surveys from the JGOFS and WOCE (Joint Global Ocean Flux Study and World Ocean Circulation Experiment) programs have given us a first truly global survey of the physical and biogeochemical properties of the ocean. These new, high quality data provide the opportunity to better quantify the present oceans reservoirs of carbon and the changes due to fossil fuel burning. In addition, diverse process studies and time-series observations have clearly revealed the complexity of interactions between nutrient cycles, ecosystems, the carbon-cycle and the physical environment. While much new understanding has been achieved, and new views of ocean biogeochemistry have been formed, so too have many new questions emerged and the overarching questions noted above have not been unequivocally answered.

This book is the outcome of the NATO Advanced Study Institute entitled "The Ocean Carbon Cycle and Climate", held in Middle East Technical University, Ankara, Turkey, 6-15 August 2002. This initiative provided an opportunity to discuss with students and young researchers backgrounds in a range of disciplines, reflecting the nature of the topic itself as well as to present them an up to date overview of current understanding of the ocean carbon cycle and its relation to the climate. This book emerges from a series of lectures and workshops at the Advanced Study Insitute, and consists of 11 chapters. Each chapter stands alone, covering some facet of the broader topic of the ocean carbon cycle and climate connection, and linked by the overarching questions and many significant areas are touched upon. They have been written in the style of scientific review papers and accepted after an international peer-reviewing process. Such a book as this cannot be fully comprehensive in scope. Indeed this volume is, by necessity, not as wide in scope as the lectures of the Advanced Study Insitute from which it derives.

The solubility pump of carbon in the ocean is the reservoir which reflects the fact that the deep waters are formed at high latitudes, where cooler, carbon-rich surface waters sink to form the deep waters of the ocean. Here, in Chapter 1, James Murray provides a concise but detailed account of the basic carbonate chemistry which controls the equilibrium dissolved inorganic carbon concentration in seawater and the timescale of equilibration with the atmosphere.

The biological pumps are those mechanisms by which the deep ocean dissolved inorganic carbon concentration is enriched due to biological production in the surface ocean, sinking and transport of organic matter, and subsequent respiration. In Chapter 2 David Kirchman reviews the current understanding of the role and actions of heterotrophic organisms and the cycling of dissolved organic carbon; important themes where there is still much to be understood and quantified. Also on a biological theme, Tom Anderson and Ian Totterdell provide in Chapter 3 an overview of modelling efforts of biological mechanisms and processes. Themes include the interaction of ecosystems and biogeochemical cycles, and the significance of elemental ratios in organic matter. The ocean's biological carbon pumps are intimately linked with, and modulated by, the global cycles of other elements; nutrients such as phosphorus, nitrogen and iron. It has long been known that the ratio of dissolved inorganic nitrogen and phosphorus in the oceans is close to that in biomass yet the residence time of nitrogen in the ocean is much shorter than that of phosphorus. Nicolas Gruber (Chapter 4) discusses the dynamic balance between nitrogen fixation and denitrification which controls the relative oceanic distributions of nitrate and phosphate, and possible feedbacks between climate, nitrogen, iron and carbon cycles.

Our knowledge of the distributions and interactions of elements in the ocean has been obtained through traditional ship-based observations, further supported recently by the development of novel autonomous and remote methods. Tommy Dickey and Gennady Korotaev discuss the techniques, current applications and future prospects of moorings and autonomous vehicles (Chapter 5) and remote sensing (Chapter 6), respectively, of observing the physical and biogeochemical state of the oceans. These emerging techniques have the potential to provide much better spatial and temporal data coverage in the future.

Mathematical models of all kinds, from the highly idealized three box ocean biogeochemistry models, to fully coupled circulation and biogeochemistry models, are being used to understand and simulate the key mechanisms of the carbon cycle. They cover a broad range from diagnostic models used to interpret observed data to conceptual models to provide a framework for examining hypotheses and performing thought experiments, and three- dimensional prognostic models which attempt to accurately simulate (and even predict) the ocean. Mick Follows and Richard Williams (Chapter 7) use highly simplified models to consider how a combination of biological and solubility processes controls the distribution and fluxes of carbon in the North Atlantic basin. Irina Marinov and Jorge Sarmiento (Chapter 8) give an overview of a hierarchy of global models. They discuss the insights these models provide into the atmosphere- ocean partitioning of carbon and the oceanic uptake of fossil fuel CO_2.

Studies of the interannual and decadal variability of the ocean carbon cycle using historical data records, and glacial- interglacial changes using ice core and sedimentary data provide insight into, and test our understanding of the relationship between carbon cycle and climate. In Chapter 9, Richard Matear discusses the observation of decadal changes in oceanic oxygen concentrations in the Southern Ocean, perhaps indicating changes in the biological pumps. He discusses how numerical models have helped to interpret these observations. On longer timescales, while there are many published hypotheses to explain the recorded glacial- interglacial changes in atmospheric carbon dioxide and the oceans role, none has yet recieved unanimous acceptance. Hezi Gildor reviews and discusses some of the predominant theories in Chapter 10. Our inability to reproduce the observed changes in high end ocean models calls into question either the underlying hypotheses or the veracity of current models. This still provides a grand challenge for the climate and biogeochemistry communities. Finally, in Chapter 11, Richard Matear provides a concise review of methods for the blending of ocean observations and models. Data assimilation techniques can be used both to constrain the model state to be consistent with observed data, and to optimize unknown model parameter values. This more intimate connection of data and models is sure to be an important methodology for future studies of ocean biogeochemistry.

While not a comprehensive survey or primer on the ocean carbon cycle and its relationship to climate, we hope that the chapters in this volume will provide useful starting points for those beginning research careers in this exciting interdisciplinary field, or those moving towards it. The papers collected in this volume constitute the most up-to-date synthesis of the state-of-the-art research in ocean carbon cycle and climate related science.

Finally, we thank those who enabled and supported this Advanced Study Institute. The primary support was given by NATO under the auspices of the Scientific Affairs Division, and we thank Dr. Alain H. Jubier, Director of the Environmental and Earth Sciences & Technology Programme. Further support was given by SCOR, IOC, NOAA, NSF and US-JGOFS. Special thanks are due to all the contributors to this volume who provided an exciting forum for discussion and the excellent series of lectures that make this book a stimulating record of events. We are also very grateful to the Middle East Technical University for warmly hosting the Institute and providing wonderful facilities and a friendly environment to foster interactions and future collaborations between participants from NATO and Partner Countries.

MICK FOLLOWS AND TEMEL OGUZ, EDITORS.

List of Contributors

Thomas R. Anderson
Southampton Oceanography Centre, George Deacon Division
Waterfront Campus, European Way
Southampton SO14 3ZH, United Kingdom
tel: +44 23 80596337, fax: +44 23 80596247

Tommy D. Dickey
Ocean Physics Laboratory, University of California,
Santa Barbara, 6487 Calle Real, Suite A, Goleta CA, 93117, USA

Mick Follows
Department of Earth, Atmosphere and Planetary Sciences, 54-1514
Massachusetts Institute of Technology,
77 Massachusetts Avenue, Cambridge, MA 02130, USA
tel: 617 253 5939; fax: 617 253 4464

Hezi Gildor
Dept. of Environmental Sciences and Energy Research
Weizmann Institute of Science, Rehovot 76100, ISRAEL
tel: 972-8-9344238, fax: 972-8-9344124

Nicolas Gruber
Institute of Geophysics and Planetary Physics and Department of
Atmospheric Sciences, 5853 Slichter Hall, University of California,
Los Angeles, CA 90095-4996, USA
tel: (310) 825 4772, fax : (310) 206 3051

David Kirchman
Acting Associate Dean and
Maxwell P. and Mildred H. Harrington Professor of Marine Studies
University of Delaware, Graduate College of Marine Studies
700 Pilottown Rd., Lewes, Delaware 19958
tel: (302) 645-4375, fax: (302) 645-4028

Gennady K. Korotaev
Marine Hydrophysical Institute, Ukranian Academy of Sciences
Kapitanskaya St., 2, Sevastopol, 99000, Ukraine

Richard J. Matear
CSIRO Marine Research and the Antarctic Cooperative Research Center,
GPO Box 1538, Hobart Tasmania, Australia

James W. Murray
University of Washington School of Oceanography
Room 413 Ocean Sciences Building Box 355351
Seattle WA 98195-5351, USA
tel: (206) 543-4730, fax (206) 685-3351

Temel Oguz
Institute of Marine Sciences, Middle East Technical University
P.O.Box 28, 33731, Erdemli, Icel,Turkey
tel: +90 324 521 2406, fax: +90 324 521 2327

Jorge L. Sarmiento
Atmospheric and Oceanic Sciences (Biogeochemistry)
Co-Director of the Carbon Modeling Consortium
AOS Program, Princeton University, Sayre Hall, Forrestal Campus
PO Box CN710, Princeton, NJ 08544-0710, USA

OCEAN CARBONATE CHEMISTRY: THE AQUATIC CHEMISTRY FUNDAMENTALS

JAMES W. MURRAY
School of Oceanography, University of Washington Seattle WA 98195-5351
jmurray@u.washington.edu

INTRODUCTION

Some of the main questions we consider when studying the global carbon cycle are the exchange of CO_2 between the ocean and atmosphere and internal redistribution of carbon species within the ocean due to uptake and remineralization of organic matter and calcium carbonate. A schematic representation of these processes is shown in Figure 1. When CO_2 evades from or invades to the ocean it changes the equilibrium that exists between the species of dissolved inorganic carbon (H_2CO_3, HCO_3^- and CO_3^{2-}) already there. Some of the CO_2 is taken75 up by phytoplankton to make algal protoplasm. Some CO_3^{2-} reacts with Ca^{2+} and is taken up in the form of calcium carbonate shells. Both organic matter and $CaCO_3$ are contained in the particles that sink from the surface ocean into the deep sea. The biologically driven fluxes are called the "Biological Pump". These particles are remineralized through respiration of the organic carbon and solubilization of the $CaCO_3$. The reactions control the distribution of carbonate species and pH in the deep sea. The pH of seawater is a master variable that reflects the net effect of all acid and base producing aquatic geochemical processes. Some of these particles reach the sediments where their input drives an extensive set of reactions termed sedimentary diagenesis. All of these processes involve various species of the ocean dissolved inorganic carbon (DIC) system, thus the fundamental aquatic

M. Follows and T. Oguz (eds.), The Ocean Carbon Cycle and Climate, 1–29.

chemistry aspects of these equilibrium reactions of the carbonate system need to be understood.

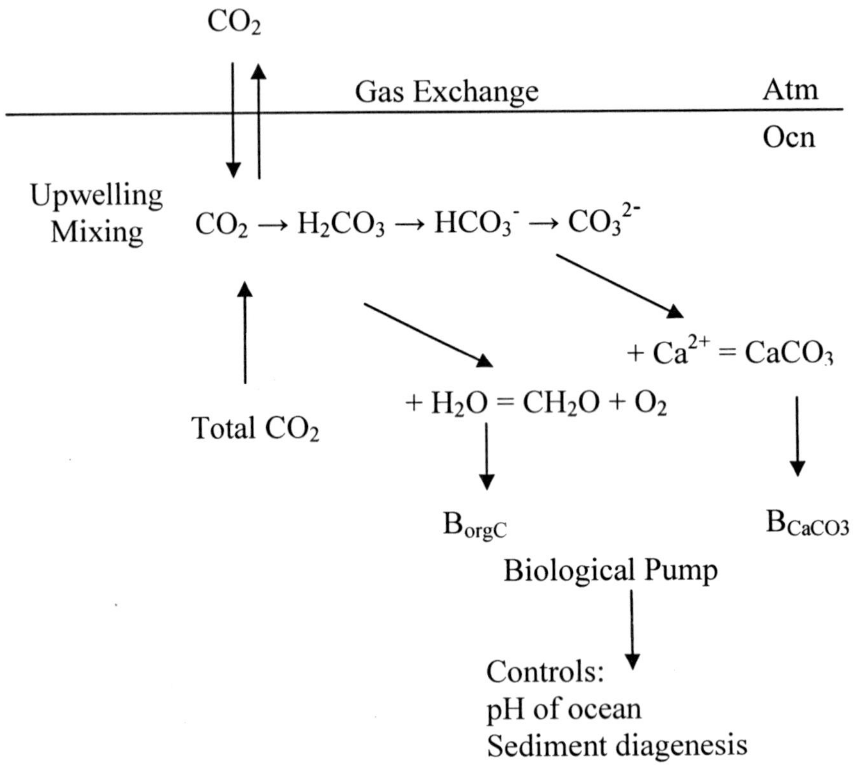

Figure 1. Schematic diagram of processes and reactions involving species of dissolved inorganic carbon (DIC) or Total CO_2 in surface seawater.

1. CARBONIC ACID

Carbonic Acid is the classic example of a diprotic acid that also has a gaseous form. Thus the system can be expressed as open or closed with respect to equilibrium with a gas phase (e.g. the atmosphere). The mass balance equations are expressed differently for open and closed systems.

There are six chemical species that we need to solve for to completely define the ocean carbonate system. These are:

$CO_2(g)$ Carbon Dioxide Gas
$H_2CO_3^*$ Carbonic Acid ($H_2CO_3^*$ equals the sum of $CO_2(aq)$+ H_2CO_3)
HCO_3^- Bicarbonate ion
CO_3^{2-} Carbonate ion
H^+ Hydrogen ion (or proton)
OH^- Hydroxide ion

There is a true carbonic acid (H_2CO_3) but it is not simple to distinguish this from CO_2 (aq) thus in practice they are summed together and called $H_2CO_3^*$. Most of $H_2CO_3^*$ is actually CO_2 (aq) rather than H_2CO_3 and the ratio of CO_2 (aq) / H_2CO_3 is approximately 300 (Stumm and Morgan, 1996).

To solve for these six unknowns we need six equations.

1.1 EQUATIONS

The first four equations are the chemical reactions defined by the equilibrium constants; K_H, K_1, K_2 and K_w. These constants are a function of temperature and pressure.

These equilibrium constants are written in terms of activities of the corresponding reactants and products. When written this way the equilibrium constants can be calculated from the standard free energy of formation (ΔG_f^o) and the standard free energy of reaction (ΔG_r^o). We refer to such equilibrium constants as being defined on the infinite dilution activity scale where concentrations are expressed as activities which are essentially apparent or effective concentrations. Here activities are written in parentheses ( ) while total concentrations are in brackets [ ].

1. $CO_2(g) + H_2O = H_2CO_3^*$ (Henry's Law)

$$K_H = (H_2CO_3^*) / f_{CO2}$$

Note that concentrations in a gas phase are usually given as partial pressure which is the product of the mole fraction in the dry gas phase (X_{CO2}) times the total dry pressure (P_T), thus $P_{CO2} = X_{CO2}\ P_T$. But in equilibrium relations gas concentrations are formally expressed as fugacity or f, which is the term for the activity of a gas in the gas phase; e.g. atmospheric f_{CO2} = $10^{3.5}$. The fugacity differs from the partial pressure by the non-ideal nature of the gas phase. Because activity corrections are so small in the atmosphere the differences between P_{CO2} and f_{CO2}.are very small ($\leq$ 0.5%) and can usually be ignored.

2. The first acidity constant of carbonic acid:

$$H_2CO_3^* = H^+ + HCO_3^-$$

$$K_1 = (HCO_3^-)(H^+) / H_2CO_3^*)$$

3. The second acidity constant of carbonic acid:

$$HCO_3^- = H^+ + CO_3^{2-}$$

$$K_2 = (H^+)(CO_3^{2-}) / (HCO_3^-)$$

4. The acidity constant for water:

$$H_2O = H^+ + OH^-$$

$$Kw = (H^+)(OH^-)$$

1.2 ACTIVITY SCALES OF EQUILIBRIUM CONSTANTS

Whenever possible, oceanographers use equilibrium constants defined in terms of concentrations rather than activities. Equilibrium constants defined this way are called apparent or operational equilibrium constants. Such constants are formally defined on the constant ionic medium activity scale, in this case the medium is seawater. We use the symbol K' to distinguish these constants from K. Such apparent equilibrium constants are written in terms of concentrations rather than activities. The exception is H^+, which is usually written as the activity of (H^+), but can be expressed in other forms as well, such as the total H^+ concentration or the seawater H^+ concentration (see section on pH measurements) (Dickson, 1984; Millero, 1995).

There are pros and cons for each approach. The pro for using K is that we can calculate K from ΔG_r^o (ΔG_r^o are easily calculated from ΔG_f^o and these values are readily tabulated and available., e.g. Stumm and Morgan., 1996) and one K can be used for all problems, in all solutions (e.g., one K fits all). The disadvantage of this approach is that all concentrations must be expressed as activities, which are obtained by multiplying the total concentration times the total activity coefficient (γ_T). This is written as:

activity or a_i or ( ) = total concentration or c_i or [ ] times the total activity coefficient or γ_T.

The activity is essentially equal to the concentration at very low concentrations (infinite dilution) and is less than the concentration except at very high salt content. We may think of the activity as the effective concentration. When ions interact with each other in more concentrated salt solutions their effective concentration is reduced due to the ion-ion interactions. The total activity coefficient is composed of two terms which account for 1) general electrostatic interactions between ions and 2) specific interactions (e.g., ion pairs or complexes) that occur between ions. We correct for electrostatic interactions using the free ion activity coefficients (γ_i). The specific interactions are expressed as the %free (f_i).

$$\gamma_T = \gamma_i \times \%\text{free}$$

There are both theoretical (e.g. the Debye-Huckel equation) and empirical (e.g. the Davies and Pitzer equations) approaches for making these corrections. In all cases the corrections vary as a function of ionic strength and include assumptions about the nature of the interactions and thus include considerable uncertainty.

Oceanographers prefer to use values of K' when these are available. Fortunately these constants have been determined experimentally for most of the important acid - base and solubility reactions in seawater. The approach for determining these constants is to measure the concentrations of reactants and products at equilibrium in seawater of known salinity (S), temperature (T) and pressure (P). If this is done for a matrix of values of S, T and P, regressions can be derived that can be used to calculate the values of K' for specific sets of oceanographic conditions of interest. The advantage is that once determined this way, the values of K' are usually more precise than the corresponding value of K. In addition, when using K' we can calculate equilibrium directly using concentrations and we don't need to estimate values for the activity corrections (γ_i and f_i). The disadvantage of this approach is that some experimentalist needs to have measured the values of K'. Fortunately these constants have been determined very precisely for the marine carbonate system.

For the ocean carbonate system the apparent constants may be written as:

$$K'_H = [H_2CO_3^*] / f_{CO2}$$

$$K'_1 = [HCO_3^-](H^+) / [H_2CO_3^*]$$

$$K'_2 = (H^+)[CO_3^{2-}] / [HCO_3^-]$$

Note that H^+ is written here in terms of activity but other pH scales may be used (see Millero, 1995).

1.2.1 Values of K and K'. The equations of Millero (1995) are currently the best available for calculating values of K' for any salinity (S) and temperature (T), pressure (P) conditions. The complete equations for calculating K' at 1 atm are given in Appendix 1. Using these equations we have calculated representative values of the various K' for S = 35, T = 25°C and P = 1 atm (Table 1). We compare the respective values of K' with values of K calculated from free energies of reaction (ΔG°_r). These constants (K) are sometimes referred to as thermodynamic constants, but this is not meant to imply that these constants are any more or less valid in a thermodynamic sense than the values of K'. They are just defined on a different activity scale (the infinite dilution versus seawater activity scale).

Table 1. Carbonate system equilibrium constants. The seawater constants are for S = 35, T = 25°C and P = 1atm.

Constant	Thermodynamic Constant (K)	Apparent Seawater Constant (K')
K_H	$10^{-1.47}$	$10^{-1.53}$
K_1	$10^{-6.35}$	$10^{-6.00}$
K_2	$10^{-10.33}$	$10^{-9.10}$
K_w	$10^{-14.0}$	$10^{-13.9}$

1.2.2. The Difference Between K and K'. The magnitude of the difference between K and K' can be explained by the following simple example. We can write the thermodynamic constant in terms of activities or concentrations and activity coefficients. Here it is illustrated for a generic monoprotic acid (HA) and a reaction written as:

$$HA = H^+ + A^-$$

$$K = (H^+)(A^-) / (HA) = (H^+)[A^-]\gamma_{T,A-} / [HA]\,\gamma_{T,HA}$$

Rearranging we get:

$$K = \frac{(H^+)[A^-]\gamma_{T,A^+}}{[HA]\gamma_{T,HA}} = K'\frac{\gamma_{T,A^+}}{\gamma_{T,HA}}$$

Thus, the difference between the magnitudes of K and K' is equal to the ratio of the total activity coefficients of the base species to the acid species.

We could turn this around and say that if you know the values of both K and K' , you can learn something about the activity corrections. We can illustrate this using the first acidity constant for carbonic acid.

For:

$$H_2CO_3^* = HCO_3^- + H^+$$

$$K_1 = (HCO_3^-)(H^+) / (H_2CO_3^*)$$

or

$$K_1 = \frac{[HCO_3-]\ \gamma_{T,HCO3}\ (H^+)}{[H_2CO_3^*]\ \gamma_{T,H2CO3}}$$

or

$$K_1 = \frac{[HCO_3](H^{\pm})\ \gamma_{T,HCO3}}{[H_2CO_3^*]\ \ \gamma_{T,H2CO3}} = 10^{-6.3}$$ from tables of $\Delta G_r°$ at 25°C and 1 atm

The value of K' has been determined for the same reaction. At S = 35, 25°C and 1 atm

$$K_1' = \frac{[HCO_3^-]\,(H^{\pm})}{[H2CO3]} = 10^{-6.0}$$

If you combine the two equations for K and K' you get:

$$1.K_1 = K_1 \ni \gamma_{T,HXO3} / \gamma_{T,\ H2XO3}$$

We can solve this for the ratio of the total activity coefficients or:

$$\gamma_{T,HCO3} / \gamma_{T,\ H2CO3} = K_1 / K_1' = 10^{-6.3} / 10^{-6.0} = 10^{-0.3} = \mathbf{5.0 \times 10^{-1}}$$

We can compare this ratio with that obtained from the Garrels and Thompson (1962) speciation model of surface seawater where:

$$\frac{\gamma_{T,HCO_3}}{\gamma_{T,H_2CO_3}} = \frac{f_{HCO_3}\gamma_{HCO_3}}{f_{H_2CO_3}\gamma_{H_2CO_3}} = \frac{(0.69)(0.68)}{(1.00)(1.13)} = \mathbf{4.1 \times 10^{-1}}$$

Not too bad. Not too good. These two estimates differ by about 20%.

Is that good enough? The answer depends on the application, but it indicates that the different approaches are not totally internally consistent. When you do the same calculation for K_2 and K_2', the differences are even larger. As the experimental values of K' are well known, the uncertainties probably mostly reside with the values of the activity coefficients for the carbonate species.

1.2.3. The Acids in Seawater. There are other acids in addition to carbonic acid. The following table summarizes the main acid producing reactions in seawater. The total concentrations of the elements are given for S = 35, and the best values of K' are given for S = 35, T = 25°C and P = 1 atm in both linear and log notation. Organic acids should be included but we don't know their concentrations or equilibrium constants well enough at present (Bradshaw and Brewer, 1988).

Table 2. Reactions, concentrations and acidity constants for acids in Seawater

Element	Reaction	mol kg^{-1}	-logC	pK'
H2O	H2O = H+ + OH			13.9
C	H2CO3 = HCO3- + H+	2.4 x 10-3	2.6	6.0

Element	Reaction	mol kg^{-1}	-logC	pK'
	HCO3- = CO32- + H+			9.1
B	B(OH)3 + H2O = B(OH)4- + H+	4.25 x 10-4	3.37	8.7
Mg	Mg2+ + H2O = MgOH+ + H+	5.32 x 10-2	1.27	12.5
Si	H4SiO4 = SiO(OH)3- + H+	1.5 x 10-4	3.82	9.4
P	H3PO4 = H2PO4- + H+	3.0 x 10-6	5.52	1.6
	H2PO4- = HPO42- + H+			6.0
	HPO42- = PO43- + H+			8.6
S(VI)	HSO4- = SO4- + H+	2.82 x 10-2	1.55	1.5
F	HF = F- + H+	5.2 x 10-5	4.28	2.5
Ca	Ca2+ + H2O = CaOH+ + H+	1.03 x 10-2	1.99	13.0
And in anoxic systems				
N	NH4+ = NH3 + H+	10 x 10-6	5.0	9.5
S(-II)	H2S = HS- + H+	10 x 10-6	5.0	7.0
	HS- = S2- + H+			13.4

2. CARBONATE SYSTEM CALCULATIONS

The concentrations of the individual carbonate species can be determined using algebraic, graphic or computer programming approaches. Here I will review each briefly. For more detail go to Sillen (1959), Butler (1964 a, b, 1991) or Stumm and Morgan (1996).

2.1 ALGEBRAIC APPROACH

The algebraic approach is simply to combine the equations by sequential substitution to derive relationships to solve for the concentration of each species as a function of pH (e.g. H^+). In the case of a simple monoprotic acid, like acetic acid, this would proceed as follows.

Let acetic acid (CH_3COOH) = HA.

The base form or acetate (CH_3COO^-) = A^-

We need to determine the concentrations of 4 species = HA, A^-, H^+ and OH^-. The 4 key equations are:

1. The reaction HA = H^+ + A^-
   Acid Hydrogen Anion (or base)

2. Equilibrium Constants $K_{HA} = (H^+)(A^-) / (HA)$ In terms of activities
   $Kw = (H^+)(OH^-)$

3. Mass balance on A $C_A = [HA] + [A^-]$ In terms of concentrations

4. Charge Balance $[H^+] = [A^-] + [OH^-]$ In terms of equivalents

By combining equations 2 and 3 given above we can write algebraic expressions to solve for the main species of acetic acid (HA) and acetate (A^-) assuming we know a value of pH.

$$[HA] = C_A\,(H^+ / K + H^+) = C_A\,(\alpha_0)$$

$$[A^-] = C_A\,(K / K + H^+) = C_A\,(\alpha_1)$$

Note that the alpha values (α_0 and α_1) are a function of only the pH and K values. The use of these alpha values is a very convenient shorthand way to set up such equilibrium derivations. The denominators are always the same for a given acid type, in this case, $K + H^+$. Some examples with applications of this will be shown later.

If pH is not known, but we do know the total concentration (C_A), we can solve for the pH using the third order polynomial:

$$[H^+]^3 + [H^+]^2 K_{HA} - [H^+](C_A K_{HA} + K_w) - KK_w = 0$$

The derivations for a diprotic acid like carbonic acid (H_2CO_3) (with two acidity constants, K_1 and K_2, and a total concentration of C_T) would be similar. The unknown species are $H_2CO_3^*$, HCO_3^-, CO_3^{2-}, H^+ and OH^-. The concentrations of the acid and the two base forms as a function of pH would be:

$$H_2CO_3^* = C_T\,(H^+)^2(1 + K_1 + K_1K_2)^{-1} = C_T\,\alpha_0$$

$$HCO_3^- = C_T\,(H^+)K_1\,(1 + K_1 + K_1K_2)^{-1} = C_T\,\alpha_1$$

$$CO_3^{2-} = C_T\,(K_1K_2)\,(1 + K_1 + K_1K_2)^{-1} = C_T\,\alpha_2$$

If pH is not known it can be solved for using the 4th order polynomial in the form:

$$(H^+)^4 + (H^+)^3K_1 + (H^+)^2\,(K_1K_2 - C_TK_1 - K_w) - (H^+)\,K_1(2C_TK_2 + K_w) - K_1K_2K_w = 0$$

2.2 GRAPHICAL APPROACH

The equations presented above can be displayed graphically to show the distributions over a large pH range for a given total concentration (see Sillen, 1959). Examples are shown for acetic acid as an example of a monoprotic acid (Fig. 2) and carbonic acid as a diprotic acid (Fig. 3). Such diagrams are drawn using -log[conc] versus pH (= -log (H_+) scales. These graphs have three general regions which are where pH = pK (system point), pH > pK and pH < pK.

The steps for their construction using a sharp pencil, a ruler and graph paper are as follows.

1. Label the x-axis with pH and the y-axis with –log concentration.

2. Draw a horizontal line corresponding to the total concentration. The sum of the individual species at any pH is never larger than this line.

3. The system points are identified on the total concentration line where pH = pK. There is one system point for the monoprotic acid and two for the diprotic case.

4. When pH = pK the concentration of the acid (HA) and its conjugate base (A^-) are equal. For example:

$$K_A = (H^+)(A^-)/(HA)$$

In log form:

$$\log K_A /(H^+) = \log (A^-)/(HA)$$

When $(H^+) = K_A$, the concentrations of (A^-) and (HA) are equal and both are equal to ½ the total concentration, thus:

$$(A^-) = (HA) = \tfrac{1}{2} C_A$$

As the log of 2 is 0.3, the crossover point of the lines on the graph is 0.3 log units below the line for the total concentration at the system point. These are indicated in the graphs shown.

5. The lines have slopes of 0, ±1 and ±2 depending on how many H^+ are involved in the reaction. For example in acid solutions HA is the main species and is equal to the total concentration C_A (except within about one pH unit of the system point). So in the region where pH < pK_A the slope of log (HA) is 0. For the pH region where pH > pK_A, the concentration of log (HA) has a slope of -1 as shown by the following derivation.

If:

$$K_A = (H^+)(A^-)/(HA)$$

Then:

$$\log (HA) = \log (H^+) + \log (A^-) - \log K_A$$

Or:

$$\log (HA) = -pH + \log (A^-) - \log K_A$$

The slope is the derivative of this equation versus pH and is equal to -1:

$$\partial \log (HA)/ \partial \, pH = -1$$

The slopes of the other lines can be obtained by similar derivations.

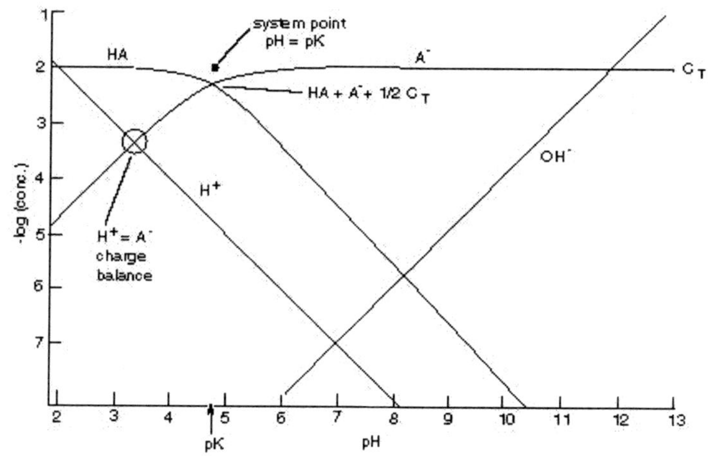

Figure 2. A distribution diagram of acetic acid (HA) with CA = 10^{-2} M and pKA = 4.7

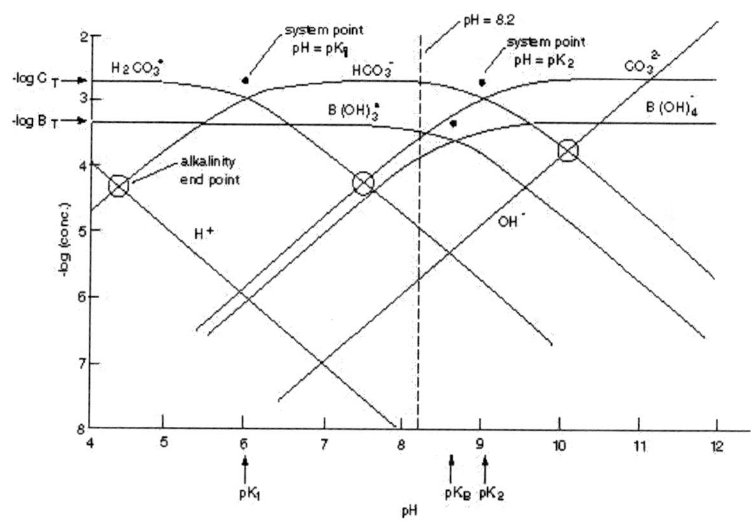

Figure 3. A distribution diagram of carbonic acid with CT = 2.0 x 10-3 (-log CT = 2.7). In this case the apparent seawater constants of $pK'_1 = 6.0$ and $pK'_2 = 9.1$ were used. The species for boric acid are also shown (pK'B = 8.7; -log CB = 3.3).

6. Such distribution diagrams can be constructed using either thermodynamic (Fig. 2 for acetic acid) or apparent equilibrium constants (Fig. 3 for carbonic acid in seawater)

Such graphs are useful for visualizing the concentrations over a large pH range. For the carbonic acid example (Fig. 3) you can see the following:

The main species at the pH of seawater (pH ≈ 8) is HCO_3^-.

When pH < pK_1, H_2CO_3 is the main species.

When pH > pK_2, CO_3^{2-} is the main species.

For this set of conditions ($C_T = 2.0 \times 10^{-3} = 10^{-2.7}$), the concentrations of HCO_3^- and CO_3^{2-} are equal at pH = 7.6. Thus, at the pH of surface seawater (see vertical line at pH = 8.2) the concentrations decrease in the order: $HCO_3^- > CO_3^{2-} > H_2CO_3$. The concentrations of CO_3^{2-} and H_2CO_3 are about 10% and 1% of HCO_3^-, respectively, at this pH. Thus, for some problems we can neglect the concentration of H_2CO_3 and approximate the Total CO_2 in seawater as: $C_T \approx [HCO_3^-] + [CO_3^{2-}]$

2.3 THE PROTON BALANCE

The proton balance has special significance in acid base systems: It determines the value of the pH. The proton balance is written as a balance between the species that have excess protons versus those that are deficient in protons relative to a defined starting point or reference level. Here are examples of the three reference levels for the carbonic acid system. If a species has lost 2 protons relative to the reference level, it's concentration is multiplied by 2.

Reference Level	excess protons	=	deficient protons
H_2CO_3 / H_2O	H^+	=	$HCO_3^- + 2\,CO_3^{2-} + OH^-$
HCO_3^- / H_2O	$H^+ + H_2CO_3$	=	$CO_3^{2-} + OH^-$
CO_3^{2-} / H_2O	$H^+ + 2H_2CO_3 + HCO_3^-$	=	OH^-

These proton conditions define three equivalence points on the graph in Fig. 3 (see circles). The first proton condition will be satisfied when $H^+ = HCO_3^-$, the second when $H_2CO_3 = CO_3^{2-}$ and the third when $HCO_3^- = OH^-$. You can see from the graph that for these conditions the other species in the proton balance have negligibly low concentrations and can be safely neglected. For the first reference level, where both reference species (H_2CO_3 and H_2O) are uncharged, the proton condition is equal to the charge balance.

You can approach each equivalence point from either the acid or base direction, depending on what the initial pH value is. Thus, these equivalence points are used to define 6 capacity factors for solutions. If you add strong acid (e.g. HCl) it would be represented in the proton balance as C_A on the right side to represent the amount of Cl^- added. The Cl^- would be a conservative tracer for the H^+ added. H^+ is already indicated on the left side.

Strong base (e.g. NaOH) is represented as C_B on the left side (e.g. for the Na^+ added).

For example, for a pure solution of H_2CO_3 (H_2CO_3/H_2O reference level) we would include the terms for strong acid and base as follows (but you would only add one at a time, of course):

$$C_B + H^+ = HCO_3^- + 2CO_3^{2-} + OH^- + C_A$$

The capacity factor $C_B - C_A$ is defined as the alkalinity and $C_A - C_B$ as the H^+-Acidity.

Then:

$$\text{Alkalinity} = C_B - C_A = HCO_3^- + 2CO_3^{2-} + OH^- - H^+$$

$$H^+\text{-Acidity} = C_A - C_B = H^+ - HCO_3^- + 2CO_3^{2-} + OH^-$$

In the case of seawater at pH = 8 you would be at a higher pH than the first equivalence point at about pH = 4.4 and you could titrate your solution with strong acid (C_A) to reach it. The amount of acid that you had to add would be called the alkalinity. The protons that you add would titrate all bases present with $pK > pK_1$ for H_2CO_3, so if you only want the carbonate part of the alkalinity you need to first subtract the concentration of the other bases present like $B(OH)_4^-$. To illustrate this, boric acid ($B(OH)_3$) is also graphed in Fig. 3. If you continued to add HCl past the equivalence point, the concentration of H^+-Acidity would steadily increase, as in this pH region H^+-acidity=H^+

If you had some acid rain with pH = 3.0, adding acid would only lower the pH. But if you added strong base (C_B) until you reached the H_2CO_3/H_2O equivalence point, that amount of base added would neutralize the strong acid present and would be called the H^+-acidity. These capacity factors can be averaged when solutions with different values are mixed.

The capacity factors defined relative to the HCO_3^-/H_2O and CO_3^{2-}/H_2O equivalence points (CO_2-acidity / p-alkalinity and acidity / caustic alkalinity) are used less frequently in oceanography than in engineering so they won't be discussed here, but see Stumm and Morgan (1996) for more detail.

2.4 OPEN CARBONATE SYSTEM

For an open system the gas phase exerts a major control on the expression of the mass balance. The exchange or chemical equilibrium of a gas between gaseous and liquid phases can be written as:

$$A(g) = A(aq)$$

At equilibrium we can define

$$K = [A(aq)] \; / \; [A(g)]$$

2.4.1 Henry's Law. We can express the gas concentration in terms of partial pressure using the
ideal gas law:

$$PV = nRT$$

so that [A(g)] is equal to the number of moles n divided by the volume

$[A(g)] = n/V = P_A / RT$ where P_A is the partial pressure of A

Then

$$K = [A(aq)] \; / \; P_A / RT$$

Or

$$[A(aq)] = (K/RT) \; P_A$$

$\mathbf{[A(aq)] = K_H \; P_A}$ units for K are mol kg^{-1} atm^{-1}; for P_A are atm and A(aq) is in mol kg^{-1}

Henry's Law states that the solubility of a gas (the concentration in the aqueous phase) is proportional to the partial pressure in the overlying gas phase. The mole fraction of CO_2 in dry air (f_G) (where f_G = moles gas i / total moles) is approximately equal to 3.75×10^{-4} (or 375 ppm) (and is increasing steadily due to anthropogenic CO_2). The Henry's Law constant S = 35 equals $10^{-1.21}$ at 0° C and $10^{-1.50}$ at 25° C. In general, the solubility of gases decreases with increasing temperature and salinity.

For the open carbonate system the concentration of carbonic acid is fixed by equilibrium with the gas phase:

$$[H_2CO_3] = K_H \; P_{CO2}$$

Remember that $[H_2CO_3] = C_T \; \alpha_o$

Then the concentration of total CO_2 for an open system can be expressed as:

$$C_T = K_H \; P_{CO2} / \alpha_o$$

The DIC will vary with pH following the pattern of α_o.

2.4.2 Graphical Display. The distributions of carbonate species as a function of pH for an open system in equilibrium with a gas phase with $P_{CO2} = 10^{-3.5}$ are shown in Fig. 4. You can see that allowing the system to be open results in a very different set of distributions than seen for a closed system (Fig. 3). Now the concentration of H_2CO_3 is constant for all pH values and equals $K_H \; P_{CO2} = 10^{-5}$. The concentration of log $[HCO_3^-]$ and log $[CO_3^{2-}]$ have slopes of +1 and +2, respectively. The concentration of Total CO_2 is constant at 10^{-5} below about pH = 6, then increases following the concentrations of the most concentrated species. The alkalinity equivalence point is still identified as the intersection of H^+ and HCO_3^-. If you make a solution with a basic pH, CO_2 will invade from the atmosphere to make very high concentrations of C_T.

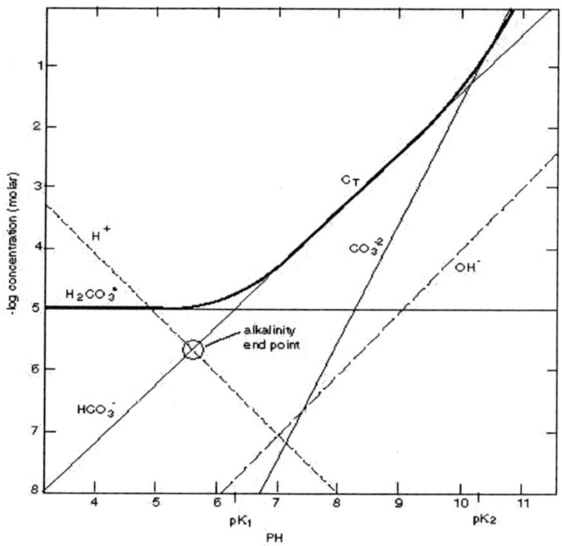

Figure 4. Distributions for an open system of CO_2 with a fixed $PCO_2 = 10^{-3.5}$

2.5 WHAT CAN YOU MEASURE?

We can not measure most of the carbonate species directly. What we can measure are pH, Total CO_2, Alkalinity and P_{CO2}. The distribution of these properties within the ocean tell us a lot about the fluxes of organic matter and carbonate minerals. We can combine these measurements to calculate the concentrations of the individual species.

pH: pH is more complicated because there are several different ways that it can be defined, measured and standardized. Historically pH was typically defined in terms of the activity of H^+ or as $pH = -\log (H^+) = -\log \gamma_H[H^+]$. But recent advances have led to definitions on the total proton scale where

$$pH_T = -\log m(H^+)_T = -\log \{[H^+] + [HSO_4^-]\} \text{ (Dickson, 1993)}$$

or the seawater scale where

$$pH_{sws} = -\log\{H_{sws}\} = -\log \{[H^+] + [HSO_4^-] + [F^-]\} \text{(Millero, 1995).}$$

Which scale you use determines which standards are used to calibrate the pH measurement. pH was historically measured using electrochemical approaches using ion-selective glass electrodes calibrated against standardized buffer solutions (historically called National Bureau of Standards (NBS) then later National Institute of Standards and Technology (NIST) buffers). These approaches were susceptible to accuracy errors, especially those due to the unknown differences in electrochemical liquid junction potentials when the electrodes were transferred from the standards to the solution of interest (e.g. MacInnes1961).

More recently a number of colorimetric dyes have been developed and the accuracy and precision of the pH measurement has improved enormously. In these new methods the ratio of the acid and base form of a pH sensitive dye are measured by absorbance (Byrne, 1987). For example, pH dependent reaction between the acid and base forms of sulfonephthalein dyes such as phenol red and bromocresol green (Byrne et al, 1988), cresol red (Byrne and Breland, 1989) or m-cresol purple (Clayton and Byrne, 1993) can be written as:

$$HL^- = L^{2-} + H^+$$

The relative concentrations of the acid and base forms are related to pH through the equation:

$$pH = pK_2 + \log [L^{2-}]/[HL^-]$$

Each species of the dye absorbs light at a different wavelength. Either single, double or scanning wavelength approaches can be used but the multiple wavelength approach is more precise. If you know the apparent acidity constant and can measure both forms you can determine the pH very precisely and with good accuracy (to 0.001 pH unit)

Total CO_2: Total CO_2 (expressed as C_T or DIC or ΣCO) is defined as the sum of the total concentrations of the three carbonate species (in mol kg^{-1}):

$$C_T = [H_2CO_3] + [HCO_3^-] + [CO_3^{2-}]$$

It is determined by acidifying a precisely known volume of seawater to low pH using phosphoric acid. This converts all the carbonate species to $H_2CO_3^*$ (essentially equivalent to CO_2(aq)). The evolved CO_2 is stripped off with an inert carrier gas (e.g. He), trapped in an absorbent and titrated coulometrically in an electrochemical cell (Johnson et al., 1985; Chipman et al., 1993; Johnson et al, 1998). For the purpose of quality control Certified Reference Materials (CRMs) have been prepared by Andrew Dickson at Scripps Institute of Oceanography. Routine analyses of shipboard CRMs help verify measurement accuracy. A precision of about ~2 μmol kg^{-1}, or better, can be obtained for DIC analyses.

Alkalinity: The alkalinity is defined as the amount of acid necessary to titrate all the weak bases (proton acceptors) in seawater (e.g. HCO_3^-, CO_3^{2-}, $B(OH)_4^-$) to the alkalinity endpoint which occurs where $(H^+) = (HCO_3^-)$ (this is about pH = 4.3 for seawater values of total CO_2; see Fig. 3) (Dickson et al., 1981; Bradshaw and Brewer, 1988). At this pH, essentially all the carbonate species (HCO_3^- and CO_3^{2-}) have been converted to H_2CO_3. Alkalinity is determined using an acid titration and for precise analysis the data are treated by a linearization procedure called the Gran Plot (Gran, 1952) (see Butler, 1991, chapter 3 for a clear derivation and example). The concentration of alkalinity is expressed as equivalents kg^{-1}, rather than moles kg^{-1}, because each species is multiplied by the number of protons it consumes (see previous discussion of proton conditions). For example, when acid is added HCO_3^- consumes one proton when it is converted to H_2CO_3. CO_3^{2-} consumes two protons, thus its concentration is multiplied by two.

$$HCO_3^- + H^+ \rightarrow H_2CO_3$$

$$CO_3^{2-} + 2H^+ \rightarrow H_2CO_3$$

An acid titration is blind to the identity of the base thus all weak acids present in seawater (with $pK > pK_1$ for carbonic acid) are titrated. Borate is the most important of these.

$$B(OH)_4^- + H^+ \rightarrow B(OH)_3^\circ$$

Thus, the total alkalinity of seawater (TALK) is defined as:

$$\text{Alkalinity} = [HCO_3^-] + 2[CO_3^{2-}] + [B(OH)_4^-] +.[SiO(OH)_3^-] + [HPO_4^{2-}] + 2[PO_4^{3-}] + [NH_3] + [HS^-] + [OH^-] - [H^+] - [HSO_4^-] - [HF] - [H_3PO_4])$$

For carbonate system calculations the total alkalinity needs to be corrected to the carbonate alkalinity (CA), defined as:

$$CA = [HCO_3^-] + 2[CO_3^{2-}]$$

Precise seawater alkalinity measurements are made by potentiometric titration using a titrator and a potentiometer. The titrating acid is hydrogen chloride made up in a sodium chloride solution to help maintain activity coefficient approximately constant during the titration. Total Alkalinity (TALK) is usually determined using a full titration curve with Gran analysis (Brewer et al., 1986) but a single point method (Perez and Fraga, 1987) is sometimes used. In principal, both total alkalinity and total CO_2 can be computed from the same titration curve, however, the coulometric method is more accurate for routine analyses of DIC. The CRMs have also been certified for TALK calibration. With care a precision of ±5 μmol kg+ (or better) can be obtained.

P_{CO2}: The P_{CO2} in a sample is the P_{CO2} that a water sample would have if it were in equilibrium with a gas phase with a fixed P_{CO2}. It is determined by equilibrating a known volume of water with a known volume of gas at a constant temperature and measuring the CO_2 in the gas phase, by non-dispersive infrared (NDIR) detection or by quantitatively converting the CO_2 to CH_4 which is then analyzed by a gas chromatograph with flame ionization detection. Precisions better than 0.5% can be obtained.

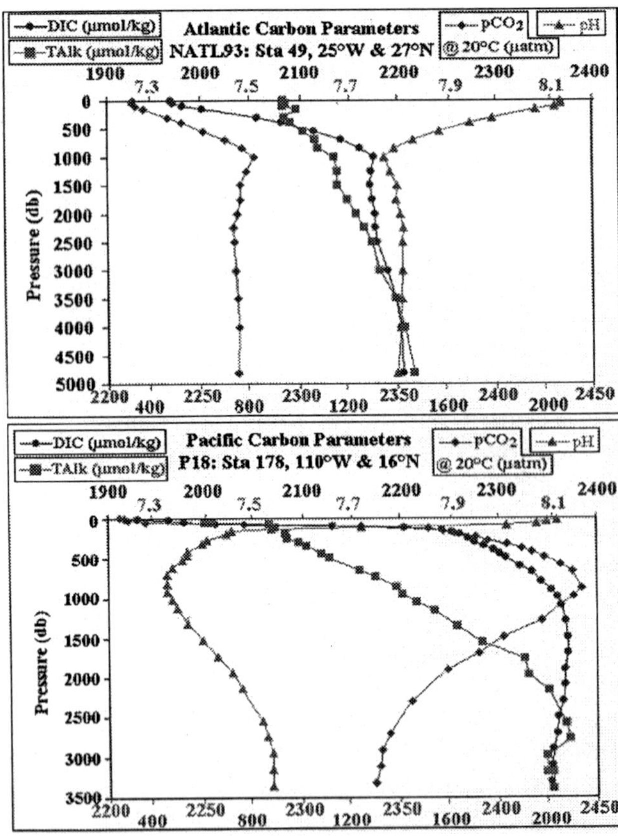

Figure 5. Depth profiles of pH, Talk, Total CO_2 and PCO_2 in the North Atlantic and North Pacific (from R. Feely, NOAA/PMEL)

3. OCEAN DISTRIBUTIONS

Here we summarize briefly the broad patterns of the distributions of the measured properties versus depth and between oceans. An example of the depth profiles of pH, Total CO_2, TAlk, and P_{CO2} from the western basin of the North Atlantic (25°W; 27°N) and the eastern North Pacific (110°W; 16°N) are shown in Fig. 5 (R. Feely, personal communication). The same concentration scales are used for both oceans to facilitate comparison.

pH - The surface values in both oceans are just slightly higher than pH=8.1. This is close to the value expected for water of seawater alkalinity in equilibrium with the atmosphere with $P_{CO2} = 10^{-3.5}$. pH then decreases to a minimum, however the minimum is more intense in the Pacific (to about pH= 7.3) than the Atlantic (pH = 7.75). The depth of this pH minimum corresponds to the depth of the oxygen minimum thus it is caused by production of CO_2 due to respiration of organic matter. In the deep sea the pH increases slowly, but at all depths, the pH in the Pacific (pH $\approx$ 7.5) is less than that in the Atlantic (pH $\approx$ 7.8).

DIC - The Total CO_2 is about 1950 μmol kg^{-1} in both the surface Atlantic and Pacific. It then increases with depth. The increase is steep in the upper 1000m and then is more gradual in deeper water. All subsurface DIC concentrations in the deep Pacific (about 2350 μmol kg^{-1}) are higher than in the deep Atlantic (about 2200 μmol kg^{-1}).

Alkalinity -The values plotted here are for total alkalinity so it includes the concentrations of all titratable bases. Alkalinity in the surface Atlantic is about 2300 μeq kg^{-1}, while the surface Pacific is slightly lower at 2250 μeq kg^{-1}. Alkalinity increases less steeply with depth than does DIC. The deep values are lower in the Atlantic (2350 μeq kg^{-1}) than the Pacific (2425 μeq kg^{-1}).

$\mathbf{P_{CO2}}$ - In most regards the distribution of P_{CO2} is a mirror image of pH. The surface values in both oceans are about 350 μatm, which is about the value of the atmosphere at the time of the cruises. P_{CO2} increases to a maximum of about 800 μatm in the Atlantic but over 2000 μatm in the Pacific.

4. CONTROLS ON OCEAN DISTRIBUTIONS

The distributions of carbonate system parameters within the ocean are primarily controlled by photosynthesis and respiration and calcium carbonate precipitation and dissolution. Here is a brief summary of how the different measured variables vary with the different processes.

a) Photosynthesis/Respiration

Organic matter (approximated as simply CH_2O for this example) is produced and consumed as follows:

$$CH_2O + O_2 \leftrightarrow CO_2 + H_2O$$

Respiration consumes dissolved oxygen and produces CO_2.

The CO_2 then reacts with water to produce carbonic acid, which dissociates to make HCO_3^- and CO_3^{2-} :

$$CO_2 + H_2O \rightarrow H_2CO_3$$

$$H_2CO_3 \rightarrow H^+ + HCO_3^-$$

$$HCO_3^- \rightarrow H^+ + CO_3^{2-}$$

As CO_2 is produced during respiration we should observe the following trends:

$$pH \downarrow DIC \uparrow \ Alk \leftrightarrow P_{CO2} \uparrow$$

The pH goes down because we are adding an acid, H_2CO_3. DIC goes up because we are adding one of the three species of DIC. P_{CO2} goes up because the pH goes down and DIC goes up. It is curious that alkalinity is unaffected by the carbonate species produced during respiration. But for each HCO_3^- that forms a corresponding H^+ is produced as well and they counter each other in the definition of alkalinity. The net effect is no change. In reality, there will actually be a small decrease in alkalinity as NH_4^+ from the organic matter is produced then oxidized to NO_3^-, apparently as HNO_3 (Brewer et al., 1975; Brewer and Goldman, 1976; Goldman and Brewer, 1980).

The trends will be the opposite for photosynthesis.

b) $CaCO_3$ dissolution/precipitation

The dissolution/precipitation of $CaCO_3$ can be written in a couple of different ways.

The standard solubility relationship is:

$$CaCO_3(s) \leftrightarrow Ca^{2+} + CO_3^{2-}$$

This can also be written as:

$$CaCO3(s) + CO2 + H2O \leftrightarrow Ca2+ + 2\ HCO3-$$

Oceanographers prefer the second reaction as it is written in terms of the main carbonate species in seawater (e.g. HCO_3^-) and because it is written in terms of CO_2. In this form it can be used to explain the coupling between production of CO_2 from respiration and dissolution of $CaCO_3$.

As the reaction shifts to the right and $CaCO_3(s)$ dissolves, CO_3^{2-} is added to solution.

We should observe the following trends.

$$pH \uparrow \quad DIC \uparrow \quad Alk \uparrow \quad P_{CO2} \downarrow$$

Note that alkalinity is affected by carbonate dissolution reactions. pH goes down because we are adding as base. DIC and Alk go up because we are adding species in their definitions. P_{CO2} goes down because the pH goes up.

5. WHAT CONTROLS THE PH OF SEAWATER?

In any given sample of seawater, the pH is controlled by alkalinity and DIC and can be calculated from these two parameters as shown below. Because H^+ and OH^- are negligible, we can approximate carbonate alkalinity as:

$$Alk \approx HCO_3^- + 2\ CO_3^{2-}$$

Bicarbonate and carbonate can be expressed in the form of Total CO_2 (i.e. C_T) and pH (e.g. α values which are defined in the next section)

$$Alk \approx C_T\ \alpha_1 + 2\ C_T\ \alpha_2$$

$$Alk = C_T\ (H^+K_1' + 2\ K_1'\ K_2')\ /\ (H^{+2} + H^+\ K_1' + K_1'K_2')$$

Rearranging, we can calculate H^+ (thus pH) from Alkalinity and C_T.

$$(H^+) = \{-K_1'\ (Alk\text{-}C_T) + [(K_1')^2\ (Alk\text{-}C_T)^2 - 4\ Alk\ K_1'\ K_2'\ (Alk - C_T)]\ \}\ /\ 2\ Alk$$

So the question of what controls pH, boils down to what controls alkalinity and total CO_2.

Internal variations of pH in the ocean are controlled by internal variations in Total CO_2 and alkalinity which are controlled by photosynthesis, respiration and $CaCO_3$ dissolution and precipitation.

The long term controls on alkalinity and DIC are the balance between the sources and sinks of alkalinity and DIC to the ocean. These are the weathering or the continents (sources) and burial in the sediments (sinks) of silicate and carbonate rocks and organic matter.

6. CARBONATE SYSTEM CALCULATIONS

The measured properties can be used to calculate the concentrations of the individual carbonate species. Because there are 6 unknowns for the open carbonate system (P_{CO2}, H_2CO_3, HCO_3^-, CO_3^{2-}, H^+ and OH^-) and we have 4 equilibrium constants (K_1, K_2, K_H, K_W) we can use any two of the four measured parameters for these calculations. Two examples are shown here. Calculations using other variables can be conducted using similar derivations.

pH and C_T

A useful shorthand is the alpha notation, where the alpha (α) values express the fraction each carbonate species as a function of the Total CO_2. These α values are a function of pH only for a given set of acidity constants.

Thus:

$$[H_2CO_3] = \alpha_o C_T$$

$$[HCO_3^-] = \alpha_1 C_T$$

$$[CO_3^{2-}] = \alpha_2 C_T$$

The value of alpha 0 (α_o) is derived as follows:

$$\alpha_o = H_2CO_3 / C_T = H_2CO_3 \ / \ (H_2CO_3 + HCO_3^- + CO_3^{2-})$$

$$= 1 / (\ 1 + HCO_3^- / H_2CO_3 \ + \ CO_3^{2-}/H_2CO_3)$$

$$= 1 / (\ 1 + K_1/H \ + \ K_1K_2/H^2)$$

$$= H^2 / (\ H^2 \ + \ HK_1 \ + \ K_1K_2)$$

The values for α_1 and α_2 can be derived in a similar manner to obtain:

$$\alpha_1 = HK_1 / (H^2 \ + \ H\,K_1 \ + \ K_1K_2)$$

$$\alpha_2 = K_1K_2 / (\ H^2 + H\,K_1 + K_1K_2)$$

Note that the denominators are the same for all three α values.

With these equations we can calculate the concentrations of the carbonate species if we know the Total CO_2 and pH.

For example:

Assume pH = 8, $C_T = 10^{-3}$, $pK_1' = 6.0$ and $pK_2' = 9.0$

$$[H_2CO_3^*] = 10^{-5} \text{ mol kg}^{-1}$$

(note the answer is in concentration because we used K')

$$[HCO_3^-] = 10^{-3} \text{ mol kg}^{-1}$$

$$[CO_3^{2-}] = 10^{-4} \text{ mol kg}^{-1}$$

As we saw in the graphical approach (Fig. 3) there is about a factor of 10 difference between the concentrations of $HCO_3^- > CO_3^{2-} > H_2CO_3$.

Alkalinity and Total CO_2 or P_{CO2}

Again if we neglect H^+ and OH^- we can write alkalinity as:

$$Alk = HCO_3^- + 2\ CO_3^{2-}$$

Then

$$Alk = C_T\,\alpha_1 + 2\ C_T\,\alpha_2$$

$$Alk = C_T\,(\alpha_1 + 2\alpha_2)$$

We can use this equation if we have a closed system and we know 2 of the 3 variables (Alk, C_T or pH).

For an open system we can express C_T in terms of P_{CO2} as shown earlier:

$$C_T = K_H\ P_{CO2} / \alpha_o$$

Now:

$$Alk = (K_H\,P_{CO2} / \alpha_o)\,(\alpha_1 + 2\alpha_2)$$

$$Alk = K_H\,P_{CO2}\ (\ (\alpha_1 + 2\ \alpha_2) / \alpha_o)$$

$$Alk = K_H\,P_{CO2}\,(\ (HK_1 + 2\ K_1K_2) / H^2)$$

If we assume:

$$Alk = 10^{-3}$$

$$P_{CO2} = 10^{-3.5}$$

$$pK_1' = 6.0$$

$$pK_2' = 9.0$$

Then we calculate a pH = 8.3

Computer Programs

Many computer programs are available that can be used to solve for the distributions of carbonate species. Examples include MINEQ+, WATEQ, HYDRAQL and PHREEQ. A google web search will provide access.

One useful program for carbonate system calculations called CO2SYS.EXE with necessary documentation is available to download from the web (Lewis and Wallace, 1997) at:

http://cdiac.esd.ornl.gov/oceans/co2rprt.html

7. REVELLE FACTOR

Burning of fossil fuels, making concrete and cutting down forests has caused a significant increase in atmospheric CO_2. As atmospheric CO_2 increases some of it goes into the ocean due to solubility reactions. The dissolution of atmospheric CO_2 in surface seawater is enhanced by the ocean's capacity to buffer the associated changes in seawater chemistry and this can be calculated using principles of aquatic chemistry.

The Revelle buffer factor (Revelle and Suess, 1957) defines how much CO_2 can be absorbed by homogeneous reaction with seawater. Homogeneous in this case means that we assume no $CaCO_3$ dissolution due to the added CO_2, which would enhance the uptake of CO_2. CO_2 exchange across the air/sea interface does not cause any $CaCO_3$ dissolution (or change in alkalinity) as long as the water remains supersaturated with respect to $CaCO_3$. This is still the case in the present ocean. The Revelle factor is defined as the fractional change in P_{CO2} divided by the fractional change in DIC in equilibrium with the atmosphere.

$$R = (\Delta P_{CO2}/P_{CO2}) / (\Delta DIC / DIC)$$

Formally this should be at constant alkalinity. Thus:

$$R = (DIC/PCO_2) \times (\partial PCO_2/\partial CT)\ alk = \frac{DIC(\partial P_{CO_2} / \partial H)alk}{P_{CO_2}(\partial C_T / \partial H)alk}$$

The full derivation is straightforward but messy. It is presented in Butler (1991, p. 143-146) where he neglects terms for boric acid in the alkalinity. After calculating the partial derivative terms, neglecting small terms and substituting you get the simple expression:

$$R = DIC / (H_2CO_3 + CO_3^{2-})$$

For typical seawater with pH = 8, Alk = $10^{-2.7}$, DIC = $10^{-2.7}$, pK_1' = 6.0 and pK_2' = 9.1 you get $H_2CO_3 = 10^{-4.7}$ and $CO_3^{2-} = 10^{-3.8}$; then the value of R = 11.2

Sundquist et al (1979) derived the Revelle factor including the contribution of borate to alkalinity and this makes a significant difference. With borate:

$$R= DIC/ \{(H_2CO_3+ CO_3^{2-} + ((\gamma HCO_3^- - 4 (CO_3^{2-})^2)/(HCO_3^- + 4 CO_3^{2-}+ \gamma)))\}$$

where

$$\gamma = (C_B K_B' (H^+)) / (K_B' + (H^+))^2$$

where K_B' is the seawater acidity constant for boric acid, C_B is the total concentration of dissolved borate in seawater. The value of R at 25°C with the borate term included is about 9.0.

Sundquist et al (1979) used the field data from GEOSECS to calculate the Revelle factor for seawater as a function of temperature which they compared with the line based on this equation (Fig. 6).

A value of about 10 for the Revelle factor means that a change of 10% in atm CO_2 is required to produce a 1% change in Total CO_2 content of seawater. It means that about 10% of the increase in atmospheric CO_2 goes into the water. A 10% increase in atmospheric CO_2 by ~30ppm or 15 mol m^{-2} would produce an increase of about 1% in DIC (9 mol m^{-2}). The buffer factor also increases with increasing P_{CO2} and increasing DIC. This means that an increase in the CO_2 content of seawater will decrease its capacity for taking up more (because CO_3^{2-} decreases)!

Temperature is not the only factor however. For example, in the surface water of the Southern Ocean temperature is low but the alkalinity is also low. Waters of lower alkalinity have less capacity to accumulate anthropogenic CO_2 for a given increase in atmospheric CO_2. Thus there are lower amounts of anthropogenic CO_2 in the southern ocean and the values increase to the north (McNeil et al., 2003).

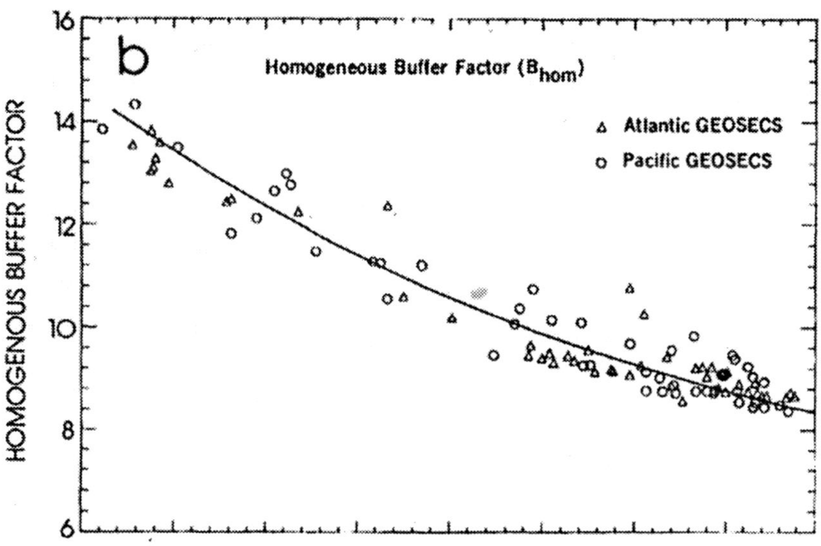

Figure 6. . The Revelle Factor (R) in the Atlantic and Pacific, here called the homogeneous buffer factor Bhom, as a function of temperature calculated using GEOSECS data (Sundquist et al., 1979).

APPENDIX:

Equations to calculate the thermodynamic equations in seawater as a function of temperature (T) and salinity (S) at 1 atm total pressure (from Millero, 1995).

Carbonic Acid: Carbonic acid constants of Roy et al (1993) converted to the seawater scale combined with the smoothed results of Goyet and Poisson (1989) give overall standard errors of 0.0012 in pK'_1 and 0.0025 in pK'_2. Millero (1995) recommended the following combined equations on the seawater pH scale.

$$\ln K'_1 = 2.18867 - 2275.0360\, T^{-1} - 1.468591 \ln (T) = (-0.138681 - 9.33291\, T^{-1})\, S^{0.5} + 0.0726483S - 0.00574938\, S^{1.5}$$

$$\ln K'_2 = -0.84226 - 3741.1288\, T^{-1} - 1.437139 \ln (T) + (-0.128417 - 24.41239\, T^{-1})\, S^{0.5} + 0.1195308\, S - 0.00912840\, S^{1.5}$$

Boric Acid: Boric acid in seawater at temperature (T) and salinity (S) with standard error of 0.0042 in $\ln K'_B$ on the seawater pH scale..

$$\ln K'_B = (-8966.90 - 2890.51\, S^{0.5} - 77.942\, S + 1.726\, S^{1.5} - 0.0993\, S^2)\, T^{-1} + (148.0248 + 137.194\, S^{0.5} + 1.62247\, S) + (-24.4344 - 25.085\, S^{0.5} - 0.2474\, S) \ln T + 0.053105\, S^{0.5}\, T$$

Water: The dissociation of water (H_2O) in seawater on the seawater pH scale can be described by:

$$\ln K'_W = 148.9802 - 13847.26\, T^{-1} - 23.6521 \ln T + (-5.977 + 118.67\, T^{-1} + 1.0495 \ln T)\, S^{0.5} - 0.01615\, S$$

Phosphoric Acid: The three dissociation constants for phosphoric acid at various T and S on the seawater pH scale come from Millero (1995) with standard errors of 0.09, 0.03 and 0.20 respectively in pK'_{P1}, pK'_{P2} and pK'_{P3}.

$$\ln K'_{P1} = 115.54 - 4576.752\, T^{-1} - 18.453 \ln T + (0.69171 - 106.736\, T^{-1})\, S^{0.5} + (-0.01844 - 0.65643\, T^{-1})\, S$$

$$\ln K'_{P2} = 172.1033 - 8814.715\, T^{-1} - 27.927 \ln T + (1.3566 - 160.340\, T^{-1})\, S^{0.5} + (-0.05778 + 0.37335\, T^{-1})\, S$$

$$\ln K'_{P3} = -18.126 - 3070.75\, T^{-1} + (2.81197 + 17.27039\, T^{-1})\, S^{0.5} + (-0.09984 - 44.99486\, T^{-1})\, S$$

Silicic Acid: The dissociation constant of silicic acid has not been measured in seawater so Millero (1995) recommended values in NaCl at the same ionic strength of seawater (I = 0.7). This regression has a standard error of 0.02 in pK_{Si}. For the seawater pH scale.

$$\ln K_{Si} = 117.40 - 8904.2\ T^{-1} - 19.334 \ln T + (3.5913 - 458.79\ T^{-1})\ I^{0.5} + (-1.5998 + 188.74\ T^{-1})\ I + (0.07871 - 12.1652\ T^{-1})\ I^2$$

Hydrogen Sulfide: The first dissociation constant of hydrogen sulfide in seawater on the Total pH scale can be estimated from:

$$\ln K'_{1S} = 225.838 - 13275.3\ T^{-1} - 34.6435 \ln T + 0.3449\ S^{0.5} - 0.0274\ S$$

Ammonium: The dissociation constant for NH_4^+ can be obtained from the following equation with a standard error of 0.01 in pK_{NH4}.

$$\ln K_{NH4} = -6285.33\ T^{-1} + 0.0001635\ T - 0.25444 + (0.46532 - 123.7184\ T^{-1})\ S^{0.5} + (-0.01992 + 3.17556\ T^{-1})\ S$$

Solubility of Calcium Carbonate: The results of Mucci (1983) can be represented by:

$$\log K_{calcite} = -171.9065 - 0.077993\ T + 2839.319\ T^{-1} + 71.595 \log T + (-0.77712 + 0.0028426\ T + 178.34\ T^{-1})\ S^{0.5} - 0.07711\ S + 0.0041249\ S^{1.5}$$

$$\log K_{aragonite} = -171.945 - 0.077993T + 2903.293\ T^{-1} + 71.595 \log T + (-0.068393 + 0.0017276\ T + 88.135\ T^{-1})\ S^{0.5} - 0.100185\ S + 0.0059415\ S^{1.5}$$

Concentration of Ca^{2+} in seawater: Millero (1982) gives:

$$[Ca^{2+}] = 0.01028\ (S/35)$$

REFERENCES

Bradshaw, A.L. and Brewer, P.G. (1988) High precision measurements of alkalinity and total carbon dioxide in seawater by potentiometric titration 1. Presence of unknown protolytes. Marine Chemistry 23, 69-86

Bradshaw, A.L. and Brewer, P.G. (1988) High precision measurements of alkalinity and total carbon dioxide in seawater by potentiometric titration 2. Measurements on standard solutions. Marine Chemistry 24, 155-162.

Brewer, P.G., Wong, G.T.F., Bacon, M.P. and Spencer D.W. (1975) An oceanic calcium problem? Earth and Planetary Science Letters, 26, 81-87.

Brewer, P.G. and Goldman, J. C. (1976) Alkalinity changes generated by phytoplankton growth. Limnology and Oceanography, 21, 108-117.

Brewer, P.G., Bradshaw, A. and Williams, R. (1986) Measurements of total carbon dioxide and alkalinity in the North Atlantic Ocean in 1981. In The Changing Carbon Cycle, A Global Analysis, edited by J. Trabalka and D. Reichle, pp 358-381, Springer-Verlag, New York

Butler, J.N. (1964a) Ionic Equilibrium, a Mathematical Approach, Addison-Wesley, Reading, Mass.

Butler, J.N. (1964b) Solubility and pH Calculations. Addison-Wesley, Reading, Mass

Butler, J.N. (1991) Carbon Dioxide Equilibria and their Applications, Lewis Publishers, Chelsea, Michigan, 259pp.

Byrne R.H. (1987) Standardization of standard buffers by visible spectrometry. Analytical Chemistry 59, 1479-1481.

Byrne, R.H., Robert-Baldo, G., Thompson S.W., and Chen, C.T.A. (1988) Seawater pH measurements: an at-sea comparison of spectrophotometric and potentiometric methods. Deep-Sea Research 35, 1405-1410.

Byrne, R.H. and Breland, J.A. (1989) High precision multiwavelength pH determinations in seawater using cresol red. Deep-Sea Research 36, 803-810.

Chipman, D.W., Marra, J. and Takahashi, T. (1993) Primary production at 47°N and 20°W in the North Atlantic Ocean: A comparison between the 14C incubation method and the mixed layer carbon budget. Deep-Sea Research, 40, 151-169.

Clayton, T.D. and Byrne, R.H. J. (1993) Spectrophotometric seawater pH measurements: total hydrogen ion concentration scale calibration of m-cresol purple and at-sea results. Deep-Sea Research 40, 2315-2329.

Dickson, A.G. (1981) An exact definition of total alkalinity and a procedure for the estimation of alkalinity and total inorganic carbon from titration data. Deep-Sea Research 28, 609-623.

Dickson, A.G. (1984) pH scales and proton-transfer reactions in saline media such as seawater. Geochimica et Cosmochimica Acta, 48, 2299-2308.

Dickson, A.G. (1993) The measurement of seawater pH. Marine Chemistry 44, 131-142.

Dickson, A.G. (1993) pH buffers for seawater media based on the total hydrogen ion concentration scale. Deep-Sea Research 40, 107-118.

Garrels, R.M. and Thompson, M.E. (1962) A chemical model for seawater at 25°C and one atmosphere total pressure. American Journal of Science, 260, 57-66.

Goldman, J.C. and Brewer, P.G. (1980) Effect of nitrogen source and growth rate on phytoplankton-mediated changes in alkalinity. Limnology and Oceanography, 25, 352-357.

Goyet C. and Poisson, A. (1989) New determination of carbonic acid dissociation constants in seawater as a function of temperature and salinity. Deep-Sea Research 36, 1635-1654.

Gran, G. (1952) Determination of the equivalence point in potentiometric titrations. Part II. Analyst, 77, 661-671

Johnson, K.M., King, A.E. and Sieburth, J.M. (1985) Coulometric TCO2 analyses for marine studies: An Introduction. Marine Chemistry, 16, 61-82.

Johnson, K.M., Dickson, A.G., Eischeid, G., Goyet, C., Guenther, P., Key, R.M., Millero, F.J., Purkerson, D., Sabine, C.L., Schottle, R.G., Wallace, D.R.W., Wilke R.J., and Winn, C.D. (1998) Coulometric total carbon dioxide analyses for marine studies: Assessment of the quality of total inorganic carbon measurements made during the US Indian Ocean CO2 Survey 1994-1996. Marine Chemistry 63, 21-37.

Lewis, E. and Wallace, D. (1997) CO2SYS.EXE A Program Developed for CO2 system calculations. Web site available at: //cdiac.esd.ornl.gov/oceans/co2rprt.html.

MacInnes, D.A. (1961) The Principles of Electrochemistry. Dover Pubs., New York 478pp

McNeil, B.I., Matear, R.J., Key, R.M., Bullister, J.L. and Sarmiento, J.L. (2003) Anthropogenic CO2 uptake by the ocean based on the global chlorofluorocarbon data set. Science, 299, 235-239.

Millero, F.J. (1995) Thermodynamics of the carbon dioxide system in the oceans. Geochemica et Cosmochimica Acta 59, 661-677.

Mucci, A. (1983) The solubility of calcite and aragonite in seawater at various salinities, temperatures and one atmosphere total pressure. American Journal of Science 283, 780-799.

Perez, F.F. and Fraza, F. (1987) A precise and rapid analytical procedure for alkalinity determination. Marine Chemistry 21, 169-182.

Revelle, R. and Suess, H.E. (1957) Carbon dioxide exchange between atmosphere and ocean and the question of an increase of atmospheric CO2 during the past decade. Tellus, 9, 18-27

Roy, R.N., Roy, L.N., Vogel, K.M., Moore, C.P., Pearson, T., Good, C.E., Millero F.J., and Campbell, C.D. (1993) Determination of the ionization constants of carbonic acid in seawater. Marine Chemistry 44, 249-268.

Sillen, L.G. (1959) Graphic Presentation of Equilibrium Data. In Treatise on Analytical Chemistry Part 1, Vol., 2, I.M. Kolthoff and P.J. Elving, Eds., Interscience, New York.

Stumm, W. and Morgan, J.J. (1996) Aquatic Chemistry, Third Edition. Wiley, New York, 780pp.

Sundquist, E. T., Plummer, L.N. and Wigley, T.M.L. (1979) Carbon Dioxide in the Ocean Surface: The Homogeneous Buffer Factor. Science, 204, 1203-1205

A PRIMER ON DISSOLVED ORGANIC MATERIAL AND HETEROTROPHIC PROKARYOTES IN THE OCEANS

DAVID L. KIRCHMAN
College of Marine Studies, University of Delaware, Lewes, DE 19958, USA
kirchman@udel.edu

INTRODUCTION

Dissolved organic material (DOM) and microbes are well-recognized to be important components of carbon cycles and other biogeochemical cycles in the oceans. The main component of DOM, dissolved organic carbon (DOC), is one of the largest pools of carbon in the biosphere, equaling atmospheric CO_2 in size (ca. 700 X 10^{15} tons) (Hedges and Oades 1997). The flux of carbon through the DOC pool is also substantial. Various approaches indicate that about 50% of primary production is some how routed through bacteria and single-cell eukaryotes (protists) that graze on bacteria, i.e. the microbial loop. Mineralization of DOM to CO_2 and other inorganic nutrients by microbial loop microbes is one of three fates for primary production, the other two being grazing by protists and macrozooplankton (Fig. 1). In the open oceans, most of the carbon produced by phytoplankton (primary production) is mineralized by protist grazers or the microbial loop, leaving <10% for export to the deep ocean via sinking of large phytoplankton, the waste products (fecal pellets) of large zooplankton, or amorphorous aggregates of unclear origin. Carbon can also be exported as DOC in some oceanic regimes, as discussed below.

This chapter attempts to review the basics of DOM and the microbial processes impacting DOM in the oceans. Both topics are thoroughly reviewed by Hansell and Carlson (2002) and Kirchman (2000) who discuss

M. Follows and T. Oguz (eds.), The Ocean Carbon Cycle and Climate, 31–63.

DOM and microbes, respectively, in much more detail than possible here. The goal of this chapter is to introduce some aspects of DOM and microbes that may be useful for those interested in examining the role of DOC-related processes in global carbon cycles.

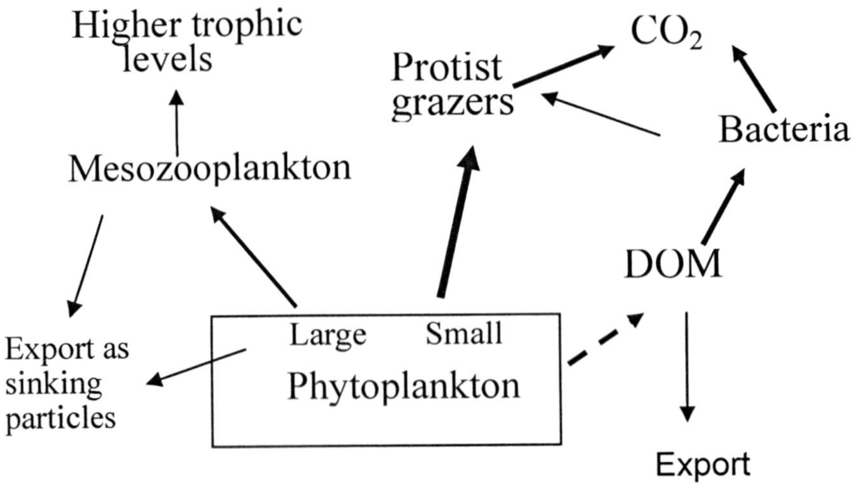

Figure 1. The role of DOM and the microbial loop (DOM–>bacteria and archaea–> grazers) in biogeochemical cycles. The relative size of the flux is indicated by the thickness of the arrow. The arrow for DOM production is dashed because many organisms in addition to phytoplankton produce DOM.

1. WHAT IS DOM?

The DOM pool is operationally defined as any organic material that passes through a GF/F filter which has a nominal pore size of about 0.7 µm. Consequently, "DOM" actually includes small bacteria and archaea, viruses, and detrital particles that are far from being dissolved. Fortunately, particulate carbon in the DOM fraction is usually not large compared to truly dissolved material even if the border between dissolved and particulate can be murky. In fact, DOC concentrations are usually much higher than particulate organic carbon (POC) concentrations, so that sometimes "total organic carbon" (TOC) is measured to avoid contamination caused by filtration. Even ignoring things such as colloids (Wells 2002), the DOM pool is complex and consists of many compounds that differ greatly in structure, molecular size, C:N ratio, and susceptibility to degradation by microbes, i.e. just about every parameter of importance to understanding its role in biogeochemical processes. To make matters worse, DOM is found in a complex medium, seawater. Because of many analytical difficulties, we know little about the chemical identity of DOM at the molecular level. About 10% of DOM can be identified as amino acids (proteins) and neutral sugars (e.g. glucose) in polysaccharides (reviewed by Benner 2002).

There is a quasi-inverse relationship between our understanding of a DOM component's chemistry and its lability and concentration (Fig. 2). Free amino acids and sugars, whose chemistry can be precisely defined, are used by microbes very rapidly and consequently their concentrations are very low. Biopolymers such as dissolved protein and polysaccharides must be hydrolyzed to smaller free amino acids and sugars before uptake by microbes, so their turnover is a bit slower and concentrations are higher. Likewise, the chemistry of these biopolymers in seawater is not as well understood as that for free amino acids and sugars; the monomers of biopolymers can be examined after acid hydrolysis and HPLC analysis, but the linkages and matrices surrounding targeted biopolymers in seawater are largely unknown.

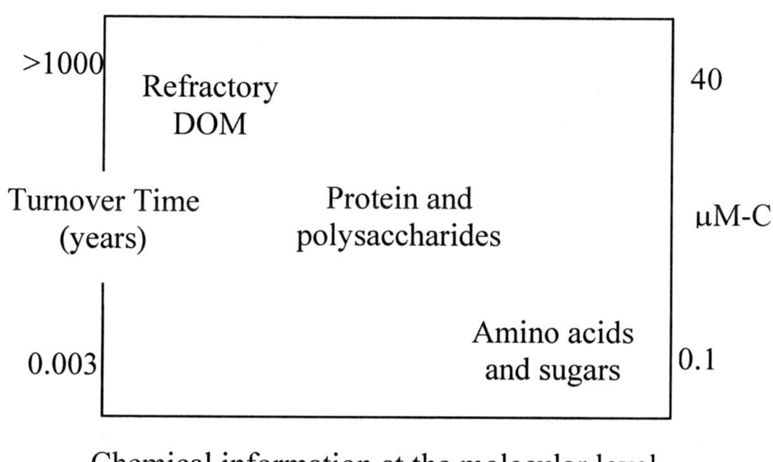

Figure 2. A schematic diagram to illustrate the relationship between turnover time, concentration and our understanding of DOM components at the molecular level. We know the most about the chemistry of the least abundant, fastest turning-over component consisting of compounds like free amino acids and sugars. At the other extreme, we know the least about the component with the highest concentration and slowest turnover time, refractory DOM.

At the other extreme, the largest component of DOM is refractory and a chemical mystery (Benner 2002).

Our fragmentary understanding of DOM chemistry greatly hinders our efforts to examine the role of DOM in biogeochemistry. Most importantly, we could obtain more precise estimates of turnover times and fluxes if armed with more insights into DOM chemistry. For example, because of our ability to measure dissolved free amino acid and neutral sugar concentrations, we can precisely estimate the flux of these compounds; they can account for large fractions of bacterial growth and, ignoring photochemistry, of the total DOM flux (Kirchman 2003).

We have some ideas about concentrations and chemical identify of DOM components such as dissolved protein and polysaccharides, but not enough to devise effective tracers and to estimate fluxes.

Much has been learned about the role of DOM in the oceanic carbon cycle from data on DOC concentrations over time and space. For example, DOC concentrations usually are higher in the euphotic zone and then decrease with depth (Fig. 3A), indicating the impact of biology on DOM processes. Depth profiles of concentrations can be used to divide the DOC pool into three components (Kirchman et al. 1993), based on lability (Table 1). The smallest pool, the labile pool, consists of those compounds that can be chemically identified at the molecular level and are used rapidly by microbes. Although concentrations of the labile pool are quite low (< 1 μM-C), fluxes can be high due to high turnover. The other extreme is the refractory pool whose concentration is taken to be that of DOC in the deep ocean (>1000 m). Williams and Druffel (1987) used ^{14}C dating to estimate that deep DOC turns over on the order of thousands of years. Since this turnover is longer than the turnover of the oceans, Williams and Druffel (1987) reasoned that the most refractory component of DOC must be uniform with depth (Fig. 3B). Consequently, concentrations of DOC found in the deep ocean provide an estimate of refractory DOC in the surface layer. Then, the semi-labile pool of the surface layer is the difference between total DOC concentrations in the surface layer and the deep DOC with the labile pool contributing only trivial amounts. There are few good estimates of the turnover time for the semi-labile pool, but weeks-to-months is probably the right time scale (Table 1).

I would like to emphasize two caveats to keep in mind when using the three component model. As with any simplification, it is difficult to fit all DOM into just three components, and the boundaries can be blurry. For example, dissolved polysaccharides are generally considered to be labile, yet their concentrations can be quite high (ca. 20% of surface layer DOC) (Pakulski and Benner 1994) and there is some evidence that they build up over the summer (Williams 1995), two characteristics of the semi-labile pool. The other caveat is that lability is currently defined strictly by bioassays or changes in concentrations in the water column; no chemical property is a good predictor of how fast a compound will be used by microbes. For these reasons, it is difficult and probably futile to try to assign particular compounds into one of the three compartments.

Table 1. The three compartment model of DOC in the upper layer of the oceans. The boundaries separating these three compartments are not strict, and there is currently no chemical parameter that can be used consistently and accurately to define these pools. The size of semi-labile pool versus that of the refractory pool also varies among the oceans.

Name	% of total DOC	Turnover Time
Labile	<1	minutes to days
Semi-labile	50	weeks to months
Refractory	50	years to millenia

It is not clear why microbes do not degrade semi-labile DOC, allowing it to increase in concentration during the growth season. One obvious answer is that the structure of semi-labile compounds simply makes them less available to microbes. This explanation seems more likely than another one often invoked: that microbes are limited by the low supply or concentration of some other element (e.g. phosphorus). Because our understanding of DOC chemistry is so limited, we do not have much evidence in support of the first hypothesis. But a few studies offer evidence against the second. Addition of phosphate and other inorganic nutrients does not appear to stimulate the use of the semilable DOC, even in environments such as the Sargasso Sea where bacterial growth may be limited by phosphate (Carlson et al. 2002).

2. FLUXES OF DOM THROUGH THE MICROBIAL LOOP

Heterotrophic prokaryotes are responsible for most of DOM mineralization in the oceans. Prokaryotes consist of Archaea and Bacteria, with the latter dominating the surface layer of the ocean (see below). Cyanobacteria are phylogenetically "bacteria" but functionally they are the same as eukaryotic algae or phytoplankton. Like eukaryotic algae, cyanobacteria are primary producers (CO_2 fixers) that account for >80% of primary production in the open oceans.

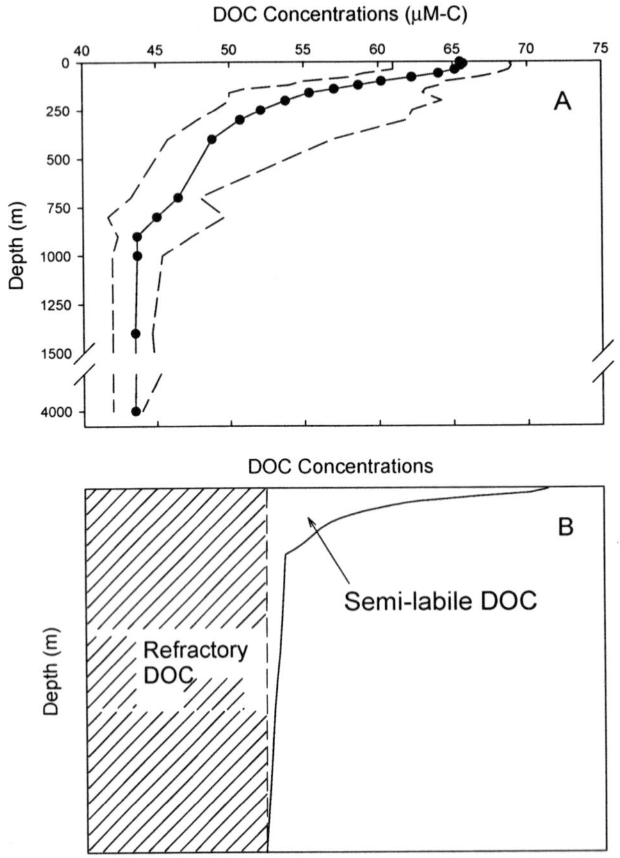

Figure 3. Dissolved organic carbon along a depth profile of the North Atlantic.

A) Concentrations of dissolved organic carbon (DOC) at the Bermuda Times Series Station in 1994 – 1998. The two dashed lines without data points follow the minimum and maximum concentrations. Data from Hansell and Carlson (2001).

B) Schematic diagram of the depth profile data, illustrating how deep DOC concentrations are used to estimate refractory DOC concentrations. The difference between deep and surface concentration is an estimate of the semi-labile pool. Concentrations of the labile pool are <1 μM-C and trivial compared to the other two DOC pools.

There is some evidence of DOM use by heterotrophic protists (single cell eukaryotes) (Sherr 1988), phytoplankton (Palenik and Morel 1990), and even invertebrate larvae. However, DOM uptake by these eukaryotes is not substantial even when it is essential for the organism. One reason for the predominance of prokaryotes in DOM uptake is that they are the most numerous organism in the biosphere (Whitman et al. 1998), and their biomass can rival that of even phytoplankton in the euphotic zone of open oceans (Ducklow 2000). Although uptake by eukaryotes can be ignored, photochemical oxidation of DOM may be an important alternative pathway to consider (Moran and Zepp 2000). In tributaries of the Amazon River, for example, photochemical oxidation of DOC to CO_2 can exceed microbial respiration (Amon and Benner 1996). Photochemistry is probably less important in the open ocean where concentrations of DOM capable of absorbing light, i.e. colored DOM, are low. For this reason, we can focus on microbially-mediated oxidation of DOM in thinking about oceanic carbon cycles.

What is the evidence that about half of primary production is somehow routed through the DOM pool and the microbial loop? Three interrelated approaches support the 50% estimate. The first two examine the decrease (uptake) over time of DOC or of oxygen (respiration) in surface water incubations of the small size fraction that includes mostly heterotrophic bacteria. These data are reviewed by Williams (2000) and Carlson (2002). Independently, these studies show the importance of DOM uptake. The data on DOC depletion and respiration are often used to estimate growth efficiency (BGE), which is an essential parameter in the third approach for estimating uptake of DOM by bacteria. Growth efficiency is defined as bacterial biomass production (BP) divided by total organic carbon uptake (U), i.e. BGE = BP/U.

The third approach estimates DOM uptake by combining estimates of growth efficiency and biomass production of heterotrophic bacteria: U = BP/BGE. Analogous to primary production, rates of heterotrophic bacterial production reflect the synthesis of new biomass before mortality (grazing and viral lysis) has taken its toll. Respiration is not included in bacterial production estimates. In the absence of mortality, bacterial biomass would increase over time initially at the rate of bacterial production. A hidden assumption of this approach is that heterotrophic bacteria rely mostly on DOM rather than particulate organic carbon. Except in rare cases, this assumption is a safe one because of the large number and high concentration of free-living bacteria and DOC relative to attached bacteria and particulate organic carbon.

A more serious problem with using bacterial production to examine DOM fluxes is that the growth efficiency can vary in ways that are still poorly understood. Growth efficiency depends on many factors that affect bacterial metabolism and varies among aquatic habitats as a function of trophic status. Freshwaters and estuaries have the highest (35%) and the open oceans (15%) the lowest BGE values reported to date (del Giorgio and Cole 2000). There are also problems associated with estimating bacterial production by current methods (incorporation of thymidine and leucine) (see

Ducklow 2000). Even ignoring the methodological problems, some investigators argue that we should concentrate on measuring bacterial respiration rather than production because respiration is the larger of the two processes. Also, because DOM can be re-cycled (in a sense) more than once, bacterial production can theoretically exceed primary production (Anderson and Ducklow 2001; Strayer 1988) and thus respiration is more directly comparable to primary production. These criticisms are quite valid, but the difficulties of directly measuring DOC depletion and respiration prevent these approaches from becoming common practice. Also, rates measured with the bacterial size fraction may not be accurate because the food web producing DOM and inorganic nutrients has been disrupted by filtration. In contrast, the ease of measuring thymidine and leucine incorporation allows much greater spatial and temporal coverage of heterotrophic bacteria-phytoplankton interactions in a nearly unperturbed state. Even given the ease of the methodology, the number of estimates of bacterial production for the open ocean is small compared to the vastness of the oceans and the sample density of primary production.

Generally, heterotrophic bacterial production correlates with primary production in the open oceans over sufficiently large and long space and time scales (Fig 4). The relationship between bacterial (BP) and primary production (PP) is often parameterized by the ratio of the two rates (BP:PP). The data in Figure 4 suggest that BP:PP values for the open oceans are generally lower than that seen in freshwaters, estuaries and the few coastal ocean samples included in the old review of Cole et al. (1988), as Ducklow (2000) first observed. Indeed, the average BP:PP in Figure 2 (the slope of the line) is 0.087 ± 0.046, over 2-fold lower than the BP:PP found by Cole et al. (1988). This issue needs to be examined by a more quantitative analysis, including an update of Cole et al. (1988).

We can use BP:PP and BGE values to estimate the fraction of primary production that is routed through the DOM pool (U:PP).

It can be shown that U:PP = (BP:PP)/BGE. Table 2 illustrates how U:PP varies with BP:PP and BGE. Taking BP:PP= 0.0867 and BGE= 0.15 (or 15%) implies that U:PP = 0.578, that is, somewhat higher than the 50% estimate quoted above. Ducklow et al. (2002) recently discussed balancing BP:PP and BGE in a carbon budget for a spring bloom in the North Atlantic. They conclude that some of the original BP:PP ratios estimated for the North Atlantic are too high to fit in our current understanding of food web dynamics and carbon cycles. Still, even the lower BP:PP ratios and our best estimate of BGE indicate that 50% or more of primary production is processed by the microbial loop via the DOM pool.

In fact, recently Hoppe et al. (2002) reported even higher BP:PP ratios and suggested that areas of the tropical Atlantic Ocean are net heterotrophic, i.e. organic carbon consumption exceeds primary production.

Table 2. Percentage of primary production that is routed through the DOM pool as a function of the ratio of bacterial production to primary production (BP:PP) and growth efficiency, expressed as a fraction. Currently, BP:PP and the growth efficiency are thought to be <0.1 and 0.15 (15%), respectively, which would imply that >50% of primary production passes through the DOM pool.

	Growth efficiency		
BP:PP	0.5	0.20	0.15
0.3	60%	150%	200%
0.2	40%	100 %	133%
0.1	20%	50%	67%

These data add to an ongoing debate about the balance between autotrophy and heterotrophy in the oceans (del Giorgio et al. 1997; Duarte et al. 2001; Williams 1998), although that question is best examined with direct measures of oxygen production and consumption (Serret et al. 2001). I side with Williams and Bowers (1999) and doubt that large oceanic regions can be heterotrophic for long. Regardless, the bacterial production data point to the importance of DOM fluxes.

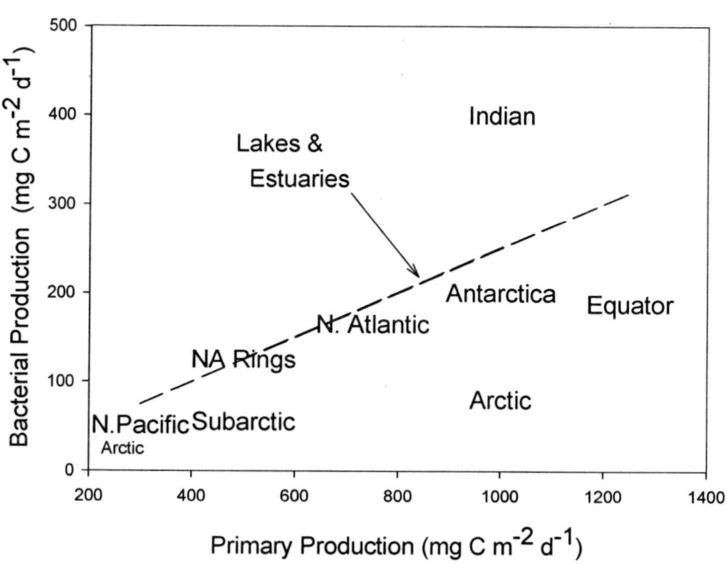

Figure 4. The mean bacterial production and ^{14}C particulate primary production for the indicated oceanic regimes, taken from Ducklow (2000) with two values from the Arctic provided by Rich et al. (1997). The dotted line labeled with "lakes and estuaries" is from Cole et al. (1988)

3. NET PRODUCTION AND EXPORT OF DOC

With few exceptions, net changes in DOC concentrations are not included in estimates of DOM fluxes (production and uptake), at least the fluxes defined above that depend on microbial activity, because net changes are generally small compared to DOM fluxes estimated by other approaches (Table 3). In fact, as a first approximation, the net daily change in total DOC is nil while there is always some bacterial growth and respiration, i.e. a measurable DOM flux. Individual labile compounds, such as free amino acids and sugars (Rich et al. 1996), can vary several fold over a couple days, but the impact of these changes on total DOC is negligible, since concentrations of labile compounds are so small (nanomoles per liter). The implication is that production of DOM is closely balanced by consumption.

Table 3. Net changes in DOC concentrations and DOM fluxes calculated from bacterial production and bacterial growth efficiency assumed to be 0.15. The bottom end of the range for net changes was set at zero because usually the net changes in DOC on a day time scale is below detection. The upper limit was taken as the range of concentration observed over the entire sampling period. The concentration range was then divided by days in the sampling period and multiplied by the euphotic zone to obtain an integrated daily rate comparable to the DOM flux.

Regime	DOM flux	Potential Net Changes in DOC	Reference
	(mmol C m^{-2} d^{-1})		
Equatorial. Pacific	58-158	0-10	Ducklow (2000), Carlson and Ducklow (1995)
Sargasso	39	0-12	Carlson et al. (1994)
Ross Sea	31	0-3	Ducklow et al. (2001), Carlson et al. (2000)

However, occasionally DOM production can greatly exceed consumption over short time scales and even relatively small daily changes in DOC can accumulate. These changes can be a substantial part of the overall carbon budget. In the few oceanic habitats with available data, it is evident that DOC can build up in the upper 100 m during the growing season in spite of continuous bacterial activity (Fig. 3A) (Hansell and Carlson 1998b). The cause of this build up is unknown (see above), but its size can be gauged by comparing it to the simultaneous depletion of nitrate, one measure of net primary production. In most of the oceans with available data, it seems that about 20% of net primary production accumulates as DOC (Table 4). The

Sargasso Sea, where DOC has been examined probably more extensively than any other open ocean, is an interesting case. During the spring bloom, 59-70% of net primary production accumulates as DOC but when calculated for the entire year, this percentage drops to <10% (Hansell and Carlson 1998b; Hansell 2002). These percentages indicate that net DOC accumulation can be substantial relative to net primary production.

Table 4. Net DOC production compared with net community (primary) production (PP). Based on the review by Hansell (2002). Net primary production was estimated from nitrate depletion.

	% of net PP	Reference
Ross Sea	8-20	Carlson et al. (2000)
Equatorial Pacific	20	Archer et al. (1997), Hansell et al. (1997)
Equatorial Atlantic	20	Thomas et al. (1995)
Iberian margin	20	Alvarez-Salgado et al. (2001)
Greenland Sea	50	Noji et al. (1999)
Sargasso Sea		Carlson et al. (1994), Hansell and Carlson (2001)
Spring bloom	59-70	
Entire year	8	

This accumulated DOM is a potential source of carbon and other essential elements (e.g. nitrogen and phosphorus) for bacterial growth in the surface layer. The oft-observed lag between bacterial and primary production can be explained by use of DOM produced and accumulated during previous periods of high primary production (Ducklow et al. 2002). This uncoupling is taken to its extreme by those arguing for large net heterotrophic regions in the oceans. Accumulated DOM, produced in a highly productive location, has been called upon to support excess heterotrophic bacterial activity in other, more oligotrophic locations (del Giorgio et al. 1997). There are two problems with this mechanism. First, to make up the oxygen and carbon deficit of some presumed heterotrophic oceanic regions, accumulated DOC produced in productive areas would have to be carried away by surface currents larger than the Gulf Stream (Williams and Bowers 1999), currents that have escaped the attention of physical oceanographers to date. Another problem is that the DOM has to be refractory enough to accumulate and survive during transport but then it is has to be labile enough to support substantial heterotrophic bacterial activity in the oligotrophic regions. Some DOM has the right degree of lability to be exported both laterally and vertically (see below), but probably not to the extent argued by some investigators.

Another fate of accumulated DOM is export from the surface layer to the mesopelagic zone and deep ocean (Hansell and Carlson 2001). As Hansell (2002) points out, DOM export requires biology to produce DOM and physics to move water with its dissolved constituents to deeper layers of the oceans. One could add that in addition to biology and physics, chemistry is needed to make DOC refractory enough to survive immediate degradation by heterotrophic bacteria and to accumulate in the surface layer. Substantial export of DOC is known to occur with the formation of North Atlantic Deep Water (NADW) (Hansell and Carlson 1998a) and North Pacific Intermediate Water (Hansell et al. 2002).

One way to estimate DOC export is to examine the relationship between apparent oxygen utilization (AOU) and DOC concentrations in various density strata (e.g. Fig. 5). The extent to which the variation in AOU is explained by DOC concentrations is a measure of how much oxygen use (total respiration) is fueled by DOC. Since little DOC is produced below the euphotic zone, DOC used by respiration must have been exported from the surface layer. The approach cannot be used to examine DOC-supported respiration in the upper 100 m where DOC-AOU correlations are not statistically significant because atmospheric exchange prevents any deficit in oxygen concentrations. At the other extreme, DOC-AOU correlations are again insignificant in the deepest ocean where DOC concentrations are relatively constant. Between those two extremes in depth, however, there are several depth strata with significant DOC-AOU correlations. At these locations, DOC appears to explain 15-47% of AOU, with each oceanic regime varying about two-fold with depth (Table 5). Hansell and Carlson (pers. comm.) estimated that global DOC export to intermediate, deep and bottom water is about 9.8 $\times 10^{12}$ mol C y^{-1} or about 10% of total export (oxygen utilization) to depths >500 m. Export to shallower waters is higher, as summarized in Table 5.

Table 5. DOC export, estimated from the correlation between DOC concentrations and apparent oxygen utilization (AOU). Based on the review by Hansell (2002).

Region	% AOU explained By DOC	Depth (m)	Reference
North Pacific Gyre	40	75-125	Emerson et al. (1997)
South Pacific	21-47	<500	Doval and Hansell (2000)
Indian Ocean	18-43	<500	Doval and Hansell (2000)
Gulf of Mexico	20-30*	0-250	Guo et al. (1994)
Western North Pacific	33	400-1000	Hansell et al. (2003)

*Upper limit, as the covariance between AOU and DOC was not corrected for mixing effects.

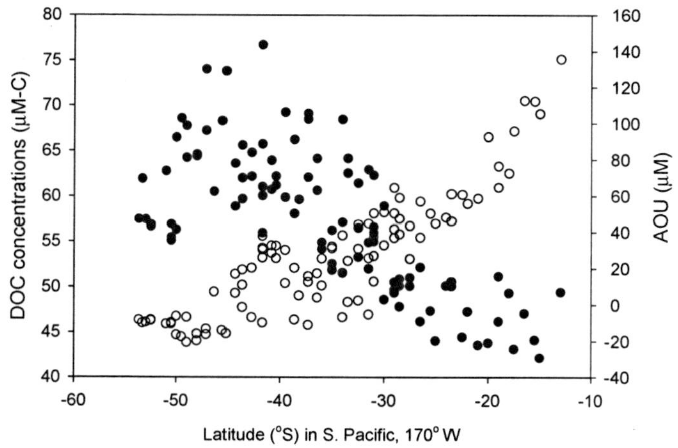

Figure 5. Example of data on DOC and apparent oxygen demand (AOU) at a density of ı ·l =26-26.5. Modified from Doval and Hansell (2000). The filled-in circles are DOC concentrations whereas the open circles are AOU.

4. CARBON STORAGE AND DOM UPTAKE IN THE DEEP OCEAN

A major issue in climate change studies is to understand the mechanisms by which carbon can be sequestered out of contact with the atmosphere on various time scales. DOC can be viewed as one such sequestration mechanism. Because of its concentration and the volume of the ocean, oceanic DOC is a major store of carbon in the biosphere. The largest fraction, which is also the most refractory, would seem not to impact atmospheric CO_2 on time scales <1000 years. The model of Anderson and Williams (1999) indicates that changes in refractory DOC can be ignored when examining global change over 200 years. However, any changes in refractory DOC, especially in the deepest depths where most of the oceanic volume lies, could have large impacts on atmospheric CO_2 on longer time scales.

It would be extremely difficult to directly measure changes over time in deep oceanic DOC, given that its turnover exceeds 1000 years. Indeed, Hansell (2002) points out that there is no evidence that deep oceanic DOC changes over time, at least during six years of work in the Sargasso Sea and over a year of study in the Arabian Sea. But DOC concentrations do appear to decrease along the conveyor belt of deep oceanic currents that cover the world's ocean (Hansell and Carlson 1998). DOC is 43-48 μM-C in deep waters of the North Atlantic (1000-4800 m), near sites of North Atlantic Deep Water (NADW) formation, and it decreases to 34-39 μM-C at 900-5000 m in the North Pacific Ocean. This deep North Pacific Water is NADW with some influence from Antarctic Bottom Water. The drop in DOC parallels the increase in concentrations of inorganic nutrients such as phosphate along the conveyor belt.

Hansell (2002) points out that we do not know if the decrease in deep oceanic DOC is due to microbial activity or to some abiotic mechanism, such as the scavenging of DOC on sinking particles. He reasoned that the amount of useable DOC is too small and the time frame too long to select for much bacterial activity. However, Williams (2000) argued that it is energetically worthwhile for a microbe to attempt to take up even a single glucose molecule dissolved in several oceans. In fact, there is enough microbial activity at depth to explain the gradient in deep oceanic DOC. If we take the microbial activity estimates (8.33×10^{-5} μmol-C liter^{-1} d^{-1}) of Nagata et al. (2000) and assume deep oceanic DOC to be 40 μM-C, the implied deep oceanic DOC turn over is on the order of 1300 years, faster than the turnover estimated from ^{14}C-dating. The model of Bendtsen et al. (2002) indicates that the observed gradient in DOC in the deep North Atlantic can be explained by temperature-dependent bacterial activity.

Bacterial activity is high in the deep oceans relative to the vertical flux of particulate organic carbon, even though deep sea activity is 100-fold (or more) lower than rates in the surface layer. Measured rates of bacterial production (mainly leucine incorporation) and assumed growth efficiencies can be used to estimate the amount of carbon used by bacteria in the mesopelagic zone of the oceans. In the North Pacific gyre (Cho and Azam 1988), other North Pacific regimes (Nagata et al. 2000; Simon et al. 1992) and the Arabian Sea (Ducklow 1993), the estimated carbon use by heterotrophic bacteria is a large fraction (30-100%) of total organic carbon supplied by sinking particles as measured by sediment traps. The numbers and activity of truly attached bacteria are low at all depths in the oceans, so this carbon mineralization in deep waters is by unattached bacteria using labile DOM. The DOC fueling deep bacteria could be from DOM exported directly from the surface layer, a possibility that has not been examined thoroughly. More likely, sinking particles disintegrate to DOM because of abiotic hydrolysis and microbial activity (Smith et al. 1992). We know little about this process in spite of its obvious importance for understanding carbon fluxes.

We need more work on DOC and microbial activity in the deep ocean. Williams (2000) pointed out that most studies of microbial activity are restricted to the upper 100 m, probably because of the emphasis on primary production, leaving the vast interior of the ocean grossly undersampled.

Unlike deep DOC, it is possible to measure changes in surface layer DOC. Church et al. (2002) reported that surface layer DOC concentrations in the North Pacific Gyre increased at a rate of 0.3 mol- C m^{-2} y^{-1} from 1993-1999. While this DOC accumulation is only about 2% of primary production, it is 15% of total organic carbon export in the North Pacific, given that export is about 2 mmol-C m^{-2} y^{-1} in this region (Emerson et al. 1997). Concentrations of dissolved organic nitrogen (DON) also increased, but dissolved organic phosphorus did not (Church et al. 2002). The implications of these changes are still not clear, but the changes seem large enough to be considered in carbon budgets and models.

5. WHAT MECHANISMS PRODUCE LABILE AND REFRACTORY DOM?

The mechanisms producing labile DOM probably differ greatly from those producing refractory DOM, and as a first approximation we can consider the two DOM pathways to be completely separate. We know the most about how labile DOM is produced.

5.1 LABILE DOM

While only one microbial group, heterotrophic bacteria (and perhaps archaea), are the main users of DOM, several organisms and processes potentially produce DOM. Direct release or excretion by phytoplankton has been examined more extensively than any other DOM production mechanism, in part because of the ease of using ^{14}C to trace carbon from the inorganic carbon pool through phytoplankton and into DOM and heterotrophic bacteria. There are many generalizations that are often quoted (e.g. DOC is released by nutrient stress nearing the end of phytoplankton blooms) but with conflicting supporting data at best (Williams 1990). In a review of the topic, Nagata (2000) concludes that phytoplankton release only about 12% of ^{14}C-primary production as labile DOM (Fig. 6). The ^{14}C-based studies used by Nagata (2000) for his 12% estimate may have underestimated release of polysaccharides and other polymers normally associated with cell walls or bound to the outside of phytoplankton (and perhaps bacteria), because these polymers would not be labeled adequately in the typical short incubation with ^{14}C. Polysaccharides are large DOM constituents (Pakulski and Benner 1994), which can build up over the growing season (Kirchman et al. 2001). In any event, the 12% estimate seems a reasonable average for direct release by phytoplankton.

Nagata (2000) then assumes that total DOM uptake and release account for 49% of primary production, implying that direct phytoplankton release is roughly 20% of total DOM production (10%/50%). The contribution by viral lysis is set at 5% of primary production or 20% of total DOM production, the most arbitrary and least-constrained part of Nagata's budget. The remaining 65% of total DOM production has to be due to release of DOM by grazers, an estimate that is supported by direct measurements (Strom et al. 1997). Nagata (2000) estimates that protist grazers account for most of this release (53% of total DOM production) with DOM release by mesozooplankton contributing the last 6%, reflecting the fact that protists graze more of phytoplankton production than do mesozooplankton (by nearly 20-fold in Nagata's budget) in the open ocean, which again is supported by direct measurements of grazing rates (Landry et al. 1995).

That grazers, not phytoplankton, are the main source of DOM has many implications for understanding carbon cycles and trophic dynamics. It could explain why bacterial production often is not correlated with primary production and more specifically why bacterial activity often lags behind phytoplankton. Whether grazers or phytoplankton (or viruses) produce the most DOM is likely to impact virtually all biogeochemical properties of DOM. For example, we may expect DOM from grazers to be partially degraded and thus less labile than DOM directly from phytoplankton.

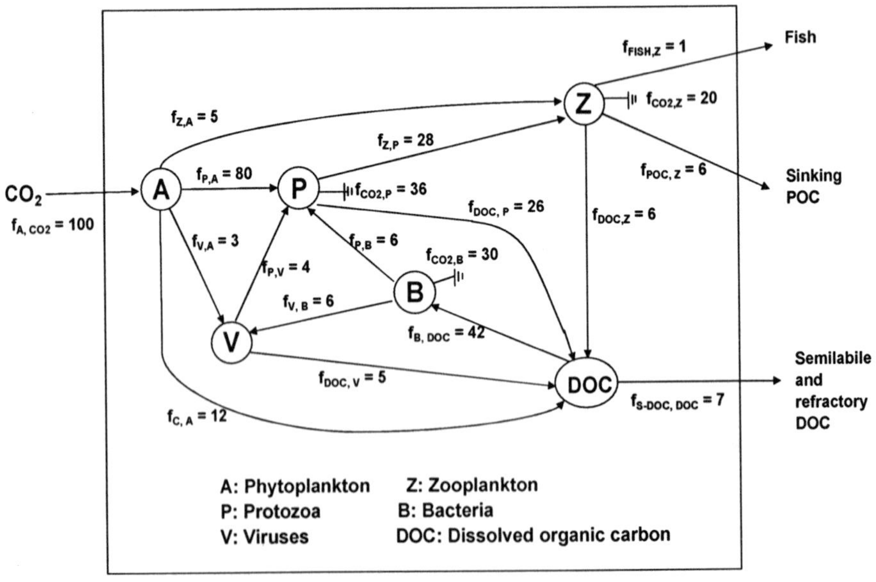

Figure 6. A budget of carbon compartments and flow used to account for DOC production. The symbols in the form of $f_{x,y}$ denote the flux from compartment y to compartment x. All fluxes are normalized to primary production ($f_{A,CO2}$) or the flux from CO_2 to phytoplankton (A). Modified from Nagata (2000).

5.2 REFRACTORY DOM

We know little about how refractory DOM is produced in the oceans. ^{13}C data indicate that it does not come from terrestrial sources, and the concentrations of known terrestrial compounds, the most important one being lignin, are very low in oceanic waters (Benner 2002). The implication is that organic material originally produced by phytoplankton is somehow transformed to refractory DOM. What complicates matters is that only a very small fraction (<1%) of this phytoplankton-produced organic carbon needs to be transformed to refractory DOM; the vast majority is mineralized back to CO_2 and inorganic nutrients mostly in the upper layer where organic material is produced.

Refractory DOM is often said be "humic material", formed by the abiotic condensation of labile, low molecular weight (LMW) DOM. Borrowing from studies on soil formation, humic material is operationally defined as the organic material that can be isolated by sorption onto XAD resins after acidification. While the same term is used for material isolated by this method for soils and seawater, it has been clear since the initial studies that seawater humic substances are not the same as those compounds isolated from soils. Regardless, the classic condensation hypothesis did seem to explain the basic dilemma of turning labile into refractory DOM; compounds synthesized abiotically are generally less available to enzymatic hydrolysis than organic material synthesized by enzymes. The hypothesis, however, always suffered from the problem that known condensation reactions required higher concentrations than observed for LMW DOM. More critically, the classic condensation model leads to predictions that are not supported by the data.

One prediction of the condensation hypothesis is that high molecular weight (HMW) DOM, if resulting from abiotic condensation of labile LMW precursors, should be more refractory than LMW DOM. In fact, the opposite seems to be the case. HMW DOM (>1000 Da) isolated by ultrafiltration is used much more quickly than its LMW DOM counterpart (Amon and Benner 1994). It remains true that biopolymers larger than about 500 Da must be hydrolyzed by ectoenzymes before the LMW byproducts can be transported into the cell and catabolized. However, it is quite misleading to assume that LMW byproducts of biopolymer hydrolysis are representative of LMW DOM, since concentrations of byproducts such as free amino acids and sugars are very low and are a trivial fraction of LMW DOM found in seawater. In contrast, biopolymers such as polysaccharides make up a large fraction of HMW DOM (Benner 2002), which is one big reason why it is more labile than LMW DOM. Benner (2002) summarizes the data indicating the LMW DOM is more refractory than HMW DOM.

Another prediction of the condensation hypothesis is that the biochemical composition of LMW DOM should be most similar to that of organisms, and HMW DOM should be composed of more unknown compounds. In fact, again the opposite is true. There is a general decrease in the fraction of material that can be identified as typical biochemicals along the size gradient from particulate material (plankton) to LMW DOM. About 85% of the organic carbon in plankton can be chemically-identified at the molecular level whereas this percentage decreases to 15% and 4% for HMW and LMW DOM, respectively (calculated from data presented by Benner 2002). So, the most refractory material is in the smallest molecular size fraction that contains the lowest percentage of known biochemicals. There appears to be a correlation between lability and the fraction of organic material identifiable at the molecular level, and perhaps this is no mere coincidence. As compounds move out of the analytical windows of our chemical methods, they also seem to become less accessible to enzymatic attack. Microbial enzymes and our analytical methods both seem focused on the compounds directly synthesized by organisms.

Another set of studies using NMR techniques have shed some light on the composition of HMW DOM (reviewed by Benner 2002). In brief, these studies indicated that the chemical bonds found in HMW DOM are similar to those of the plankton and are not those expected from condensation reactions. In particular, McCarthy et al. (1997) reported ^{15}N-NMR results indicating that most of the nitrogen in HMW DOM is in amide bonds, which make up common biopolymers, including protein, peptidoglycan (found in bacterial cell wall) and chitin (common in invertebrates and selected algae). The structural compounds, peptidoglycan and chitin, are most likely the major components of HMW DOM, because they are thought to be more resistant to microbial attack than protein (but see below). These results do not rule out all abiotic reactions, but they do indicate that HMW DOM is not completely from the condensation or coagulation of LMW DOM.

It has been implied that phytoplankton are the source for the refractory DOM. Several lines of evidence indicate, however, that refractory compounds in fact come from bacteria. Tanoue et al. (1995) reported one of the first sightings of a bacterial biomarker, a membrane protein (porin) unique to bacteria, in the HMW fraction of both surface and deep DOM. Subsequently, McCarthy et al. (1998) found more evidence, high amounts of D-amino acids, that bacterial cell wall remains make up a large fraction of HMW DOM. Abiotic racimerization or the switching of L-amino acids (the form found in proteins) to the D form is too slow to explain the high D/L ratios measured by McCarthy et al. (1998). The most likely source of the D-amino acids is the peptide bridge found in peptidoglycan, a biopolymer synthesized only by bacteria. These results are consistent with the ^{15}N-NMR results (McCarthy et al. 1997), suggesting peptidoglycan as one possible source of the amide bonds found in HMW DOM.

More recent data complicate the peptidoglycan story. First, concentrations of the amino sugar, muramic acid, which is unique to peptidoglycan, are not as high as one may expect if peptidoglycan were common in the HMW DOM fraction (Benner and Kaiser 2003). Second, turnover of peptidoglycan does not appear to be any slower than protein turnover, at least in short term incubations (Nagata et al. 2003). However, the experiments of Nagata et al. (2003) revealed one crucial difference between peptidoglycan and protein degradation. Unlike protein degradation, LMW byproducts were released during the degradation of peptidoglycan. What is even more interesting is that the composition of these byproducts could not be identified by the same chemical methods that successfully characterized the starting material. In short, it seems that material reminiscent of refractory DOM was produced during peptidoglycan degradation (Nagata et al 2003).

While the precise precursor may still be in doubt, it is clear that bacteria can be a source, if not the sole mechanism for producing refractory DOM. The role of heterotrophic bacteria was clearly demonstrated by Ogawa et al. (2001) who fed glucose to a bacterial size fraction and measured glucose and DOC concentrations over time. They found that within two days glucose dropped to unmeasurable levels while concentrations of DOC, which initially was nearly all glucose, also decreased, but less than expected, considering how much glucose had disappeared. The difference between the expected and the actual decrease in DOC concentrations indicates the formation of unknown organic material in the DOM pool; after the glucose disappeared, <15% of the DOM could be identified. Ogawa et al. (2001) calculated turnover times exceeding a year for this unknown DOM.

The mechanism by which glucose was somehow transformed into refractory DOM is not clear. Ogawa et al. (2001) recognized that viruses and protists may have been involved, but the authors favor another mechanism: nonspecific activity by bacterial enzymes produce compounds which escape further bacterial attack and detection by standard chemical analysis. Regardless of the precise mechanism, this experiment rules out several mechanisms and indicates that light, phytoplankton, zooplankton and sediments are all unnecessary in the production of refractory DOM.

Perhaps most important, the experiment indicates that refractory DOM can be produced quite quickly; organic compounds with turnover times exceeding a year were produced within a couple days. Similarly, Norrman et al. (1995) observed production of refractory DOC over a few days during a phytoplankton bloom in a mesocosm. In both cases, the production rates are undoubtedly abnormally high because of high DOC concentrations (glucose in the experiments of Ogawa et al. (2001)) or of excessive phytoplankton activity (in the eutrophic mesocosm of Norrman et al. (1995)). The oceans would quickly fill up with refractory DOM if produced at the rate suggested by these two studies. Even so, these data suggest that refractory DOM can be produced on much shorter time scales than it is degraded. This apparent uncoupling between production and mineralization of refractory DOM seems quite important in thinking about DOM-microbe interactions and in including DOC in carbon budgets.

6. CHARACTERIZING BACTERIA COMMUNITIES IN THE OCEAN

Just as the DOM pool is a complex mixture of several compounds that differ greatly in biogeochemically-relevant properties, so too is the prokaryotic community a complex assemblage of many organisms that potentially differ in their biogeochemical roles. Again analogous to DOM, we still do not know much about the taxonomic composition ("community structure") of prokaryotic communities in the oceans, in spite of extensive work over the last 15 years. The fundamental problem is, few marine microbes can be isolated and cultured in the lab, which is the first step in traditional methods for identifying microbes. Consequently, microbial ecologists have to turn to molecular methods and other techniques for deducing the role of uncultured microbes in biogeochemical processes such as the uptake and production of DOM.

Giovannoni and Rappé (2000) and Kirchman (2002) provide more complete reviews than possible here about microbial community structure in the oceans. This chapter will discuss some observations that I think are most relevant to thinking about microbial processes in carbon cycle models.

Starting with the highest phylogenetic level, prokaryotes consist of two domains, bacteria and archaea, which share some morphological traits (they cannot be distinguished with simple microscopy) but differ in several other ways. The first archaea to be characterized were extremophiles (e.g. hyperthermophiles and halophiles) able to survive and grow under extreme conditions using unique metabolisms, such as the production of methane (methanogens). Recent work, however, indicates that most marine archaea are not extremophiles and are not just in environments with high temperatures, pH or salinity. They can be found virtually in all marine habitats, but archaea appear to be particularly abundant in the deep ocean (Karner et al. 2001). The number of crenarchaeota, one of the three main groups of archaea, is about equal to bacterial abundance in the North Pacific below 1000 m (Karner et al. 2001). Another main archaeal group, euryarchaeota, is relatively rare throughout the water column (Karner et al. 2001). While archaea are abundant in the deep ocean, bacteria still greatly outnumber archaea in the surface layer.

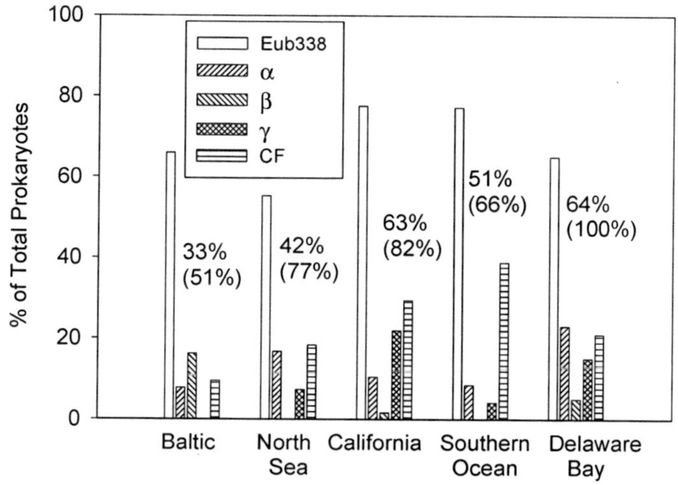

Figure 7. The major bacterial groups in selected oceanic regimes and the Delaware Bay. The oceanic data are summarized by Kirchman (2002) and the Delaware Bay data are from Cottrell and Kirchman (2003). The first percentage is the fraction of total bacterial abundance (DAPI direct counts) accounted for by the four major groups. The percentage in parenthesis is the fraction of Eub338-positive microbes (Eub338 targets all bacteria) accounted for by the four groups.

We know little about the metabolic capacities of these marine archaea and whether they differ greatly from bacteria. Marine archaea do appear capable of using DOM (Ouverney and Fuhrman 2000), but much more work on this topic is needed.

Of the over 50 divisions or phyla of bacteria found in the biosphere (Rappé and Giovannoni 2003), only three are usually abundant in surface waters of the oceans (Giovannoni and Rappé 2000). One of these is the cyanobacterial group consisting of three genera: *Synechococcus*, *Prochlorococcus*, and *Trichodesmium*. The other two divisions are the proteobacteria and the Bacteriodetes. The latter division includes *Cytophaga-Flavobacterium*. Proteobacteria include several subdivisions of which two, alpha- and gamma-proteobacteria, are most abundant in the oxic waters of the oceans (Fig. 7). Beta-proteobacteria are common in lakes and freshwater reaches of estuaries, but are usually <10% of abundance in marine waters. Epsilon-proteobacteria, which include sulfate reducers, are abundant only in anoxic habitats. Usually *Cytophaga-Flavobacterium* and alpha-proteobacteria are the two most abundant bacterial groups in marine surface waters (Kirchman 2002), as determined by one molecular approach, fluorescence in situ hybridization (FISH) with oligonucleotide probes. We are still in the early days of describing the variation in these bacterial groups and understanding what factors determine this variation.

The simplicity implied by stating that the oceans are dominated by a few bacterial groups is misleading. Each of these groups contains many different bacteria with potentially different biogeochemical roles. As measured by 16S rRNA sequences, the phylogenetic marker of choice, the difference within a proteobacterial subdivision or the *Cytophaga-Flavobacterium* cluster is greater than the difference among the common phytoplankton groups, such as diatoms and coccolithophorids or among common zooplankton, such as copepods and euphausids. The number of bacterial genera and species in the ocean is also not clear, nor is it obvious whether bacterial communities are dominated by a few species or are evenly divided among several species. A recent study using fluorescence in situ hybridization (FISH) suggests that one alpha proteobacterial group, SAR11, is abundant and ubiquitous in the oceans (Morris et al. 2002), but even this group is quite diverse (Giovannoni and Rappé 2000). And there is even more diversity at even finer phylogenetic levels than indicated by commonly used 16S rRNA gene sequences. It is well known that 16S rRNA is not a good phylogenetic marker for closely related organisms. For example, different ecotypes of the abundant cyanobacterial genus *Prochlorococcus* are known to inhabit different depths and light regimes in oligotrophic oceans, but these ecotypes have virtually the same 16S rRNA signature (<3% difference) (Moore et al. 1998). In short, while only a few of the many bacterial divisions are found in the oceans, the diversity within bacterial divisions is incredibly high with unclear implications for understanding the role of bacteria in biogeochemical cycles. The different levels of bacterial phylogeny are summarized in Figure 8.

I think the critical question microbial ecologists have to answer is, what phylogenetic level is most appropriate for describing bacterial activity in the oceans and other aquatic environments? The answer probably varies with the particular problem being addressed. Consider two extremes. Microbial ecologists have long discussed DOM uptake by "bacteria", i.e. the action of the entire bacterial domain (not to mention archaea), without apparently needing to know which bacterial group was doing what. At the other extreme, we know that some strains of a bacterial species are pathogenic while others are not; many strains of *E.coli* are common, harmless residents of the human gut microflora whereas others cause dysentery and even death.

It remains to be seen whether we need to consider finer phylogenetic levels (the extreme being strains) in describing DOM uptake.

Phylogenetic Level: Example

Domain (3): Bacteria

Division (>50): Proteobacteria

Subdivision (5): alpha proteobacteria

Clades (45): Sar11

"Species" (??): ??

Figure 8. Summary of the major phylogenetic levels used to organize bacterial diversity and an example for each level. The arrow indicates the progression from the broadest level to the finest. "Species" is not a well defined concept in bacterial taxonomy. The numbers in parenthesis refer to subgroups at each level. There are three domains of which Bacteria is one. There are about 52 subdivisions of Bacteria of which Proteobacteria is one, and so on. Based on Giovannoni and Rappé (2000) and Rappé and Giovannoni (2003).

7. UPTAKE OF SELECT DOM COMPONENTS BY AQUATIC BACTERIAL GROUPS

Current models of carbon cycles in the ocean depict bacteria by a single compartment (Fasham et al. 1999)–if bacteria are included at all. The implied assumption of a single compartment is that all bacteria are using all DOM components equally. Unfortunately, there are few data available to test this assumption. Laboratory-bound microbiologists would immediately point to the wealth of data showing how different bacterial species differ in using organic compounds, and some of this information is useful for understanding natural bacterial assemblages. However, it is quite difficult to apply information about cultured bacteria to the oceans because so few marine bacteria can be cultured and because those that can be cultured differ greatly from those that cannot (Giovannoni and Rappé 2000). Furthermore, even if we could culture the major oceanic bacteria, it would be difficult to extrapolate from laboratory experiments to the ocean because of the difficulties in reproducing oceanic conditions in the lab.

Only a few studies provide data that bear on the question of whether all uncultured bacteria take up DOM equally. Cottrell and Kirchman (2000a) combined microautoradiography and FISH (Micro-FISH) to examine uptake of selected ^3H-labeled organic compounds by bacteria identifiable by FISH probes for the proteobacterial subdivisions and the *Cytophaga-Flavobacterium*. Cottrell and Kirchman (2000a) found that not even free amino acids were taken up equally by the bacterial groups they examined. Their data suggest that the alpha-proteobacteria seem to dominate use of LMW DOM (free amino acids and N-acetylglucosamine) whereas *Cytophaga-Flavobacterium* dominated uptake of HMW organic compounds (protein and chitin). This proficiency in using biopolymers is consistent with laboratory experiments with *Cytophaga-Flavobacterium* (Kirchman 2002).

Other approaches can be used to examine DOM uptake by specific bacterial groups or at least to make inferences about specific DOM-bacteria interactions. One approach is to follow the incorporation of radiolabeled or stable isotopically-labeled compounds into taxon-specific RNA or phospholipids (Boschker et al. 1998). Another approach is to follow bacterial community structure following the addition of various DOM components (e.g. Carlson et al. 2002); the increase of a bacterial group in response to the addition implies, although does not necessarily prove, that these bacteria are keying in on the added organic carbon. Covert and Moran (2001) isolated the LMW and HMW fractions of DOM, added back an inoculum of a natural bacterial assemblage and examined with a DNA fingerprint method the bacterial communities that grew up in each incubation. The bacterial groups that grew on these DOM fractions differed from those expected from the results of Cottrell and Kirchman (2000a), for many possible reasons. The two studies used different techniques to examine the microbes in two different locations. Also, the compounds used by Cottrell and Kirchman (2000) may not model the LMW and HMW material examined by Covert and Moran (2001). Perhaps even more problematically, Covert and Moran (2001) found that a second experiment did not yield the same communities as in the first. Differences in starting LMW and HMW DOM could explain this lack of consistency between experiments. More troubling, the results may indicate the need to look at finer phylogenetic levels than examined by Covert and Moran (2000) and certainly by Cottrell and Kirchman (2000a).

8. NEW TYPES OF BACTERIAL METABOLISMS– EVEN MORE BACTERIAL COMPONENTS FOR MODELS?

Many other members of the bacterial community are well known to participate in several biogeochemical processes important in the oceans

(Capone 2002). One example worth mentioning is the group of chemoautotrophic microbes that oxidize ammonium eventually to nitrate (nitrification). Much less is known about two new types of microbes that are potentially involved in uptake, mineralization, and perhaps production of DOM. Both types harvest light energy while using DOM to supplement their energy and carbon needs, as illustrated in Figure 9.

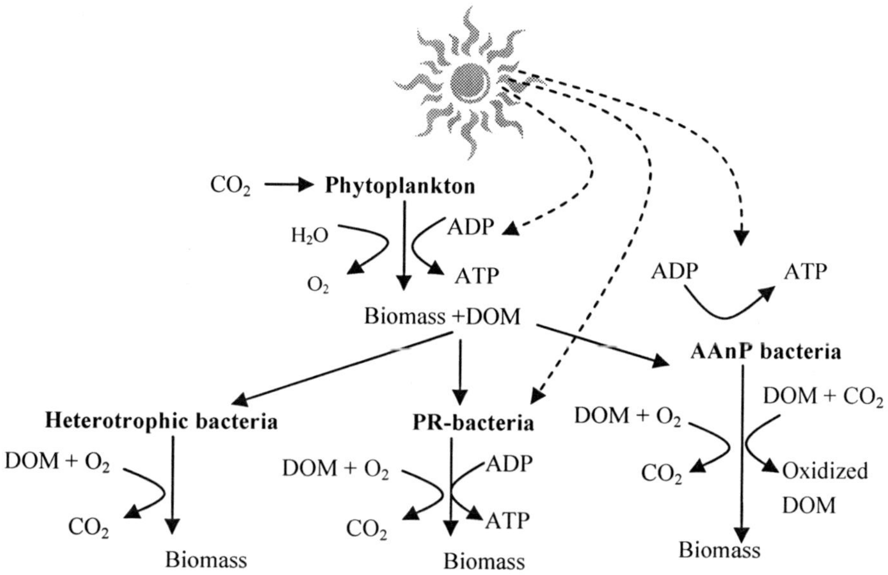

Figure 9. A schematic diagram illustrating the use of light by microbes, including two newly discovered bacterial groups, aerobic anoxygenic photosynthesizing (AAnP) bacteria and bacteria containing proteorhodopson (PR). Heterotrophic bacteria also synthesize ATP during the oxidation of DOM, which is not depicted in this diagram. Based on a more detailed diagram in Karl (2002).

One new microbial group, aerobic anoxygenic photosynthesizing (AAP) bacteria, is really not new in some sense, since culturable representatives had been isolated from sediments and the water column from a couple marine habitiats (Yurkov and Beatty 1998). In culture, these microbes use bacteriochlorophyll a (bchl a) to harvest light energy to carry out CO_2 fixation, analogous to oxygenic photosynthesizing cyanobacteria and eukaryotic algae and plants. Unlike oxygenic photosynthetic organisms, however, AAP bacteria do not use H_2O as the reductant for the synthesis of NADH needed for reducing CO_2; hence AAP bacteria do not evolve O_2 (they are anoxygenic) even though they tolerate aerobic conditions and even require O_2 (they are aerobic). Bacteria belonging to the alpha-proteobacterial subdivision with cultured AAP representatives had been detected in the oceans (Giovannoni et al. 1990), but initial attempts to find bchl were unsuccessful, suggesting that oceanic AAP was insignificant.

More recently, Kolber et al. (2001) and Kolber et al. (2000) used biophysical techniques to detect AAP activity in the surface ocean and isolated AAP bacteria from a variety of oceanic habitats. The isolates were all closely related to *Erythrobacter*, a subgroup of alpha-proteobacteria. Using a molecular approach Beja et al. (2002) found *puf* genes, which direct the synthesis of the AAP light-harvesting complex, in uncultured marine bacteria more closely related to beta-proteobacteria than to alpha-proteobacteria. The reductant for these AAP bacteria remains unknown, but DOM is one possibility (Yurkov and Beatty 1998).

The other new microbial group, proteorhodopson-containing bacteria, is also not so new. Beja et al. (2000) completely sequenced a 130,000 base clone from a library constructed with DNA isolated from uncultured bacteria taken from the Pacific Ocean; they concentrated on this clone because screening for 16S rRNA genes indicated that it contained DNA from the SAR86 subgroup of gamma-proteobacteria. SAR86 had been found previously (Giovannoni et al. 1990), but more sequence data than just from the 16S rRNA gene were needed to indicate the presence of proteorhodopson. Rhodopson is a well characterized membrane protein found in the eyes of animals and in extreme halophilic archaea. In archaea, rhodopson can function as chloride ion pumps, photo-receptors, or light-driven proton pumps, the latter being capable of driving ATP synthesis. Likewise, proteorhodopson is thought to enable SAR86 bacteria to use light energy to synthesize ATP. Prior to the work of Beja et al. (2000), rhodopson had not been found in bacteria. In addition to the gene, Beja et al. (2001) found evidence of active proteorhodopson in surface seawater samples and suggested that the light absorption properties of proteorhodopson vary among oceanic habitats in response to the light environment.

The biogeochemical significance of these two new phototrophic microbes is still unclear. CO_2 fixation by AAP is probably much less than 10% of primary production (Kolber et al. 2001) and bchl a is <2% of chlorophyll a present in oxygenic primary producers (Goericke 2002). Kolber et al. (2001) estimated the abundance of AAP bacteria to be about 10% of total prokaryotic abundance, based on biophysical data, although molecular analysis indicates that the number could be much higher. Using FISH, Beja et al. (2001) found SAR86 to also comprise about 10% of total prokaryotic abundance in the Pacific Ocean, although Cottrell and Kirchman (2000b) were not able to detect it in coastal waters off California.

We need more information about both microbial groups because they have radically new, to the euphotic zone of the ocean, types of metabolisms. CO_2 fixation by AAP is at most a small fraction of total primary production, and light harvesting by both groups is probably trivial compared to that of oxygenic phototrophic cyanobacteria and eukaryotic phytoplankton. What is potentially much more significant is their use of DOM. For starters, it is possible that our current methods for measuring heterotrophic bacterial production (thymidine and leucine incorporation) do not include these two new metabolic types, if for no other reason than bacterial production is measured in dark incubations. All other measures of DOM use are also done in dark incubations. DOM uptake cannot be greatly underestimated by our current methods, because that possibility has already been discounted by analyses of bacterial production and growth efficiencies (Ducklow et al. 2002). Still, at least one form of DOM use by AAP bacteria may not be captured by current measurements and budgets. To obtain reducing power for CO_2 reduction, these bacteria may partially oxidize DOM, without carbon assimilation, oxygen consumption or CO_2 production, and release a more oxidized, perhaps more refractory organic compound into the DOM pool. In short, our concepts about many aspects of DOM processes may be affected if these light-dependent microbes are abundant and active in the oceans.

9. SUMMARY

The community structure of marine bacterial and archaeal assemblages remains poorly understood in spite of much work over the past 15 years. In addition to more widespread use of current molecular methods, the application of novel techniques in genomics should greatly help us understand these uncultured microbes and reveal new types of microbes, as already demonstrated by the work on AAP and proteorhodopsin. It is now technically possible (although expensive) to completely sequence the microbial assemblage from a water sample and to reconstruct the genome of each microbe. Whole genome sequences from an increasing number of microbes in pure cultures will also help field-orientated microbial ecologists as will other genomic-based technology.

Once armed with this wealth of molecular data, microbial ecologists interested in DOM face two challenges. The first challenge is to use the molecular data to understand DOM uptake and in particular the detailed interactions between microbes and DOM. While molecular microbial ecologists have been describing the composition of microbial communities, molecular organic geochemists have been elucidating the composition of DOM and the mechanisms by which it is produced. Combining data from these two fields will not be easy but it will undoubtedly yield many insights into microbe-DOM interactions. As these interactions become clear, the second challenge will become more important to face: how do we use this information to better understand carbon cycles and other biogeochemical processes at basin and global scales?

ACKNOWLEDGMENTS

Hugh Ducklow, Tom Anderson, Craig Carlson, and Dennis Hansell critically read an early draft of this paper and provided many useful comments. Carlson and Hansell generously provided their published data for constructing some of the figures presented here. I also thank the organizers of this symposium, Mick Follows and Temel Oguz. DLK was supported by NSF and DOE during preparation of this paper.

REFERENCES

Alvarez-Salgado, X. A., J. Gago, B. M. Miguez, and F. F. Perez. 2001. Net ecosystem production of dissolved organic carbon in a coastal upwelling system: the Ria de Vigo, Iberian margin of the North Atlantic. Limnol. Oceanogr. 46:135-147.

Amon, R. M. W. and R. Benner. 1994. Rapid cycling of high-molecular-weight dissolved organic matter in the ocean. Nature 369:549-552.

Amon, R. M. W. and R. Benner. 1996. Photochemical and microbial consumption of dissolved organic carbon and dissolved oxygen in the Amazon River system. 60:1783 - 1792.

Anderson, T. R. 2002. The magnitude of spring bacterial production in the North Atlantic Ocean. Limnol. Oceanogr. 47: 1684-1693.

Anderson, T. R. and H. W. Ducklow. 2001. Microbial loop carbon cycling in ocean environments studied using a simple steady-state model. Aquat. Microb. Ecol. 26: 37-49.

Archer, D., E.T. Peltzer and D. L. Kirchman. 1997. A timescale for dissolved organic carbon production in equatorial Pacific surface waters. Global Biogeochem. Cycles 11:435-452.

Beja, O., L. Aravind, E. V. Koonin, M. T. Suzuki, A. Hadd, L. P. Nguyen, S. Jovanovich, C. M. Gates, R. A. Feldman, J. L. Spudich, E. N. Spudich and E. F. DeLong. 2000. Bacterial rhodopsin: Evidence for a new type of phototrophy in the sea. Science 289:1902-1906.

Beja, O., E. N. Spudich, J. L. Spudich, M. Leclerc and E. F. DeLong. 2001. Proteorhodopsin phototrophy in the ocean. Nature 411:786-789.

Beja, O., M. T. Suzuki, J. F. Heidelberg, W. C. Nelson, C. M. Preston, T. Hamada, J. A. Eisen, C. M. Fraser and E. F. DeLong. 2002. Unsuspected diversity among marine aerobic anoxygenic phototrophs. Nature 415:630-633.

Bendtsen, J., C. Lundsgaard, M. Middelboe and D. Archer. 2002. Influence of bacterial uptake on deep-ocean dissolved organic carbon. Global Biogeochem. Cycle 16: 1127, doi: 10.1029/2002GB001947.

Benner, R. 2002. Chemical composition and reactivity. In Biogeochemistry of Marine Dissolved Organic Matter. D. A. Hansell and C. A. Carlson (eds.), pgs. 59-90. San Diego Academic Press.

Benner, R. and Kaiser, K. 2003. Abundance of amino sugars and peptidoglycan in marine particulate and dissolved organic matter. Limnol. Oceanogr. 48:118-128.

Boschker, H. T. S., S. C. Nold, P. Wellsbury, D. Bos, W. de Graff, R. Pel, R. J. Parkes and T. E. Cappenberg. 1998. Direct linking of microbial populations to specific by geochemical processes by ^{13}C-labeling of biomarkers. Nature 392:801-805.

Capone, D.G. 2002. The marine microbial nitrogen cycle. In Microbial Ecology of the Oceans, D. L. Kirchman (ed.), pgs. 455-493, New York: Wiley-Liss, Inc.

Carlson, C. A. and H. W. Ducklow. 1995. Dissolved organic-carbon in the upper ocean of the central Equatorial Pacific-Ocean, 1992 - daily and finescale vertical variations. Deep-Sea Res. II 42:639-656.

Carlson, C. A., H. W. Ducklow and A. F. Michaels. 1994. Annual flux of dissolved organic-carbon from the euphotic zone in the northwestern Sargasso Sea. Nature 371:405-408.

Carlson, C.A., D.A. Hansell, E.T. Peltzer and W.O. Smith, Jr. 2000. Stocks and dynamics of dissolved and particulate organic matter in the Southern Ross Sea, Antarctica. Deep-Sea Res. II 47:3201-3225.

Carlson, C.A. 2002. Production and removal processes. In Biogeochemistry of Marine Dissolved Organic Matter, D. A. Hansell and C. A. Carlson (eds.), pgs. 91-151, San Diego: Academic Press.

Carlson, C. A., S.J. Giovannoni, D.A. Hansell, S.J. Goldberg, R. Parsons, M.P. Otero, K. Vergin and B.R. Wheeler. 2002. Effect of nutrient amendments on bacterioplankton production, community structure, and DOC utilization in the northwestern Sargasso Sea. Aquat. Microb. Ecol. 30: 19-36.

Cho, B. C. and F. Azam. 1988. Major role of bacteria in biogeochemical fluxes in the ocean's interior. Nature 332:441-443.

Church, M. J., H. W. Ducklow and D. M. Karl. 2002. Multiyear increases in dissolved organic matter inventories at station ALOHA in the North Pacific Subtropical Gyre. Limnol. Oceanogr. 47:1-10.

Cole, J. J., S. Findlay and M. L. Pace. 1988. Bacterial production in fresh and saltwater ecosystems: a cross-system overview. Mar. Ecol. Prog. Ser. 43:1-10.

Cottrell, M. T. and D. L. Kirchman. 2003. Contribution of major bacterial groups to bacterial biomass production (thymidine and leucine incorporation) in the Delaware Estuary. Limnol. Oceanogr. 48: 168-178.

Cottrell, M. T. and D. L. Kirchman. 2000a. Natural assemblages of marine proteobacteria and members of the *Cytophaga-Flavobacter* cluster consuming low- and high- molecular-weight dissolved organic matter. Appl. Environ. Microbiol. 66:1692-1697.

Cottrell, M.T. and D.L. Kirchman. 2000b. Community composition of marine bacterioplankton determined by 16S rDNA clone libraries and fluorescence in situ hybridization. Appl. Environ. Microbiol. 66:5116-5122.

Covert, J. S. and M. A. Moran. 2001. Molecular characterization of estuarine bacterial communities that use high- and low-molecular weight fractions of dissolved organic carbon. Aquat. Microb. Ecol. 25:127-139.

del Giorgio, P. A., J. J. Cole and A. Cimbleris. 1997. Respiration rates in bacteria exceed phytoplankton production in unproductive aquatic systems. Nature 385:148-151.

del Giorgio, P.A. and J.J. Cole. 2000. Bacterial energetics and growth efficiency. In Microbial Ecology of the Ocean, D. L. Kirchman (ed), pgs. 289-325. New York: Wiley-Liss.

Doval, M. D. and D. A. Hansell. 2000. Organic carbon and apparent oxygen utilization in the western South Pacific and the central Indian Oceans. Mar. Chem. 68:249-264.

Duarte, M. C., S. Agusti, J. Aristegui, N. Gonzalez and R. Anadon. 2001. Evidence for a heterotrophic subtropical northeast Atlantic. Limnol. Oceanogr. 46:425-428.

Ducklow, H. W. 1993. Bacterioplankton distributions and production in the northwestern Indian Ocean and Gulf of Oman, September 1986. Deep-Sea Res. II 40: 753-771.

Ducklow, H. 2000. Bacterial production and biomass in the oceans. In Microbial Ecology of The Oceans, D. L. Kirchman (ed.), pgs. 85-120, New York: John Wiley & Sons.

Ducklow, H., C. Carlson, M. Church, D. Kirchman, D. Smith and G. Steward. 2001. The seasonal development of the bacterioplankton bloom in the Ross Sea, Antarctica 1994-1997. Deep-Sea Res. II. 48:4199-4221.

Ducklow, H. W., D. L. Kirchman and T. R. Anderson. 2002. The magnitude of spring bacterial production in the North Atlantic Ocean. Limnol. Oceanogr. 47: 1684-1693.

Emerson, S., P. Quay, D. Karl, C. Winn, L. Tupas and M. Landry. 1997. Experimental determination of the organic carbon flux from open-ocean surface waters. Nature 389:951-954.

Fasham, M. J. R., P. W. Boyd and G. Savidge. 1999. Modeling the relative contributions of autotrophs and heterotrophs to carbon flow at a Lagrangian JGOFS station in the Northeast Atlantic: The importance of DOC. Limnol. Oceanogr. 44:80-94.

Giovannoni, S. J., T. B. Britschgi, C. L. Moyer and K. G. Field. 1990. Genetic diversity in Sargasso Sea bacterioplankton. Nature 345:60-63.

Giovannoni, S. J. and M. S. Rappé. 2000. Evolution, diversity and molecular ecology of marine prokaryotes. In Microbial Ecology of the Ocean, D. L. Kirchman (ed.), pgs. 47-84, New York: Wiley Interscience.

Goericke, R. 2002. Bacteriochlorophyll a in the ocean: Is anoxygenic bacterial photosynthesis important? Limnol. Oceanogr. 47:290-295.

Guo, L., C. H. Coleman, Jr. and P. H. Santschi. 1994. The distribution of colloidal and dissolved organic carbon in the Gulf of Mexico. Mar. Chem. 45:105-119.

Hansell, D. A. and C. A. Carlson. 1998a. Deep-ocean gradients in the concentration of dissolved organic carbon. Nature 395:263-266.

Hansell, D.A. and C.A. Carlson. 1998b. Net community production of dissolved organic carbon. Global Biogeochem. Cycles 12: 443-453.

Hansell, D. A. and C. A. Carlson. 2001. Biogeochemistry of total organic carbon and nitrogen in the Sargasso Sea: control by convective overturn. Deep-Sea Res. II. 48:1649-1667.

Hansell, D. A., C. A. Carlson, N. R. Bates and A. Poisson. 1997. Horizontal and vertical removal of organic carbon in the equatorial Pacific Ocean: A mass balance assessment. Deep-Sea Res. II. 44:2115-2130.

Hansell, D.A. 2002. DOC in the global ocean carbon cycle. In Biogeochemistry of Marine Dissolved Organic Matter. D. A. Hansell and C. A. Carlson (eds), 685-715, San Diego: Academic Press.

Hansell, D.A., C.A. Carlson and S. Suzuki. 2002. Dissolved organic carbon export with North Pacific Intermediate Water formation. Global Biogeochem. Cycles 16: 77-84.

Hansell, D. A. and C. A. Carlson. 2002. Biogeochemistry of Marine Dissolved Organic Matter, 774 pgs. Academic Press.

Hedges, J. I. and J. M. Oades. 1997. Comparative organic geochemistries of soils and marine sediments. Org. Geochem. 27:319-361.

Hoppe, H. G., K. Gocke, R. Koppe and C. Begler. 2002. Bacterial growth and primary production along a north-south transect of the Atlantic Ocean. Nature 416:168-171.

Karl, D. M. 2002. Microbiological oceanography - Hidden in a sea of microbes. Nature 415:590-591.

Karner, M. B., E. F. DeLong and D. M. Karl. 2001. Archaeal dominance in the mesopelagic zone of the Pacific Ocean. Nature 409:507-510.

Kirchman, D. L. 2000. Microbial Ecology of the Oceans. New York: John Wiley & Sons.

Kirchman, D.L. 2003. The contribution of monomers and other low molecular weight compounds to the flux of DOM in aquatic ecosystems. In Aquatic Ecosystems - Dissolved

Organic Matter, Findlay, S. and Sinsabaugh, R.L. (eds), pgs. 217-241. Academic Press, New York.

Kirchman, D. L., B. Meon, H. W. Ducklow, C. A. Carlson, D. A. Hansell and G. F. Steward. 2001. Glucose fluxes and concentrations of dissolved combined neutral sugars (polysaccharides) in the Ross Sea and Polar Front Zone, Antarctica. Deep-Sea Res. II 48:4179-4197.

Kirchman, D.L. 2002. The ecology of *Cytophaga-Flavobacteria* in aquatic environments. FEMS Microb. Ecol. 39:91-100.

Kirchman, D. L., C. Lancelot, M. Fasham, L. Legendre, G. Radach and M. Scott. 1993. Dissolved organic matter in biogeochemical models of the ocean. In Towards a Model of Biogeochemical Ocean Processes. G. T. Evans and M. J. R. Fasham (eds.), pgs. 209-225 NATO ASI Series. Springer-Verlag.

Kolber, Z. S., F. G. Plumley, A. S. Lang, J. T. Beatty, R. E. Blankenship, C. L. VanDover, C. Vetriani, M. Koblizek, C. Rathgeber and P. G. Falkowski. 2001. Contribution of aerobic photoheterotrophic bacteria to the carbon cycle in the ocean. Science 292:2492-2495.

Kolber, Z. S., C. L. Van Dover, R. A. Niederman and P. G. Falkowski. 2000. Bacterial photosynthesis in surface waters of the open ocean. Nature 407:177-179.

Landry, M. R., J. Constantinou and J. Kirshtein. 1995. Microzooplankton grazing in the Central Equatorial Pacific during February and August, 1992. Deep-Sea Res. II 42:657-671.

McCarthy, M. D., J. I. Hedges and R. Benner. 1998. Major bacterial contribution to marine dissolved organic nitrogen. Nature 281:231-234.

McCarthy, M., T. Pratum, J. Hedges and R. Benner. 1997. Chemical composition of dissolved organic nitrogen in the ocean. Nature 390:150-154.

Moore, L. R., G. Rocap and S. W. Chisholm. 1998. Physiology and molecular phylogeny of coexisting *Prochlorococcus* ecotypes. Nature 393:464-467.

Moran, M. A. and R. G. Zepp. 2000. UV radiation effects on microbes and microbial processes. In Microbial Ecology of the Ocean, D. L. Kirchman (ed.), pgs. 201-228, New York: Wiley-Liss.

Morris, R. M., M.S. Rappé, S.A. Connon, K.L. Vergin, W.A. Siebold, C.A. Carlson, S.J. Giovannoni. 2002. SAR11 clade dominates ocean surface bacterioplankton communities. Nature 420: 806-810.

Nagata, T. 2000. Production mechanisms of dissolved organic matter. In Microbial Ecology of the Oceans, D. L. Kirchman (ed), pgs. 121-152, New York: Wiley-Liss.

Nagata, T., B. Meon and D. L. Kirchman. 2003. Microbial degradation of peptidoglycan in seawater. Limnol. Oceanogr. 48: 745-754.

Nagata, T., H. Fukuda, R. Fukuda and I. Koike. 2000. Bacterioplankton distribution and production in deep Pacific waters: Large-scale geographic variations and possible coupling with sinking particle fluxes. Limnol. Oceanogr. 45:426-435.

Noji, T. T., F. Rey, L. A. Miller, K. Y. Borsheim and J. Urban-Rich. 1999. Fate of biogenic carbon in the upper 200 m of the central Greenland Sea. Deep-Sea Res. II. 46:1497-1509.

Norrman, B., U. L. Zweifel, C. S. Hopkinson and B. Fry. 1995. Production and utilization of dissolved organic carbon during an experimental diatom bloom. Limnol. Oceanogr. 40:898-907.

Ogawa, H., Y. Amagai, I. Koike, K. Kaiser and R. Benner. 2001. Production of refractory dissolved organic matter by bacteria. Science 292:917-920.

Ouverney, C. C. and J. A. Fuhrman. 2000. Marine planktonic Archaea take up amino acids. Appl. Environ. Microbiol. 66:4829-4833.

Pakulski, J. D. and R. Benner. 1994. Abundance and distribution of carbohydrates in the ocean. Limnol. Oceanogr. 39:930-940.

Palenik, B. and F. M. M. Morel. 1990. Amino acid utilization by marine phytoplankton: a novel mechanism. Limnol. Oceanogr. 35:260-269.

Rappé, M. S. and S. J. Giovannoni. 2003. The uncultured microbial majority. Annu. Rev Microbiol. 57:369-394

Rich, J., M. Gosselin, E. Sherr, B. Sherr and D. L. Kirchman. 1997. High bacterial production, uptake and concentrations of dissolved organic matter in the Central Arctic Ocean. Deep-Sea Res. II 44:1645-1663.

Rich, J. H., H. W. Ducklow and D. L. Kirchman. 1996. Concentrations and uptake of neutral monosaccharides along 140 degrees W in the equatorial Pacific: Contribution of glucose to heterotrophic bacterial activity and the DOM flux. Limnol. Oceanogr. 41:595-604.

Serret, P., C. Robinson, E. Fernandez, E. Teira and G. Tilstone. 2001. Latitudinal variation of the balance between plankton photosynthesis and respiration in the eastern Atlantic Ocean. Limnol. Oceanogr. 46:1642-1652.

Sherr, E. B. 1988. Direct use of high molecular weight polysaccharide by heterotrophic flagellates. Nature 335:348-351.

Simon, M., N. A. Welschmeyer and D. L. Kirchman. 1992. Bacterial production and the sinking flux of particulate organic matter in the subarctic Pacific. Deep-Sea Res. I 39:1997-2008.

Smith, D. C., M. Simon, A. L. Alldredge and F. Azam. 1992. Intense hydrolytic enzyme activity on marine aggregates and implications for rapid particle dissolution. Nature 359:139-142.

Strayer, D. 1988. On the limits to secondary production. Limnol. Oceanogr. 33: 1217-1220.

Strom, S. L., R. Benner, S. Ziegler and M. J. Dagg. 1997. Planktonic grazers are a potentially important source of marine dissolved organic carbon. Limnol. Oceanogr. 42:1364-1374.

Tanoue, E., S. Nishiyama, M. Kamo and A. Tsugita. 1995. Bacterial membranes: possible source of a major dissolved protein in seawater. Geochimica et Cosmochimica Acta 59:2643-2648.

Thomas, C., G. Cauwet and J. F. Minster. 1995. Dissolved organic-carbon in the Equatorial Atlantic Ocean. Mar. Chem. 49:155-169.

Wells, M. L. 2002. Marine colloids and trace metals. In Biogeochemistry of Marine Dissolved Organic Matter, D. A. Hansell and C. A. Carlson (eds.), pgs. 367-404. San Diego: Academic Press.

Whitman, W. B., D. C. Coleman and W. J. Wiebe. 1998. Prokaryotes: The unseen majority. Proc. Nat. Acad. Sci. USA 95:6578-6583.

Williams, P. J. leB. 1990. The importance of losses during microbial growth: commentary on the physiology, measurement and ecology of the release of dissolved organic material. Marine Microbial Food Webs 4:175-206.

Williams, P. J. leB. 1995. Evidence for the seasonal accumulation of carbon-rich dissolved organic material, its scale in comparison with changes in particulate material and the consequential effect on net C/N assimilation ratios. Mar. Chem. 51: 17-29.

Williams, P. J. leB. 1998. The balance of plankton respiration and photosynthesis in the open oceans. Nature 394:55-57.

Williams, P. J. leB. 2000. Heterotrophic bacteria and the dynamics of dissolved organic material. In Microbial Ecology of the Oceans, D. L. Kirchman (ed.), pgs. 153-200, New York: Wiley-Liss.

Williams, P. J. leB. and D.G. Bowers. 1999. Regional carbon imbalances in the oceans. Science 284:1735b.

Williams, P. M. and E. R. M. Druffel. 1987. Radiocarbon in dissolved organic matter in the central North Pacific Ocean. Nature 330: 246-248.

Yurkov, V. V. and J. T. Beatty. 1998. Aerobic anoxygenic phototrophic bacteria. Microbiol. Mol. Biol. Rev. 62:695-724.

MODELLING THE RESPONSE OF THE BIOLOGICAL PUMP TO CLIMATE CHANGE

T.R. ANDERSON & I.J. TOTTERDELL
Southampton Oceanography Centre, University of Southampton, Waterfront Campus, Southampton, United Kingdom
tra@soc.soton.ac.uk; Ian.J.Totterdell@soc.soton.ac.uk

INTRODUCTION

The concentrations of so-called "greenhouse" gases, carbon dioxide (CO_2), methane (CH_4), nitrous oxide (N_2O) and halocarbons and other halogenated compounds, are increasing in the atmosphere mainly due to human activities. These gases trap outgoing radiation from the Earth's surface, causing a warming of the lower atmosphere. The fate of carbon produced from fossil fuel emissions has received a great deal of attention both in scientific circles and in the media. The present atmospheric CO_2 increase is caused by anthropogenic emissions of this gas, about three quarters of which are due to fossil fuel burning which released 6.3 ± 0.4 Gt C yr^{-1} between 1990 and 1999 (IPCC 2001). The remainder is due to land use change. The resulting atmospheric CO_2 increase is 3.2 ± 0.1 Gt C yr^{-1}. This increase is less than the emissions because the ocean and terrestrial biosphere sequester much of the carbon. The flux from atmosphere to ocean is estimated to be 1.7 ± 0.5 Gt C yr^{-1} between 1990 and 1999 (IPCC 2001). The large volume of the ocean, in conjunction with the chemistry of the marine carbonate system, means that it potentially has the capacity to continue taking up CO_2 of anthropogenic origin for at least several centuries.

Carbon sequestration in the ocean is driven by two pumps – the "solubility pump" and the "biological pump" (Volk and Hoffert 1985). The exchange of CO_2 across the air-sea interface is driven by the solubility

M. Follows and T. Oguz (eds.), The Ocean Carbon Cycle and Climate, 65–96.

pump, which is the uptake of C by air-sea exchange due to undersaturation of the partial pressure of CO_2 (pCO_2) with respect to the atmosphere and subsequent transport of C-enriched waters to depth by physical processes (Follows et al. 1996). Biology also has an important impact on the vertical distribution of C in the ocean via the biological pump (Figure 1). Organic matter is produced by photosynthesis in the sunlit upper ocean (the euphotic zone). This material is processed by the marine ecosystem, and a part is exported to depth via sinking particles (Lampitt and Antia 1997), mixing and sinking of small particles and dissolved organic matter (Carlson et al. 1994), and transfer by vertically migrating zooplankton (Longhurst and Harrison 1988). Once exported this material can then be sequestered (in either organic or inorganic form) in the ocean interior. Some species of phytoplankton and zooplankton produce calcium carbonate tests which are also transferred to depth – the "carbonate pump". The biological pump acts to generate a gradient of dissolved inorganic carbon (DIC) in the ocean, increasing with depth. The lowered DIC, and hence pCO_2, at the ocean surface accentuates the solubility pump and associated air-sea transfer of CO_2.

In pre-industrial time, the gross flux of CO_2 from the atmosphere (and also a small flux from terrestrial sources) into the ocean was in an approximate steady state with the flux back to the atmosphere. Rising atmospheric CO_2 since the onset of the industrial revolution has given rise to undersaturation of ocean pCO_2 with respect to the atmosphere, leading to CO_2 entering the ocean via the solubility pump. It is well known that land plants exhibit elevated production in response to increasing atmospheric CO_2, such that terrestrial ecosystems may sequester significant amounts of CO_2 of anthropogenic origin (Cao and Woodward, 1998). Conventional wisdom is that biological production in the ocean is not influenced by pCO_2 due to the abundance of DIC in seawater. Algae are able to utilise the large bicarbonate (HCO_3^-) pool, as well as CO_2, as a carbon source for photosynthesis (Korb et al. 1997; Nimer et al. 1997). Production is therefore primarily controlled by other factors such as nutrients and light. There is therefore no direct stimulation of primary production by CO_2 of anthropogenic origin in the ocean.

If it is assumed that (1) the pre-industrial C cycle was in steady state and continues to operate in the same way without being affected by anthropogenic perturbations and (2) marine biology is not influenced by pCO_2, then the task of modelling the fate of anthropogenic CO_2 can be considerably simplified by using perturbation models (simulating differences between the present and pre-industrial pools and fluxes) which consider the response of the solubility pump, but not the biological pump, to climate change. Early estimates of oceanic uptake of CO_2 were made using simple box models (e.g. Oeschger et al. 1975; Siegenthaler 1983). These models could not however provide the horizontal spatial resolution required for a good understanding of the C cycle. Three dimension ocean general circulation models were therefore developed for this purpose (e.g. Maier-Reimer and Hasselmann 1987; Sarmiento et al. 1992) which require a description of carbonate chemistry in order to calculate pCO_2, and a time history of atmospheric CO_2. While predicted ocean uptake rates of anthropogenic CO_2 were broadly realistic, the authors in these studies acknowledged the need for future work looking at the whole of the oceanic C cycle and its response to climate change.

In reality climate change will impact on the natural C cycle in the ocean. Box modelling studies have shown the potential for changes in the ocean biota to have a significant impact on atmospheric CO_2 (e.g. Knox and McElroy 1984; Joos et al. 1991). Changes in circulation as a result of warming (e.g. Stocker and Schmittner 1997; Stouffer and Manabe 1999) may be expected to alter the functioning of both the solubility and biological pumps.

A full description of the DIC system in the ocean (rather than just the perturbation), including the biological pump, is then necessary in order to study the natural C cycle. Inclusion of a nutrient element, N or P, is needed in order to derive production by plankton. It is useful to distinguish between "new" production which is fuelled by newly available sources of N to the euphotic zone, e.g. due to vertical mixing from deep waters, and "regenerated" production which is driven by recycling of N within the euphotic zone (Dugdale and Goering 1967). In a simplified steady state the flux of nutrients into the euphotic zone driving new production is exported back to depth in equal proportion (Figure 2). Regenerated production via ammonium can be thought of as a rapidly spinning wheel which does not impact on the steady state fluxes of nutrient to and from the euphotic zone. Working with this scheme, the simplest way to represent the biological pump in models is to relax predicted nutrients in the euphotic zone back towards observed distributions, exporting excess nutrients back to depth. For example Najjar et al. (1992) used a phosphate restoring scheme of this kind to evaluate the relative roles of dissolved and sinking particulate material in the transport of organic matter to the ocean interior. More recently Gnanadesikan et al. (2003) used a nutrient-restoring model to examine the effects of iron fertilization on uptake of CO_2 by the ocean. Fertilization was imposed by restoring upper ocean phosphate to zero, rather than to observations. Results showed that a century after fertilizing the reduction in atmospheric CO_2 is between 2 and 44% of the initial pulse of C exported to the deep ocean, depending on how rapidly surface nutrients recover. The modelled recovery rate was however very sensitive to the representation of biological productivity and remineralization in the model.

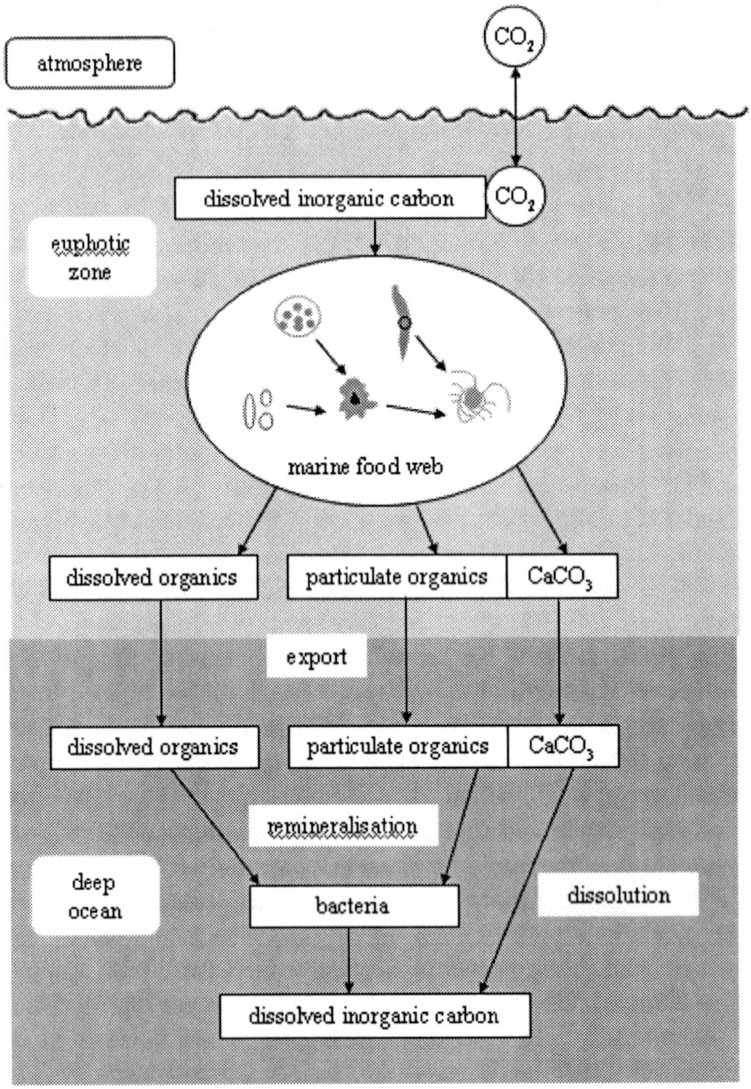

Figure 1. The biological pump in the ocean showing production of organic matter in the surface ocean and subsequent remineralization in the deep ocean.

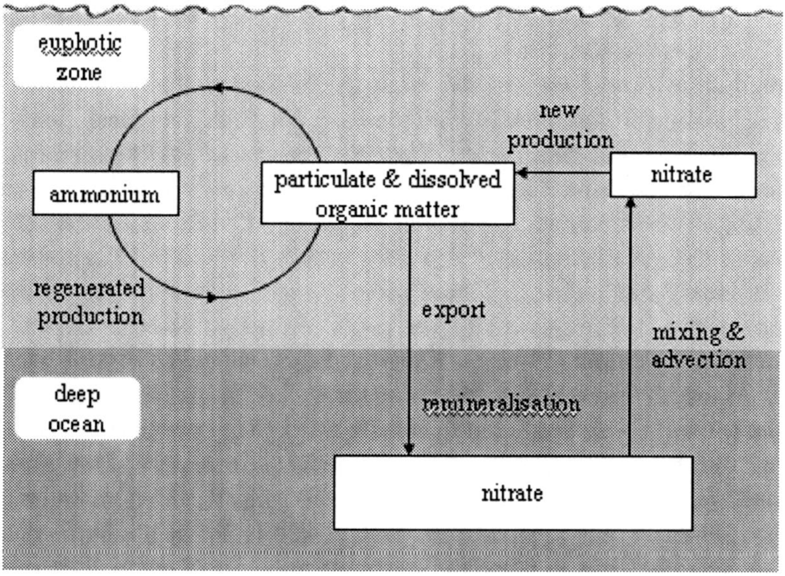

Figure 2. Schematic of nitrogen balance between euphotic zone and deep ocean. Import of nitrate from deep waters fuels "new" production, which is balanced by export of organic N back to the deep ocean.

The difficulty with nutrient restoring methods is in deciding what to do if the physical circulation, and hence transport of nutrients, is altered as a result of changing climate. One approach adopted by Sarmiento and Le Quéré (1996) when undertaking a transient run looking at changing climate was to calculate the nutrient transport terms based on nutrient restoring in their main spin-up version of the model, and then to continue to apply those same transport terms (except in areas where nutrient became depleted). Results showed reductions in cumulative 350-year oceanic uptake in simulations of 2 X CO_2 (atmospheric CO_2 increased at a rate of 1% per year until it had doubled after 70 years) and 4 X CO_2 scenarios of 14% and 28% respectively relative to a baseline run. Almost all of this effect was due to a slowing of the ocean circulation; although the import of C-rich deep waters to the ocean surface was reduced (which on its own would cause a net ingassing) the main effect was a reduction in the downwelling of waters rich in anthropogenic carbon. The result is however compromised by the assumption that the flux of biogenic material to the ocean interior, a large fraction of the total export, continues unabated. Sarmiento and Le Quéré (1996) also considered extreme cases in which the effect of climate change was to completely destroy the marine biota or to make it super-efficient at producing export. The wide variation in response led them to conclude that an improved representation of ocean biology is needed in order to reliably model climate change scenarios.

An alternative simple way to represent the biological pump in models is to model new production dynamically, although without explicit representation of biological state variables such as phytoplankton or zooplankton. For example, Bacastow and Maier-Reimer (1990) modelled new production as a Michaelis-Menten function of nutrient concentration (i.e. of the form N/(k+N), where k is a constant) multiplied by a latitudinally varying incident light factor. All new production is then exported to depth using a power law function, with a fixed proportionality assumed for N:C (or P:C) in order to calculate C fluxes. They used their model to investigate the effect of uniformly decreasing the circulation (i.e. all velocities) by 40% throughout the model on predicted C distributions in the ocean. Interestingly they found that both the solubility and biological pumps were strengthened. Less upwelling of CO_2-rich waters to the surface leads to reduced outgassing of CO_2 to the atmosphere. Slower horizontal velocities at the surface also gave the biota more time to consume nutrients and export material to depth. In a later study with a similar model, Maier-Reimer et al. (1996) explored the effects of a more realistic weakening of the thermohaline overturning on C cycling, finding that it reduced the strength of the biological pump, although the overall effect on predicted CO_2 uptake was small. Matear and Hirst (1999) undertook a series of general circulation model (GCM) runs investigating the effect of changing circulation and/or the imposition of anthropogenic CO_2 of atmospheric origin on ocean CO_2 uptake and storage. Climate change reduced the vertical supply of nutrients and DIC to the upper ocean, which produced opposing responses in the biological and solubility pumps regarding the air-sea exchange of CO_2. The decrease in nutrient supply reduced export production, increased surface pCO_2 and caused a flux of CO_2 out of the ocean. However reduced supply of DIC to the upper ocean produced a net uptake of CO_2 by the ocean. The lack of consistency between the results of these models, and the possibility that some of them are compromised by the simplistic way (necessarily so at the time) that processes were represented, highlights the need for an accurate representation of both the physical and biological processes involved. Much has nevertheless been learned from using these simple representations of the biological pump in GCMs, and there certainly may be utility in future studies of this type. For example when modelling the interannual variability of the air-sea flux of oxygen in the North Atlantic, McKinley et al. (2000) were able to reproduce the mean amplitude and phase of the surface annual cycle of O_2 in the Sargasso Sea but could not capture the full variability shown in data, which was acceptable given the goals of their work.

The relationship between nutrient supply, net community production and export of carbon in the ocean is complex. Simple correlations between primary production and flux estimates of particulate organic carbon (POC) in sediment traps tend to be weak (Lampitt and Antia 1997), indicating that plankton community structure is an important determinant of export flux. For example Boyd and Newton (1995) compared sediment trap fluxes at 3100 m in the northeast Atlantic with primary production during 1989 and 1990. Trap fluxes were significantly greater in 1989, despite similar primary productivity, which the authors conclude was due to a difference in the dominant phytoplankton species. Nevertheless a balance between simplicity and complexity needs to be struck in models (we will discuss this later), and simple empirical relationships may have an important role to play in large, computer-intensive GCM modelling studies.

The characteristics of food webs that determine the fate of biogenic C can be summarised into two key factors (Legendre and Rassoulzadegan 1996): the size spectrum of the primary producers, and the degree of coupling between the processes of production and consumption of organic matter. Marine pelagic ecosystems encompass organisms ranging in size from bacteria (~0.1 μm) to phytoplankton (0.1 to 100 μm) and zooplankton (100 to 10000 μm). The surface/volume ratio of organisms affects their ability to exchange resources with their environment, notably to take up nutrients and absorb light in the case of phytoplankton. Allometric scaling therefore predicts that small phytoplankton should outcompete large phytoplankton when nutrients are scarce, i.e. in oligotrophic environments, although cell shape is also important (Grover 1989). Generation times of organisms are generally inversely related to body size. The micrograzers which consume the smallest phytoplankton grow quickly (Gillooly 2000), resulting in a tight coupling between phytoplankton and zooplankton. Higher nutrient conditions permit larger algal species to flourish, which are then often able to outstrip their grazers in terms of growth rate, giving rise to the chlorophyll blooms which characterise the seasonal regimes of temperate latitudes.

The pelagic food chain has been traditionally thought of as being a chain from phytoplankton being grazed by zooplankton, which in turn may be grazed by higher tropic levels. This traditional chain is short and efficient, in that significant amounts of carbon may reach larger organisms via which it may be exported from the euphotic zone. In recent years however, it has been realised that a significant fraction of primary production is converted to dissolved organic matter (DOM) and processed by bacteria which are in turn consumed by micrograzers – the so-called "microbial loop" (Azam et al. 1983).

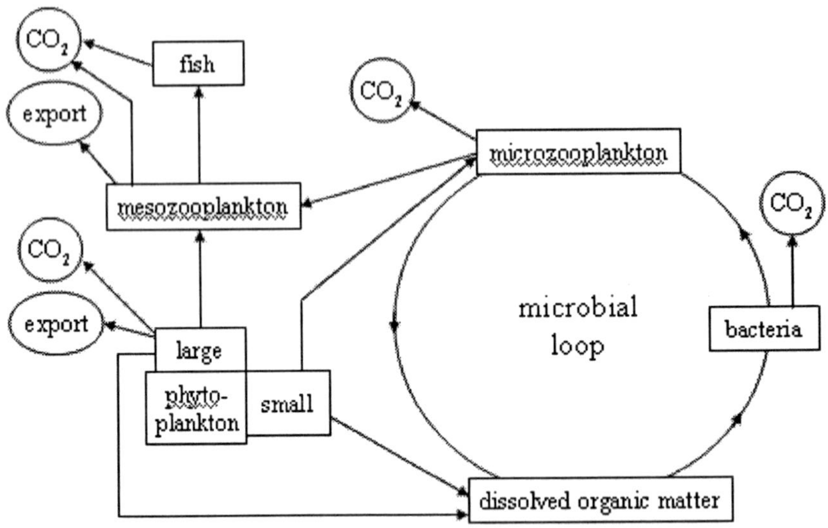

Figure 3. Simplified food web illustrating export pathways via the "traditional" food chain (left hand side of diagram), and remineralization of C to CO_2 via the microbial loop.

A large proportion of the C processed by this loop is respired to CO_2 due to the high energetic demands of bacteria and zooplankton (Anderson and Ducklow 2001). As a result little C is passed to higher trophic levels from where is may be exported as rapidly sinking particles, or directly by vertically migrating zooplankton. It should be noted however that DOM can be a significant export term (Anderson and Williams 1999). The relative dominance of these different food webs is therefore expected to vary with different environmental and hydrodynamic conditions (Margalef, 1978) as a result of the size-dependent physiological properties of the phytoplankton, and the degree of coupling with higher trophic levels. In general large phytoplankton tend to be the predominant primary producers in the weakly stratified waters of neritic areas (ocean waters between the low tide mark and a depth of about 200 m) and high latitudes, whereas smaller cells tend to be more important in the strongly stratified waters of lower latitudes and temperate summers (Malone, 1971). Indeed it is possible to divide the ocean into biogeochemical provinces, based on factors including the seasonality of mixed layer depth, irradiance and chlorophyll (Longhurst, 1995).

The extent to which changes in ocean circulation due to climate warming will impact on biological communities and hence the biological pump is not well known. In principle, tackling this issue in climate models ideally requires that biology be explicitly represented, rather than be included as an implicit nutrient restoring term. In this way can the effect of changing hydrodynamics on the functioning of marine ecosystems and the resulting export flux be fully addressed. Of course one has to bear in mind that added complexity brings with it its own intrinsic problems (discussed later in this article). Bearing in mind the many uncertainties involved in climate change modelling, the use of simple representations of the biological pump may in some instances be the most appropriate choice.

The field of marine ecosystem modelling has progressed rapidly over the years, from early models which included just nutrients, phytoplankton and zooplankton (e.g. Wroblewski et al., 1988) to models which include detritus and extra components of the recycling pathways such as dissolved organic matter and bacteria (e.g. Fasham et al., 1990; Anderson and Williams, 1998), to complex models which incorporate multiple plankton groups, multiple elements, and a wide array of processes (e.g. Moloney and Field 1991; Baretta et al. 1995; Lancelot et al., 2000). It is by no means simple to incorporate ecosystem models into GCMs, and so simpler models are understandably usually chosen for this purpose. It should also be recognised that many of the more complicated models were designed to address questions other than those concerning the global carbon cycle, such as seasonal dynamics in a particular region. The most commonly used ecosystem model structure incorporated into three dimensional hydrodynamical models is Nutrient-Phytoplankton-Zooplankton-Detritus (NPZD) (e.g. Oschlies and Garçon, 1999; McCreary et al., 2001; Palmer and Totterdell, 2001). More sophisticated models such as the Fasham et al. (1990) ecosystem model, which includes recycling pathways via ammonium and dissolved organic nutrients, have also been used in GCMs (e.g. Sarmiento et al., 1993; Ryabchenko et al., 1998). Other developments of ecosystem models employed in circulation models include the representation of long-lived dissolved organic matter (e.g. Six and Maier-Reimer 1996), and more recently multiple phytoplankton and zooplankton groups (e.g. Christian et al. 2002). However it is important to realise that increasing complexity in models does not necessarily lead to improvements in predictive capability; indeed in many instances, simpler models may perform best in this regard. To quote from Nihoul and Djenidi (1987): "The essence of modelling is the selection of a limited number of representative state variables. There must be sufficiently few of them for their evolution equations to be amenable to analysis, but enough of them to describe adequately the system's behaviour ... A simplistic model can produce very precise results entirely disconnected from reality, an intricated model may contain too many assumptions and uncertain figures to provide a satisfactory representation of it."

So what then is the ideal level of complexity for models to be incorporated into GCMs addressing possible changes in the biological pump in response to climate change? There is no simple answer to this question.

Models need to be complex enough, in terms of both processes and state variables, to capture feedbacks between biology and changing temperature, CO_2 and hydrodynamics, yet simple enough that parameterisations robustly generate realistic dynamics over the global domain. A challenge indeed. In the following two sections of this article we will identify the likely feedbacks between changing climate and the ocean's biological pump, and then discuss at length the issue of what level of complexity might be most suitable in order to capture these feedbacks in climate change studies.

1. THE BIOLOGICAL PUMP: CLIMATE FEEDBACKS

The biologically mediated flux of CO_2 between the atmosphere and ocean can be altered in response to changes in one or more of the following: (i) ocean nutrient inventories, (ii) the utilisation of nutrients in areas of the ocean which have high concentrations in surface waters, or (iii) the average elemental composition of biogenic material exported.

1.1 CHANGING NUTRIENT INVENTORIES

The effect of enhanced nutrients on carbon sequestration is well known in terrestrial systems, which are often considered to be N-limited, such that added N produces a plant growth response and an increase in storage (Vitousek and Howarth, 1991). Enhanced N deposition from fossil fuel combustion is expected to lead to carbon sequestration, particularly in forest ecosystems (Peterson and Melilo, 1985). Addition of nutrients to nutrient-limited areas of the ocean should similarly stimulate production and export. Nutrient supply is thought to have mediated variations in the sequestration of CO_2 in the oceans associated with the glacial-interglacial cycles during the past two million years. During peak glacial times, atmospheric pCO_2 was 80-100 ppm lower than that of interglacial periods. One explanation is that the whole-ocean nutrient reservoir was larger during glacial times (Ganeshram et al. 2000), strengthening the biological pump in nutrient-limited low latitude regions of the ocean. The potential impact of terrestrial N sources on sequestration of anthropogenic CO_2 in the contemporary ocean appears however to be small. Direct uptake of carbon by consumption of nutrients added to the oceans by rivers was estimated by Sabine and Mackenzie (1991) to be around 0.3 Gt C yr^{-1}. Atmospheric inputs of nitrogen are similar in magnitude to river inputs (Cornell et al. 1995).

An alternate means of increasing the N inventory in the ocean is by fixation of N_2 gas by biological organisms. Nitrogen fixation in the ocean has been estimated to be between 0.04 and 0.2 Gt N yr^{-1} (Galloway et al. 1995). The large uncertainty is due to poor knowledge of the abundance of N_2-fixing organisms due to severe undersampling of the ocean and wide variation in estimates of nitrogen fixation rates under natural conditions. Gruber and Sarmiento (1997) used geochemical methods to estimate a global N_2-fixation of 0.08 Gt N yr^{-1}. Increases in the nutrient inventory due to N_2-fixation are however generally balanced by denitrification losses on continental margins which are considerable (Galloway et al. 1995). Nevertheless whereas N_2-fixation and denitrification may be in balance over time scales of millions of years, imbalances in the N budget of the ocean in which denitrification is not equal to N_2-fixation could be occurring on shorter time scales.

It is worth noting that nutrients supplied from below by vertical mixing will in general be matched by a commensurate flux of DIC. The net transfer of carbon to the deep ocean is therefore approximately zero when these nutrients are utilised by the marine ecosystem and exported as organic matter.

1.2 NUTRIENT CONSUMPTION IN HIGH-NUTRIENT WATERS

The ocean can broadly be divided into three regimes according to nutrient status: oligotrophic (low nutrient, with associated low primary production), eutrophic with high production (large spring blooms of chlorophyll) such that nutrients are seasonally drawn down to low levels, and eutrophic areas in which nutrients remain high throughout the year and in which there is relatively low seasonality in chlorophyll. The last of these, the so-called high nutrient low chlorophyll (HNLC) areas, notably the Southern Ocean, northeast Pacific and equatorial Pacific, provide an apparently readily available source of nutrients which, if consumed by biota and exported, would cause a large sequestration of CO_2. It should be noted that nutrients are not drawn to zero during during the summer months in some of the high latitude seasonal regimes (e.g. Fasham et al. 1993). Also note that our usage of "nutrients" here refers only to macronutrients, e.g. nitrate, phosphate, silicate, which are embraced by the concept of "N" in HNLC. It does not include micronutrients such as iron, which are often not freely available in HNLC areas. Consumption of high-latitude nutrients has been hypothesised to contribute to variations in atmospheric pCO_2 between glacial and interglacial periods of the Earth's history. For example, stimulation of production through supply of iron may have caused the observed initial 40 ppm glacial-interglacial change (Watson et al. 2000).

So why are nutrients not consumed in HNLC areas, and how might this change with changing climate? Various hypotheses have been proposed for the factors controlling production in HNLC areas (e.g. see Fasham 1995). Limitation of primary production by iron has been suggested (Martin and Fitzwater 1988; Martin et al. 1990) as an important controlling factor. Artificial iron fertilization experiments carried out at sea have resulted in blooms of chlorophyll being stimulated in HNLC areas (e.g. see Boyd 2002). However we cannot be sure whether an ongoing increased supply of iron would result in significant consumption of nutrients because other limiting factors may come into play. Grazing by zooplankton may exert sufficient top-down control of phytoplankton to prevent them blooming in HNLC areas (Frost 1991; Cullen et al. 1992). Chai et al. (1996) used a model to study the causes of HNLC in the equatorial Pacific, and concluded that although supply of additional iron would enhance phytoplankton productivity, it would not eliminate the high nutrient condition because of the tight coupling between phytoplankton growth and zooplankton grazing. Other limiting factors such as light have also been suggested to be important (Mitchell et al. 1991).

It is by no means clear how climate change will impact on the biogeochemistry of HNLC areas. Atmospheric delivery of iron in dust may be important in supporting phytoplankton growth. For example Gabric et al. (2002) showed a coupling between satellite-derived chlorophyll and aerosol density in the atmosphere, for the Southern Ocean. Changes in the circulation of the atmosphere could thus impact directly on the productivity of HNLC areas. Modelling studies have indicated that climate warming will lead to the thermohaline circulation becoming more sluggish and increased stratification in the upper ocean (Bopp et al., 2001; Plattner et al. 2001), resulting in a higher light environment for surface dwelling plankton, but a lower supply of nutrients by vertical mixing. Any change in hydrodynamics as a result of changing climate will impact on the competitive interactions of different phytoplankton groups, and consequently food web structure as a whole. For example, it is well known that diatoms (non-motile) tend to thrive in turbulent nutrient-rich waters, whereas dinoflagellates, which possess flagella and hence are mobile, can regulate their position in the water column and so favours survival in stratified waters (Margalef, 1978). Diatoms are generally associated with high export regimes because of their fast sinking rates. Species shifts could affect the size distribution and tropic interactions of the planktonic food web and hence the strength of the biological pump, particularly in HNLC areas.

1.3 ELEMENTAL COMPOSITION OF EXPORT

Export of biogenic carbon from the upper ocean can be in the form of organic matter in particles, dissolved organics, or calcium carbonate. Marine stoichiometry has been widely described by the Redfield ratio of 106C:16N:1P (Redfield et al. 1963) which characterizes ratios in particulate matter in surface waters and nutrients in deep water, implying that export fluxes must also be in this ratio. Much of the particulate export is in non-living detritus, consisting of non-living phytoplankton aggregates ("marine snow": Lampitt et al. 1993) and zooplankton fecal pellets. Yet is there any *a priori* reason why these export fluxes should be in Redfield proportion? This ratio may be achieved in phytoplankton biomass under nutrient-replete conditions, but it is plastic. Geider and La Roche (2002) concluded that neither analysis of laboratory C:N:P data, nor a theoretical approach based on the biochemical composition of phytoplankton, could support the contention that the Redfield ratio reflects a physiological or biochemical constraint on the elemental composition of primary production. The C:N ratio of sinking zooplankton fecal pellets is to some extent variable (Anderson 1994), but just how it might respond to changing community structure is not clear. The C:N of particulate organic matter in surface waters often does not deviate far from Redfield (e.g. Chen et al., 1996; Kepkay et al., 1997), although there are exceptions (e.g. Daly et al., 1999). Any change in the C:N ratio of export of organic matter could have significant consequences for C sequestration in the ocean (e.g. see model sensitivity experiments of Heinze et al., 1991), but whether such changes are likely to occur is not known.

The C:N ratio of DOM may often be significantly higher than the Redfield ratio (e.g. Hopkinson et al., 1997; Søndergaard et al., 2000). This material is produced via various processes including phytoplankton exudation, zooplankton "messy feeding", viral lysis of bacteria and phytoplankton and dissolution of detritus. High C:N in marine DOM is likely due to phytoplankton extracellular release of organic C, with no associated N. This release may be a result of "photosynthetic overflow", i.e. high photosynthetic rates under nutrient-limiting conditions giving rise to the production of carbohydrates which algae cannot use for growth (Wood and Van Valen, 1990), and so highest rates are generally associated with oligotrophic conditions (Teira et al., 2001). Associated with DOM, an interesting potential export route for carbon is via transparent exopolymer particles (TEP). TEP are polymeric gel particles that form abiotically from dissolved and collidal organic matter, and are characterised by high C:N (Engel and Passow, 2001). Although TEP do not sink gravitationally, they are exported as aggregates (Passow et al., 2001). Significantly, the production of DOM and TEP are not reliant on nutrients. Shifts in ocean circulation towards increased stratification and reduced nutrient supply to the euphotic zone could favour increased production of these C-rich substrates, which could be exported to deep waters.

In addition to production of organic carbon, many marine organisms, including coccolithophorids, foramaniferans, corals and molluscs, produce calcium carbonate ($CaCO_3$). Precipitation thus occurs according to the following reaction:

$$Ca^{2+} + 2HCO_3^- \xrightarrow{\text{calcification}} CaCO_3 + CO_2 + H_2O$$

Contrast this equation with one for photosynthesis by algae:

$$6CO_2 + 12H_2O \xrightarrow{\text{primary production}} C_6H_{12}O_6 + 6O_2 + 6H_2O$$

Primary production consumes CO_2, producing organic matter which may be exported to depth. The vertical gradient of DIC thus produced acts to reduce pCO_2 in surface waters and hence encourages uptake of CO_2 by the ocean from the atmosphere. In contrast the calcification reaction produces CO_2 rather than consuming it. This process therefore favours evasion of CO_2 to the atmosphere and acts to oppose uptake of atmospheric CO_2 by the ocean, despite the efficient transfer of the biogenic carbonate to deep water via sinking particles.

So how might calcification rates respond to climate change? Invasion of anthropogenic CO_2 into the ocean may have a very significant effect on calcification. A basic knowledge of carbonate chemistry is needed in order to understand this effect. Dissolved inorganic carbon (DIC) in seawater exists as several chemical species characterised by the following reactions:

$$CO_2 + H_2O \Leftrightarrow H_2CO_3 \Leftrightarrow H^+ + HCO_3^- \Leftrightarrow 2H^+ + CO_3^{2-}$$

The equilibrium state of these reactions is strongly affected by pH ($pH = -\log_{10}[H^+]$). As CO_2 invades the ocean from the atmosphere, the carbon moves from left to right along the above equation resulting in the formation of bicarbonate (HCO_3^-), H^+ ions are generated, and pH decreases. The resulting increase in acidity gives rise to conditions which favour the presence of bicarbonate rather than carbonate (CO_3^{2-}). In other words, anthropogenic CO_2 invading the ocean will cause conditions which are less favourable for calcification. Zondervan et al. (2002) showed that the ratio of biogenic inorganic carbon to organic carbon in the coccolithophore *Emiliania huxleyi* decreases with increasing CO_2 concentration in laboratory experiments. The current rise in atmospheric CO_2 may thus slow down planktonic calcification (Riebesell et al., 2000; Zondervan et al., 2001). Coral reefs are particularly threatened: carbonate precipitation by reef-building organisms may decrease by 14 to 30 percent by the middle of this century (Kleypas et al., 1999). The consequences for other calcifying organisms may be equally severe.

Predicting feedbacks in the climate system of this kind is however far from simple, because other factors also influence the ecology of calcifying organisms. Episodes of elevated sea temperatures over the last 20 years have led to mass coral bleaching (Hoegh-Guldberg 1999). This bleaching occurs when reef-building corals lose or expel their algal symbionts (zooxanthellae) in response to stress. Another factor of potential importance relates to differences between different phytoplankton groups relating to an important enzyme in the Calvin cycle (the dark reactions of photosynthesis in C_3 plants), ribulose bisphosphate caboxylase/oxygenase (Rubisco). Large differences exist between phytoplankton groups between their specificity for Rubisco (Tortell, 2000), which operates below its maximum capability at typical CO_2 concentrations in seawater. Calcifying algae such as the coccolithophorids have relatively high specificity and so are sensitive to increasing CO_2 in seawater, potentially increasing their growth rates (Riebesell et al., 2001), and so may potentially benefit from the invasion of anthropogenic CO_2 into the ocean. Changing circulation due to climate change may also affect the relative dominance of calcifying organisms in the marine environment. For example, modelling results suggest that increased stratification of the upper water column may favour the coccolithophore *Emiliania huxleyi* to bloom due to the high light environment thus created (Tyrrell and Taylor, 1995). Accurately modelling the interacting effects of changing pH, increasing temperature and altered stratification on the growth of carbonate-forming organisms presents a considerable challenge.

2. MODELLING APPROACHES

2.1 ISSUES OF MODEL COMPLEXITY

It is clear from the preceding section that possible responses of the biological pump to climate change involve the complex interplay between ocean physics and chemistry and how these impinge on the competitive interactions of different plankton groups. This would at first sight suggest that highly complex ecosystem models, incorporating multiple plankton groups, are required in order to capture the complicated dynamics involved. There has been a general trend in recent years toward using ever more complicated ecosystem models (e.g. Baretta et al., 1995; Lancelot et al., 2000) with the realisation that marine food webs are complex, and with the availability of powerful computers to undertake the necessary number crunching. However increasing model complexity is no guarantee of improvement in predictive ability. Complex models tend to be difficult to use and understand. Outputs may be less stable than those of simpler models, particularly if underlying mechanisms have not been correctly parameterised.

The issue of model complexity has been debated in the literature for decades. Simple models, while easy to apply, may lack the realism necessary in order to capture feedback mechanisms, e.g. associated with climate change. It is well known that an inherent danger of simple empirical relationships is inadvertent extrapolation beyond the conditions in which they were derived and tested. All models are simplifications of reality. Biological systems are by their nature complex, and this has led some researchers to question whether modelling is a meaningful approach for studying such systems (Hedgpeth, 1977; Passioura, 1996). Yet simple ecosystem models incorporated into general circulation models of the ocean have apparently done a remarkably good job at reproducing global patterns of nutrients and chlorophyll (e.g. Sarmiento et al., 1993; Six and Maier-Reimer, 1996; Palmer and Totterdell 2001). Certainly much has been learned from studies of this sort about the factors controlling observed distributions of plankton in the ocean. However it is important to realise that just because a model is able to reproduce some parts of the system, this does not mean that all aspects of the model are functioning in a fully realistic manner. For example, just because a model correctly predicts a bloom of phytoplankton, it does not mean that the resulting predicted export to the deep ocean is correctly simulated. At face value the output of a such a model may be correct, but if network flow is incorrect, predicted responses to novel conditions may be fallacious. Complex models will nearly always do better than simple models when calibrated for an individual data set because of the extra flexibility afforded by the extra parameters. It is however one thing to calibrate a model (adjust model parameters) to fit an individual data set, and another to validate that model by testing its predictive capability against an independent data set. Complex models may not be able to robustly simulate the dynamics at alternate sites without recourse to changing the model structure or parameter values.

Error in model predictions can derive from either inadequate structure or parameter values. The conceptual relationship between these two sources of error, for two contrasting model domains, is shown in Figure 4. The potential for improved predictions increases in both cases as the structure of the model is improved (e.g. extra state variables added to represent different plankton groups). Errors associated with the parameterisation of that extra structure are however much greater if the model is applied to a global rather than local domain. Consider an extreme but illustrative example. The food web of the Black Sea is complex, with a particular omnivorous dinoflagellate, *Noctiluca*, being a dominant species during the last two decades. Oguz et al. (2001) have successfully used an ecosystem model to study the food web dynamics in the Black Sea; their model included compartments for diatoms, flagellates, microzooplankton, omnivorous mesozooplankton, *Noctiluca*, the jellyfish *Aurelia* and the ctenophore *Mnemiopsis*. In theory one could apply such a model to a global domain. If the parameterisation of all state variables was perfect, then those plankton groups of specific importance to the Black Sea would be driven to extinction in other areas where they are not important. However such a model would probably produce all sorts of unexpected behaviour if run in a global domain because the mathematical expressions used would be unable to fully capture the exact niches of the species involved.

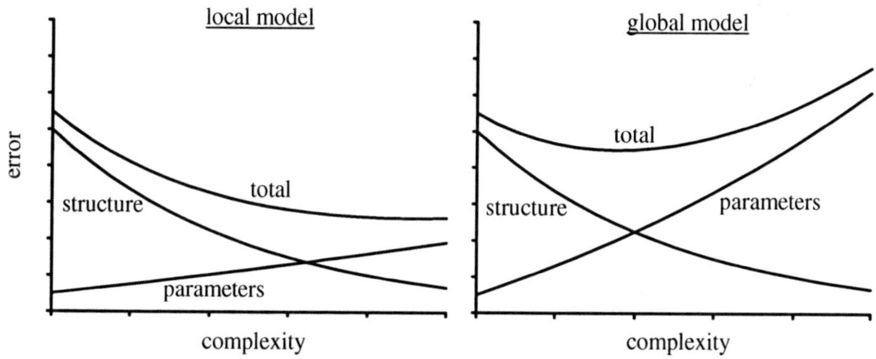

Figure 4. Conceptual sources of error as related to model complexity for models applied to local and global domains (based on Reynolds and Acock, 1985). The decrease in error associated with model structure remains the same in both cases, but the error associated with parameterization increases faster when the model is applied to a larger domain.

Climate change studies require that ecosystem models be globally robust such that they can give an accurate simulation of the annual cycle of total and new production in the euphotic zone. The ideal is to have a single model with a single parameter set.

The alternative, using different parameter sets for different areas, is unpalatable, although not necessarily impossible, because the modeller then has to take into consideration how boundaries of the regions for each parameter set might change, e.g. during a climate change scenario. There are a number of examples in the literature where researchers have tried to see if a model with a unique parameter set can successfully simulate alternate scenarios. Fasham (1993) was able to model the seasonal ecosystem dynamics at contrasting sites in the North Atlantic using a single parameter set, providing phytoplankton and zooplankton loss terms were made functions of biomass. However in a similar study (Fasham 1995) he found that he could not model the contrasting dynamics in the North Atlantic and North Pacific oceans using an ecosystem model without resorting to alternate parameter sets for each site. It may be the case that the differences in biological populations in different parts of the world will preclude the development of a world ecosystem model with a generic parameter set, but at least the attempt should be made (Fasham, 1995).

The enormous computational demand of studies involving coupled GCMs of the earth system needs to be taken into consideration when choosing the representation of biology in models. Whereas it is desirable to have as realistic a model of biological processes as possible, there is also a need for an efficient model from which understanding of a limited number of parameter uncertainties can be obtained. Such understanding requires ensemble runs of climate models investigating the uncertain parameters in these models. This process is different from trying to find the best ecosystem model for the current climate and using that model to help understand the system better. Both approaches will ultimately help us understand the response of the biological pump to climate change, but require different types of model development.

Validation of ocean models to be used on climate studies is an important issue. Of course we can never be totally sure that a model will respond in a correct manner when forced with a future climate change scenario. Confidence can however be gained if a model is able to (i) capture the seasonality and spatial variability associated with a range of ecosystem types, and (ii) simulate observed interannual or interdecadal variability in biological tracers. As an example of the former, Fasham et al. (1993) compared how well a biogeochemical model of the North Atlantic predicted the seasonal dynamics at Bermuda Station "S" (31°N 65°W) and ocean weather station "India" (59°N 19°W). Studying interannual variability poses a major challenge. Forcing fields for the years in question need to be assembled, as well as the relevant data sets with which to compare the model. A good example of the response of planktonic biomass to interannual differences in climate forcing is provided by the North Atlantic Oscillation (NAO). The NAO index is based on the average atmospheric pressure difference during winter between the Azores and Iceland, and has important effects on the circulation patterns of the North Atlantic (Curry and McCartney 2001). Modern-day GCMs are able to capture such interannual variability in the circulation (e.g. Cooper and Gordon 2002). Plankton populations similarly show interannual variability which is correlated with the NAO (Taylor et al. 1992; Edwards et al. 2001). Taylor et al. (2002) used an ecosystem model embedded within a 1-dimensional physical model for a generic, seasonally stratified water column typical of the North Sea. Model results showed correlations between simulated plankton populations and the latitude of the Gulf Stream.

Finally, it is important to note that complexity can only be effectively added to models if there are sufficient data for validation purposes. Some variables such as chlorophyll can be measured by remote sensing. However such measurements, while very useful particularly for understanding spatial and temporal variability in biological systems, are not enough on their own to fully constrain complex models. If the modelling community wishes to move towards complex models with multiple functional groups, then equivalent observations need to be made. For example, models are notoriously sensitive to zooplankton parameters. Unfortunately quantifying zooplankton biomass, including micrograzers, is a laborious task. Fasham et al. (1993), in their comparison of ecosystem dynamics at different sites in the North Atlantic, concluded by appealing for more observational data on the seasonal cycles of zooplankton biomass and growth rates. In a separate study, Fasham et al. (1999) used a size-structured ecosystem model to study primary and secondary production, and particle flux, on a cruise in the northeast Atlantic during a spring bloom period. They found that a unique model solution was not possible because measurements of dissolved organic carbon, important for fuelling bacterial production, were not made.

2.2 SECOND GENERATION ECOSYSTEM MODELS

Marine biology has commonly been afforded simple parameterisations in GCM modelling studies, although more recent models have had extra complexity added. A summary of the parameterisations of ocean biology in a variety of published GCMs arranged in chronological order is shown in Table 1. Those models which have an explicit ecosystem model are usually restricted to NPZD, perhaps also with a compartment for DOM, almost invariably couple different currencies using the Redfield ratio, and often use simple empirical functions to distribute exported biogenic material through the water column. Bearing in mind that additional complexity does not necessarily lead to improvement in predictive capability, and also the computational demands and associated costs with increasing the number of state variables to be included in GCM modelling studies, it behoves the rest of the scientific community to make a strong case to modellers for any extra complexity to be added to climate models beyond the bare minimum. Of course there is no simple answer as to how much complexity is required in models of the biological pump in the ocean. It is likely that simple NPZD-type models are unable to capture many of the feedbacks described earlier, and so there must be a case for trying so-called "second generation" ecosystem models which incorporate additional structure.

Table 1. Parameterization of ocean biology in ocean GCMs

model	domain	currencies	production OM[1]	export	remin. POM[2]	prod. $CaCO_3$ [3]	remin. $CaCO_3$
Bacastow & Maier-Reimer (1990)	global	P, C	EP = f(N,I)	POM, DOM	exponential fn.	RR = 0.15	exponential fn
Najjar et al. (1992)	global	P	nutrient restoring	POM, DOM	power fn	N/A	N/A
Fasham et al. (1993)	N. Atlantic	N	N, P, Z, D, B, LDOM	POM, (DOM)	power fn.	N/A	N/A
Sarmiento & Le Quéré (1996)	global	P, C	nutrient restoring	POM, DOM	power fn.	Alkalinity restoring	
Six & Maier-Reimer (1996)	global	P, C	N, P, Z, D, SDOM	POM, DOM	Suess fn.	RR = variable[4]	exponential fn
Yamanaka & Tajika (1996)	global	P, C	EP = f(N,I)	POM, DOM	power fn.	RR = 0.08	power fn.
Matear & Hirst (1999)	N. Pacific	P, C, O	EP = f(N,I,T)	POM	power fn.	RR = 0.08	power fn.
Oschlies & Garçon (1999)	N. Atlantic	N	N,P,Z,D	POM	explicit detritus	N/A	N/A
Bopp et al. (2001)	global	P, C, Si	N, P, Z, D, SDOM	POM, DOM	power fn.	fn. of temperature	exponential fn
Palmer & Totterdell (2001)	global	N, C	N, P, Z, D	POM	explicit detritus	RR = 0.06	below lysocline
Christian et al. (2002)	trop. Pacific	N, Fe	N, 2P ,2Z, 2D	POM	power fn.	N/A	N/A

1. Production of organic matter. EP: export production; ecosystem state variables are N: nutrient, P: phytoplankton, Z: zooplankton, D: detritus, B: bacteria, LDOM: labile DOM, SDOM: semi-labile DOM; I: light, T: temperature
2. Remineralisation of POM: power function takes for $a(z/100)^{-b}$ (e.g. Martin et al., 1987)
3. Production of $CaCO_3$. RR: rain ratio, i.e. ratio of $CaCO_3$ production to soft tissue C
4. RR negatively related to silicate

The logical approach is to gradually increase complexity, adding new components one at a time, and testing to see whether they have significant influence on model output. The scientific community is certainly adopting this approach, although most studies using complex ecosystem models have been restricted to localised areas. We now give two topical examples of how second generation ecosystem models are being developed to address biological aspects of climate change in the ocean: the development of models which represent the ecology of N_2-fixers and carbonate producers. Our aim here is not to review all the current developments in ecosystem modelling, but to highlight the complex nature of biological systems and discuss the implications for modelling studies of the biological pump in the ocean.

The explicit inclusion of nitrogen fixers in models is difficult because the ecological controls on the fixation process are complex. Fixation is energetically expensive, the organisms involved may require relatively more nutrients than other phytoplankton, and because they are relatively rich in N they may be preferentially grazed. Developing models to include these factors "will be difficult, but it should not be impossible" (Vitousek et al. 2002). Simple models of N_2-fixation are currently in use. For example, Bissett et al. (1999) parameterised N_2-fixation by *Trichodesmium* in a 1-dimensional model of the Bermuda Atlantic Time Series (BATS) site as the product of an empirical temperature-dependent number of trichomes per litre and a fixation rate per trichome. Nitrogen fixation is in reality affected by a number of factors. Iron bioavailability is thought to be an important (Falkowski 1997), although it should be noted that other factors, notably phosphorus, may be limiting in iron-replete areas of the ocean such as the North Atlantic (Sanudo-Wilhelmy et al. 2001). Open ocean N_2-fixation may be enhanced by warming and increased stratification of the water column (Karl et al. 2002).

The inclusion of the role of iron in limiting the production of phytoplankton as a whole, as well as N_2-fixers in particular, has proved particularly challenging for the modelling community. There are many poorly quantified processes relating to the cycling of iron in the ocean, notably the fraction of iron of aeolian origin which becomes soluble and bioavailable upon deposition in surface waters, and rates of biological uptake and scavenging (Wells et al., 1995; Johnson et al., 1997).

The simplest way to include iron limitation in models is to impose an empirical limitation factor (e.g. Denman and Peña, 1999; Popova et al., 2000). Such an empirical response will probably not however be responsive to climate change scenarios. The alternative is to include iron explicitly as a state variable, with representations of aeolian fluxes from the atmosphere and vertical mixing from the deep ocean (e.g. Loukos et al., 1997; Leonard et al., 1999; Moore et al., 2002). This approach is hampered by our lack of understanding of iron biogeochemistry in the ocean. For example, Moore et al. (2002) make the entirely reasonable assumptions that 2% of iron of aeolian origin enters the dissolved pool upon deposition, all dissolved iron is bioavailable, and that scavenging rates vary between 1 and 2.74% per year, but acknowledge that future versions of the model will have to be refined as our understanding of the marine iron cycle improves.

Modelling the production of carbonate-forming organisms is an equally challenging proposition. To quote Moore et al. (2002): "At present it is not possible to model dynamically or predict calcite formation in the ocean". Simple formulations have thus far been employed. One means of accounting for carbonate production is to force alkalinity profiles towards observed values (Sarmiento and Le Quéré, 1996). As with nutrient-restoring simulations, responses to changing climate cannot be studied using such an approach. The next level of sophistication is to assume that calcium carbonate is produced in a fixed ratio to organic carbon by phytoplankton (e.g Palmer and Totterdell, 2001) or is exported in a constant "rain ratio" to particulate organic carbon (e.g Bacastow and Maier-Reimer, 1990; Yamanaka and Tajika, 1996). There is considerable variation in the parameterisations of where $CaCO_3$ is remineralised in different models. For example, Palmer and Totterdell (2001) assumed that biogenic calcium carbonate is instantly remineralised between the lysocline (the depth in the ocean below which carbonate is no longer saturated) and the base of the water column, based on the premise that $CaCO_3$ is supersaturated in the upper 2000 m or so of the ocean, and so does not dissolve. In comparison, when modelling the upper 400 m of the water column at the Bermuda Atlantic Time Series (BATS) site, Anderson and Pondaven (2003) assumed that $CaCO_3$ remineralises in the upper water column at a rate of 0.05 d^{-1} (Tyrrell and Taylor, 1996), consistent with recent evidence suggesting that as much as 60-80% of this material dissolves in the upper 500-1000 m of the water column (Milliman et al., 1999).

The ecology of carbonate-forming organisms is not well understood, making the development of physiologically-based ecosystem models which include them explicitly a difficult proposition. The coccolithophore *Emiliania huxleyi*, for example, is ubiquitous throughout most of the world's oceans, although only forming blooms in a few areas, notably the North Atlantic (e.g. Ziveri et al., 2000). It would appear to favour the high light conditions which occur under strong stratification, as well as low phosphate relative to nitrate (Egge and Heimdal, 1994) because it may be able to utilise dissolved organic forms of phosphorus. There have been few attempts to explicitly include carbonate-forming organisms in ecosystem models. Coccolithophores were included in a modelling study of mesocosm experiments in the Norwegian fjords (Aksnes and Egge, 1994) and in a model applied to the North Atlantic (Tyrrell and Taylor, 1996), in both case making growth rate a function of light and phosphorus availability. These models were able to achieve reasonable representations of these systems with adequate tuning, but may not be robust when applied in predictive mode to wider domains. Tyrrell and Taylor (1996) acknowledged that the output of their model could be strongly altered by changing the values of loosely-constrained parameters within their ranges of likely values.

There are many other developments currently taking place in ecosystem modelling. A number of modellers have chosen to include diatoms in their models (e.g. Lancelot et al., 2000; Chai et al., 2002; Moore et al., 2002), necessitating the inclusion of silicate as an additional nutrient. Moore et al. (2002) included two detrital pools. Small phytoplankton were largely recycled within the surface mixed layer via a non-sinking detrital pool, while diatoms were largely exported via a detrital pool sinking at 20 m per day. Other models have split the phytoplankton based on size rather than functional groups. For example, Leonard et al. (1999) included separate size fractions for nanophytoplankton (< 10 μm) and netphytoplankton (> 10 μm). The size of particles exported from the euphotic zone is affected by aggregation processes, which have also been addressed in modelling studies (Jackson, 2001; Jackson and Burd, 2002; Ruiz et al., 2002). The division of phytoplankton into subgroups generally requires that zooplankton also need to be separated on the basis of size in order to correctly simulate top-down control of algal community structure. A typical division is to distinguish between micro- and mesozooplankton (e.g. Fasham et al. 1999; Leonard et al. 1999; Chai et al. 2002). Modelling the larger of these size classes, in particular, is not easy because mesozooplankton often have a complex life history and exhibit behaviour such as diel vertical migration. Models have been developed for particular species (e.g. Carlotti and Radach 1996), but are complex and not generally suitable for GCM modelling studies. A further topical issue in marine ecosystem modelling is the representation of dissolved organic matter and its consumption by bacteria in models (e.g. Anderson and Williams 1998, 1999). A complete discussion of the development of ecosystem models is however beyond the scope of this work.

Despite the many advances in marine ecosystem modelling, the export of organic matter and its subsequent remineralisation in deep waters have nevertheless been afforded simple representations in most studies using general circulation models of the ocean. Material sinking from the upper ocean is often instantly redistributed throughout the water column (e.g. Sarmiento and Le Quéré, 1996; Six and Maier-Reimer, 1996; Bopp et al., 2001) as empirical functions of depth (Suess, 1980; Martin et al., 1987). The often-cited Martin et al. (1987) function, derived from sediment trap data, takes the form of a power function:

$$F_z = F_{100}(z/100)^b$$

where the flux at depth z is determined from the flux at 100 m (F100) and exponent b. However Berelson (2001) compared the value of this exponent between different locations and found it to be highly variable. Other models include sinking detritus throughout the water column. Oschlies and Garçon (1999) used a detrital sinking and remineralisation rates of 5.0 m d-1 and 0.05 d-1 respectively (based on Fasham et al., 1990). In comparison, the global ocean model (HadOCC) of Palmer and Totterdell (2001) uses sinking and remineralisation rates of 10.0 m d-1 and 0.02 d-1 respectively. Further work with this model (e.g. Cox et al. 2000) uses a detrital compartment with the same sinking rate, but a variable remineralization rate with depth which effectively generates a nutrient profile described by the Martin curve. These types of model do not include any details of the organisms responsible for the decreasing flux of organic matter with depth. Particles sinking out of the sunlit upper ocean into the "twilight zone" are degraded by both bacteria (e.g. Turley and Mackie 1994; Tamburini et al. 2002) and zooplankton (e.g. see Jackson and Burd 2002). The relative importance of these organisms in transforming particle flux is however poorly understood. The twilight zone is an area in need of research, which will hopefully lead to the development of improved models of remineralization in the water column. Another promising new avenue of research is the realisation that fluxes of ballast minerals (silicate and carbonate biominerals, and dust) determine deep water fluxes of particulate organic carbon, with the subsequent development of models based on this premise (Armstrong et al., 2002).

Ecosystem models in GCMs have used Redfield stoichiometry to couple carbon to nutrient elements. As we described earlier, organic matter may in reality depart significantly from Redfield ratios. Model studies have shown that any change in the Redfield ratio will have a large effect on carbon storage in the ocean (Heinze et al., 1991). Although phytoplankton growing under optimal conditions tend to have C:N near to Redfield, it is well known that zooplankton (e.g. Gismervik, 1997) and bacteria (e.g. Goldman et al., 1987) tend to have significantly lower C:N ratios. It is relatively simple to model trophic interactions where the elemental composition of consumers differs from that of their prey (e.g. Anderson and Hessen, 1995; Anderson and Williams, 1998). Moreover different zooplankton assimilation efficiencies can be applied for carbon and nutrient elements such that the C:N (or C:P) of detritus is elevated relative to that of the food consumed; the efficiency for carbon tends to be lower than that of nitrogen (Anderson, 1994). The C:N of sinking particles may also increase with depth because of preferential remineralisation of nutrient elements relative to C (Thomas et al., 1999; Ono et al., 2001), which is also easy to include in models. Just how important it is to include these sorts of departures from Redfield stoichiometry in GCM modelling studies is not clear.

In conclusion, there are currently many exciting and new advances being made in the field of marine ecosystem modelling. Our understanding of marine biogeochemistry is improving all the time, and ever more complex ecosystem models are being developed to represent the many processes of interest. Yet although detailed models have been developed for localised systems, the extent to which the complexity of these models can be incorporated into GCM modelling studies on a global scale investigating the role of the biological pump in climate change is not known. Indeed simple representations of the biological pump without an explicit ecosystem model, such as the nutrient-restoring approach may, at least in some instances, be an entirely reasonable if not the best approach to adopt in climate change modelling work. Simply adding complexity *ad hoc* into ecosystem models to be used in GCM studies will not necessarily lead to robust performance throughout the wide range of ecosystem types that encompass the globe. Moreover, whether it is possible to create an ecosystem model with a single parameter set which is complex enough to capture system feedbacks associated with climate change, yet simple enough to robustly perform in GCMs, is a moot point. There is an ongoing need for biogeochemical modelling on a range of scales, using models of various complexities, in order to understand the factors controlling the interactions between plankton and their environment, and how these may be altered by climate change.

ACKNOWLEDGEMENTS

TRA is funded by the Natural Environment Research Council, U.K. IJT is funded by a contract from the Met Office in the U.K. We wish to thank two anonymous referees for their valuable comments on the manuscript.

REFERENCES

Aksnes, D.L., Egge, J.K., Rosland, R. and Heimdal, B.R. (1994). Representation of Emiliania huxleyi in phytoplankton simulation models. A first approach. Sarsia 79, 291-300.

Anderson, T.R. (1994). Relating C:N ratios in zooplankton food and faecal pellets using a biochemical model. J. Exp. Mar. Biol. Ecol. 184, 183-199.

Anderson, T.R. and Hessen, D.O. (1995). Carbon or nitrogen limitation in marine copepods? J. Plankton Res. 17, 317-331.

Anderson, T.R. and Williams, P.J. le B. (1998). Modelling the seasonal cycle of dissolved organic carbon at Station E1 in the English Channel. Est. Coastal Shelf Sci. 46, 93-109.

Anderson, T.R. and Williams, P.J. le B. (1999). A one-dimensional model of dissolved organic carbon cycling in the water column incorporating combined biological-photochemical decomposition. Global Biogeochem. Cycles 13, 337-349.

Anderson, T.R. and Ducklow, H.W. (2001). Microbial loop carbon cycling in ocean environments studied using a simple steady-state model. Aquat. Microb. Ecol. 26, 37-49.

Anderson, T.R. and Pondaven, P. (2003) Non-Redfield carbon and nitrogen cycling in the Sargasso Sea: pelagic imbalances and export flux. Deep-Sea Res. I 50, 573-591.

Armstrong, R.A., Lee, C., Hedges, J.I., Honjo, S. and Wakeham, S.G. (2002). A new, mechanistic model for organic carbon fluxes in the ocean based on the quantitative association of POC with ballast minerals. Deep-Sea Res. II 49, 219-236.

Azam, F., Fenchel, T., Field, J.G., Gray, J.S., Meyer-Reil, L.A. and Thingstad, F. (1983). The ecological role of water-column microbes in the sea. Mar. Ecol. Prog. Ser. 10, 257-263.

Bacastow, R. and Maier-Reimer, E. (1990). Ocean-circulation model of the carbon cycle. Climate Dynam. 4, 95-125.

Baretta, J.W., Ebenhöh, W., Ruardij, P. (1995). The European Regional Seas Ecosystem Model, a complex marine ecosystem model. Neth. J. Sea Res. 33, 233-246.

Berelson, W.M. (2001). The flux of particulate organic carbon into the ocean interior: A comparison of four U.S. JGOFS regional studies. Oceanography 14, 59-67.

Bissett, W.P., Walsh, J.J., Dieterle, D.A. and Carder, K.L. (1999). Carbon cycling in the upper waters of the Sargasso Sea: I. Numerical simulation of differential carbon and nitrogen fluxes. Deep-Sea Res. I 46, 205-269.

Bopp, L., Monfray, P., Aumont, O., Dufrense, J.-L., Le Treut, H., Madec, G., Terray, L. and Orr, J.C. (2001). Potential impact of climate change on marine export production. Global Biogeochem. Cycles 15, 81-99.

Boyd, P. and Newton, P. (1995). Evidence of the potential influence of planktonic community structure on the interannual variability of particulate organic carbon flux. Deep-Sea Res. I 42, 619-639.

Boyd, P.W. (2002). The role of iron in the biogeochemistry of the Southern Ocean and equatorial Pacific: a comparison of in situ iron experiments. Deep-Sea Res. 49, 1803-1821.

Cao M. and Woodward F.I. (1998). Net primary and ecosystem production and carbon stocks of terrestrial ecosystems and their responses to climate change. Global Change Biology 4: 185-198.

Carlotti, F. and Radach, G. (1996). Seasonal dynamics of phytoplankton and Calanus finmarchicus in the North Sea as revealed by a coupled one-dimensional model. Limnol. Oceanogr. 41, 522-539.

Carlson, C.A., Ducklow, H.W. and Michaels, A.F. (1994). Annual flux of dissolved organic carbon from the euphotic zone in the northwestern Sargasso Sea. Nature 371, 405-408.

Chai, F., Lindley, S.T. and Barber, R.T. (1996). Origin and maintenance of a high nitrate condition in the equatorial Pacific. Deep-Sea Res. II 43, 1031-1064.

Chai, F., Dugdale, R.C., Peng, T.-H., Wilkerson, F.P. and Barber, R.T. (2002). One-dimensional ecosystem model of the equatorial Pacific upwelling system. Part I: model development and silicon and nitrogen cycle. Deep-Sea Res. II 49, 2713-2745.

Chen, C.-T.A., Lim, C.-M., Huang, B.-T. and Chang, L.-F. (1996). Stoichiometry of carbon, hydrogen, nitrogen, sulfur and oxygen in the particulate matter of the western North Pacific marginal seas. Mar. Chem. 54, 179-190.

Christian, J.R., Verschell, M.A., Murtugudde, R., Busalacchi, A.J. and McClain, C.R. (2002). Biogeochemical modelling of the tropical Pacific Ocean. I: Seasonal and interannual variabililty. Deep-Sea Res. 49, 509-543.

Cooper, C. and Gordon, C. (2002). North Atlantic oceanic decadal variability in the Hadley Centre coupled model. J. Clim. 15, 45-72.

Cornell, S., Rendell, A. and Jickells, T. (1995). Atmospheric inputs of dissolved organic nitrogen to the oceans. Nature 376, 243-246.

Cox, P.M., Betts, R.A., Jones, C.D., Spall, S.A. and Totterdell, I.J. (2000). Acceleration of global warming by carbon-cycle feedbacks in a coupled climate model. Nature 408, 184-187.

Cullen, J.J., Lewis, M.R., Davis, C.O. and Barber, R.T. (1992). Photosynthetic characteristics and estimated growth rates indicate grazing is the proximate control of primary production in the equatorial Pacific. J. Geophys. Res. 97, 639-654.

Curry, R.G. and McCartney, M.S. (2001). Ocean gyre circulation changes associated with the North Atlantic Oscillation. J. Physical Oceanography 31, 3374-3400.

Daly, K.L., Wallace, D.W.R., Smith, W.O. Jr., Skoog, A., Lara, R., Gosselin, M., Falck, E. and Yager, P.L. (1999). Non-Redfield carbon and nitrogen cycling in the Arctic: Effects of ecosystem structure and dynamics. J. Geophys. Res. 104, 3185-3199.

Denman, K.L. and Peña, M.A. (1999). A coupled 1-D biological/physical model of the northeast subarctic Pacific Ocean with iron limitation. Deep-Sea Res. II 46, 2877-2908.

Dugdale, R.C. and Goering, J.J. (1967). Uptake of new and regenerated forms of nitrogen in primary productivity. Limnol. Oceanogr. 12, 196-206.

Edwards, M., Reid, P. and Planque, B. (2001). Long-term regional variability of phytoplankton biomass in the northeast Atlantic (1960-1995). ICES J. Mar. Sci. 58, 39-49.

Egge, J.K. and Heimdal, B.R. (1994). Blooms of phytoplankton including Emiliania huxleyi (Haptophyta). Effects of nutrient supply in different N:P ratios. Sarsia 79, 333-348.

Engel, A. and Passow, U. (2001). Carbon and nitrogen content of transparent exopolymer particles (TEP) in relation to their Alcian Blue adsorption. Mar. Ecol. Prog. Ser. 219, 1-10.

Falkowski, P.G. (1997). Evolution of the nitrogen cycle and its influence on the biological sequestration of CO2 in the ocean. Nature 387, 272-275.

Fasham, M.J.R., Ducklow, H.W. and McKelvie, S.M. (1990). A nitrogen-based model of plankton dynamics in the oceanic mixed layer. J. Mar. Res. 48, 591-639.

Fasham, M.J.R. (1993). Modelling the marine biota. In: The Global Carbon Cycle (ed. M. Heimann). NATO ASI Ser. Vol. I15, Springer-Verlag, Berlin Heidelberg, pp. 457-504.

Fasham, M.J.R., Sarmiento, J.L., Slater, R.D., Ducklow, H.W. and Williams, R. (1993). Ecosystem behaviour at Bermuda Station "S" and ocean weather station "India": A general circulation model and observational analysis. Global Biogeochem. Cycles 7, 379-415.

Fasham, M.J.R. (1995). Variations in the seasonal cycle of biological production in subarctic oceans: A model sensitivity analysis. Deep-Sea Res. 42, 1111-1149.

Fasham, M.J.R., Boyd, P.W. and Savidge, G. (1999). Modeling the relative contributions of autotrophs and heterotrophs to carbon flow at a Lagrangian JGOFS station in the Northeast Atlantic: The importance of DOC. Limnol. Oceanogr. 44, 80-94.

Follows MJ, Williams RG and Marshall JC (1996). The solubility pump of carbon in the subtropical gyre of the North Atlantic. Journal of Marine Research 54: 605-630.

Frost, B.W. (1991). The role of grazing in nutrient-rich areas of the open sea. Limnol. Oceanogr. 36, 1616-1630.

Gabric, A.J., Cropp, R., Ayers, G.P., McTainsh, G. and Braddock, R. (2002). Coupling between cycles of phytoplankton biomass and aerosol optical depth as derived from SeaWiFS time series in the subantarctic Southern Ocean. Geophys. Res. Lett. 29, 10.1029/2001GL013545.

Galloway, J.N., Schlesinger, W.H., Levy, H., Michaels, A. and Schnoor, J.L. (1995). Nitrogen fixation: Anthropogenic enhancement – environmental response. Global Biogeochem. Cycles 9, 235-252.

Ganeshram RS, Pedersen TF, Calvert SE, McNeill GW and Fontugne MR (2000). Glacial-interglacial variability in denitrification in the world's oceans: Causes and consequences. Paleoceanography 15: 361-376.

Geider, R.J. and La Roche, J. (2002). Redfield revisited: variability of C:N:P in marine microalgae and its biochemical basis. European J. Phycology 37, 1-17.
Gillooly, J.F. (2000). Effect of body size and temperature on generation time in zooplankton. J. Plankton Res. 22, 241-251.
Gismervik, I. (1997). Stoichiometry of some marine planktonic crustaceans. J. Plankton Res. 19, 279-285.
Goldman, J.C., Caron, D.A. and Dennett, M.R. (1987). Regulation of gross growth efficiency and ammonium regeneration in bacteria by substrate C:N ratio. Limnol. Oceanogr. 32, 1239-1252.
Gnanadesikan, A., Sarmientio, J.L. and Slater, R.D. (2003). Effects of patchy ocean fertilization on atmospheric carbon dioxide and biological production. Global Biogeochem. Cycles, 17, 1050, doi:10.1029/2002/BG001940.
Grover, J.P. (1989). Influence of cell shape and size on algal competitive ability. J. Phycol. 25, 402-405.
Gruber, N. and Sarmiento, J.L. (1997). Global patterns of marine nitrogen fixation and denitrification. Global Biogeochem. Cycles 11, 235-266.
Hedgpeth, J.W. (1977). Models and muddles. Helgoländer wiss. Meeresunters 30, 92-104.
Heinze C, Maier-Reimer E and Winn K (1991). Glacial pCO2 reduction by the world ocean: experiments with the Hamburg carbon cycle model. Paleoceanography 6: 395-430.
Hoegh-Guldberg, O. (1999). Climate change, coral bleaching and the future of the world's coral reefs. Marine Freshw. Res. 50, 839-866.
Hopkinson, C.S. Jr., Fry, B. and Nolin, A.L. (1997). Stoichiometry of dissolved organic matter dynamics on the continental shelf of the northeastern U.S.A. Cont. Shelf Res. 17, 473-489.
IPCC (2001). Climate Change 2001: The Scientific Basis. Intergovernmental Panel on Climate Change Report 2001, Cambridge University Press, 881 pp.
Jackson, G.A. (2001). Effect of coagulation on a model planktonic food web. Deep-Sea Res. I 48, 95-123.
Jackson, G.A. and Burd, A.B. (2002). A model for the distribution of particle flux in the mid-water column controlled by subsurface biotic interactions. Deep-Sea Res. II 49, 193-217.
Johnson, K.S., Gordon, R.M. and Coale, K.H. (1997). What controls dissolved iron concentrations in the world ocean? Mar. Chem. 57, 137-161.
Joos, F., Sarmiento, J.L. and Siegenthaler, U. (1991). Estimates of the effect of Southern Ocean iron fertilization on atmospheric CO2 concentrations. Nature 349, 772-775.
Karl, D., Michaels, A., Bergman, B., Capone, D., Carpenter, E., Letelier, R., Lipshcultz, F., Paerl, H., Sigman, D. and Stal, L. (2002). Dinitrogen fixation in the world's oceans. Biogeochemistry 57/58, 47-98.
Kepkay, P.E., Jellett, J.F. and Niven, S.E.H. (1997). Respiration and the carbon-to-nitrogen ratio of a phytoplankton bloom. Mar. Ecol. Prog. Ser. 150, 249-261.
Kleypas, J.A., Buddemeier, R.W., Archer, D., Gattuso, J.-P., Langdon, C. and Opdyke, B.N. (1999). Geochemical consequences of increased atmospheric carbon dioxide on coral reefs. Science 284, 118-120.
Knox, F. and McElroy, M.B. (1984). Changes in atmospheric CO2: influence of the marine biota at high latitude. J. Geophys. Res. 89, 4629-4637.
Korb, R.E., Saville, P.J., Johnston, A.M. and Raven, J.A. (1997). Sources of inorganic carbon for photosynthesis by three species of marine diatom. J. Phycol. 33, 433-440.
Lampitt RS, Hillier WR and Challenor PG (1993). Seasonal and diel variations in the open ocean concentration of marine snow aggregates. Nature 362: 737-739.
Lampitt R.S. and Antia A.N. (1997). Particle flux in deep seas: regional characteristics and temporal variability. Deep-Sea Res. I 44, 1377-1403.

Lancelot, C., Hannon, E., Becquevort, S., Veth, C. and De Baar, H.J.W. (2000). Modeling phytoplankton blooms and carbon export production in the Southern Ocean: dominant controls by light and iron in the Atlantic sector in Austral spring 1992. Deep-Sea Res. I 47, 1621-1662.

Legendre, L. and Rassoulzadegan, F. (1996). Food-web mediated export of biogenic carbon in oceans: hydrodynamic control. Mar. Ecol. Prog. Ser. 145, 179-193.

Leonard, C.L., McClain, C.R., Murtugudde, R., Hofmann, E.E. and Harding, L.W. Jr. (1999). An iron-based ecosystem model of the central equatorial Pacific. J. Geophys. Res. 104, 1325-1341.

Longhurst, A.R. and Harrison, W.G. (1988). Vertical nitrogen flux from the oceanic photic zone by diel migrant zooplankton and nekton. Deep-Sea Res. 35, 881-889.

Longhurst, A. (1995). Seasonal cycles of pelagic production and consumption. Prog. Oceanogr. 36, 77-167.

Loukos, H., Frost, B., Harrison, D.E. and Murray, J.W. (1997). An ecosystem model with iron limitation of primary production in the equatorial Pacific at 140°W. Deep-Sea Res. II 44, 2221-2249.

Maier-Reimer, E., Hasselmann, K. (1987). Transport and storage of carbon dioxide in the ocean in an inorganic ocean-circulation carbon cycle model. Climate Dynam. 2, 63-90.

Maier-Reimer, E., Mikolajewicz, U. and Winguth, A. (1996). Future ocean uptake of CO2: interaction between ocean circulation and biology. Clim. Dyn. 12, 711-721.

Malone, T.C. (1971). The relative importance of nannoplankton and netplankton as primary producers in tropical oceanic and neritic phytoplankton communities. Limnol. Oceanogr. 16, 633-639.

Margalef, R. (1978). Life-forms of phytoplankton as survival alternatives in an unstable environment. Oceanol. Acta 1, 493-509.

Martin, J.H., Knauer, G.A., Karl, D.M. and Broenkow, W.W. (1987). VERTEX: carbon cycling in the northeast Pacific. Deep-Sea Res. 34, 267-285.

Martin, J.H. and Fitzwater, S. (1988). Iron deficiency limits phytoplankton growth in the north-east Pacific subarctic. Nature 331, 341-343.

Martin, J.H., Fitzwater, S. and Gordon, R.M. (1990). Iron deficiency limits phytoplankton growth in Antarctic waters. Global Biogeochem. Cycles 4, 5-12.

Matear, RJ and Hirst, AC (1999). Climate change feedback on the future oceanic CO2 uptake. Tellus 51B, 722-733.

McCreary, J.P. Jr., Kohler, K.E., Hood, R.R., Smith, S., Kindle, J., Fischer, A.S. and Weller, R.A. (2001). Influences of diurnal and intraseasonal forcing on mixed-layer and biological variability in the central Arabian Sea. J. Geophys. Res. 106, 7139-7155.

Milliman, J.D., Troy, P.J., Balch, W.M., Adams, A.K., Li, Y.-H. and Mackenzie, F.T. (1999). Biologically mediated dissolution of calcium carbonate above the chemical lysocline? Deep-Sea Res. I 46, 1653-1669.

Mitchell, B.G., Brody, E.A., Holm-Hansen, O., McClain, C. and Bishop, J. (1991). Light limitation of phytoplankton biomass and macronutrient utilization in the Southern Ocean. Limnol. Oceanogr. 30, 1662-1677.

Moloney, C.L. and Field, J.G. (1991). The size-based dynamics of plankton food webs. I. A simulation model of carbon and nitrogen flows. J. Plankton Res. 13, 1003-1038.

Moore, J.K., Doney, S.C., Kleypas, J.A., Glover, D.M. and Fung, I.Y. (2002). An intermediate complexity marine ecosystem model for the global domain. Deep-Sea Res. II 49, 403-462.

Najjar, R.G., Sarmiento, J.L. and Toggweiler, J.R. (1992). Downward transport and fate of organic matter in the ocean: Simulations with a general circulation model. Global Biogeochem. Cycles 6, 45-76.

Nihoul, J.C.J. and Djenidi, S. (1987). Perspective in three-dimensional modelling of the marine system. In: Three-dimensional models of marine and estuarine dynamics (eds. J.C.J. Nihoul and B.M. Jamart). Elsevier Oceanography Series 45, pp. 1-33.

Nimer, N.A., Inglesias-Rodriguez, M.D. and Merrett, M.J. (1997). Bicarbonate utilization by marine phytoplankton species. J. Phycol. 33, 625-631.

Oeschger, H., Siegenthaler, U., Schotterer, U. and Gugelmann, A. (1975). A box diffusion model to study the carbon dioxide exchange in nature. Tellus 27, 168-192.

Oguz, T., Ducklow, H.W., Purcell, J.E. and Malontte-Rizzoli, P. (2001). Modeling the response of top-down control exerted by gelatinous carnivores on the Black Sea pelagic food web. J. Geophys. Res. 106, 4543-4564.

Ono, S., Ennyu, A., Najjar, R.G. and Bates, N.R. (2001). Shallow remineralisation in the Sargasso Sea estimated from seasonal variations in oxygen, dissolved inorganic carbon and nitrate. Deep-Sea Res. II 48, 1567-1582.

Oschlies, A. and Garçon, V. (1999). An eddy-permitting coupled physical-biological model of the North Atlantic 1. Sensitivity to advection numerics and mixed layer physics. Global Biogeochem. Cycles 13, 135-160.

Palmer, J.R. and Totterdell, I.J. (2001). Production and export in a global ocean ecosystem model. Deep-Sea Res. I 48, 1169-1198.

Passioura, J.B. (1996). Simulation models: Science, snake oil, education or engineering? Agron. J. 88, 690-694.

Passow, U., Shipe, R.F., Murray, A., Pak, D.K., Brzezinski, M.A. and Alldredge, A.L. (2001). Origin of transparent exopolymer particles (TEP) and their role in sedimentation of particulate matter. Cont. Shelf Res. 21, 327-346.

Peterson, BJ and Melilo, JM (1985). The potential storage of carbon caused by eutrophication of the biosphere. Tellus 37B: 117-127.

Plattner, G.-K., Joos, F., Stocker, T.F. and Marchal, O. (2001). Feedback mechanisms and sensitivities of ocean carbon uptake under global warming. Tellus B 53, 564-592.

Popova, E.E., Ryabchenko, V.A. and Fasham, M.J.R. (2000). Biological pump and vertical mixing in the Southern Ocean: Their impact on atmospheric CO2. Global Biogeochem. Cycles 14, 477-498.

Redfield, A.C., Ketchum, B.H. and Richards, F.A. (1963). The influence of organisms on the composition of sea-water. In The Sea, vol. 2 (ed. M.N. Hill). Interscience, New York, pp. 26-77.

Reynolds, J.F. and Acock, B. (1985). Predicting the response of plants to increasing carbon dioxide: A critique of plant growth models. Ecol. Modell. 29, 107-129.

Riebesell, U., Zondervan, I., Rost, B., Tortell, P.D., Zeebe, R.E. and Morel, F.M.M. (2000). Reduced calcification of marine plankton in response to increased atmospheric CO2. Nature 407, 364-367.

Riebesell, U., Zondervan, I., Bjørn, R. and Zeebe, R.E. (2001). Effects of increasing atmospheric CO2 on phytoplankton communities and the biological carbon pump. Global Change Newsletter 47, 12-15.

Ruiz, J., Prieto, L. and Ortegón, F. (2002). Diatom aggregate formation and fluxes: a modelling analysis under different size-resolution schemes and with empirically determined aggregation kernels. Deep-Sea Res. I 49, 495-515.

Ryabchenko, V.A., Gorchakov, V.A. and Fasham, M.J.R. (1998). Seasonal dynamics of biological productivity in the Arabian Sea euphotic zone as simulated by a three-dimensional ecosystem model. Global Biogeochem. Cycles 12, 501-530.

Sabine, C.L. and Mackenzie, F.T. (1991). Oceanic sinks for anthropogenic CO2. Int. J. Energy Environ. Econ. 1, 119-127.

Sanudo-Wilhelmy, S.A., Kustka, A.B., Gobler, C.J., Hutchins, D.A., Yang, M., Lwiza, K., Burns, J., Capone, D.G., Raven, J.A. and Carpenter, E.J. (2001). Phosphorus limitation of nitrogen fixation by Trichodesmium in the central Atlantic Ocean. Nature 411, 66-69.

Sarmiento, J.L, Orr, J.C. and Siegenthaler, U. (1992). A perturbation simulation of CO2 uptake in an ocean general circulation model. J. Geophys. Res. 97, 3621-3645.

Sarmiento, J.L., Slater, R.D., Fasham, M.J.R., Ducklow, H.W., Toggweiler, J.R. and Evans, G.T. (1993). A seasonal three-dimensional ecosystem model of nitrogen cycling in the North Atlantic euphotic zone. Global Biogeochem. Cycles 7, 417-450.

Sarmiento, JL, and Le Quéré, C. (1996). Ocean carbon dioxide uptake in a model of century-scale global warming. Science 274, 1346-1350.

Siegenthaler, U. (1983). Uptake of excess CO2 by an outcrop-diffusion model of the ocean. J. Geophys. Res. 88, 3599-3608.

Six, K.D. and Maier-Reimer, E. (1996). Effects of plankton dynamics on seasonal carbon fluxes in an ocean general circulation model. Global Biogeochem. Cycles 10, 559-583.

Sondergaard, M., Williams, P.J. le B., Cauwet, G., Riemann, B., Robinson, C., Terzic, S., Woodward, E.M.S. and Worm, J. (2000). Net accumulation and flux of dissolved organic carbon and dissolved organic nitrogen in marine plankton communities. Limnol. Oceanogr. 45, 1097-1111.

Stocker, T.F. and Schmittner, A. (1997). Influence of CO2 emission rates on the stability of the thermohaline circulation. Nature 388, 862-865.

Stouffer, R.J. and Manabe, S. (1999). Response of a coupled ocean-atmosphere model to increasing atmospheric carbon dioxide: sensitivity to the rate of increase. J. Clim. 12, 2224-2237.

Suess, E. (1980). Particulate organic carbon flux in the oceans – surface productivity and oxygen utilization. Nature 288, 260-263.

Tamburini, C., Garcia, J., Ragot, M. and Bianchi, A. (2002). Biopolymer hydrolysis and bacterial production under ambient hydrostatic pressure through a 2000 m water column in the NW Mediterranean. Deep-Sea Res. II 49, 2109-2123.

Taylor, A.H., Colebrook, J.M., Stephens, J.A. and Baker, N.G. (1992). Latitudinal displacements of the Gulf Stream and the abundance of plankton in the north-east Atlantic. J. Mar. Biol. Ass. UK 72, 919-921.

Taylor, A.H., Allen, J.I. and Clark, P.A. (2002). Extraction of a weak climatic signal by an ecosystem. Nature 416, 629-632.

Teira, E., Pazó, M.J., Serret, P. and Fernández, E. (2001). Dissolved organic carbon (DOC) production by microbial populations in the Atlantic Ocean. Limnol. Oceanogr. 46, 1370-1377.

Thomas, H., Ittekkot, V., Osterroht, C. and Schneider, B. (1999). Preferential recycling of nutrients – the ocean's way to increase new production and to pass nutrient limitation? Limnol. Oceanogr. 44, 1999-2004.

Tortell, P.D. (2000). Evolutionary and ecological perspectives on carbon acquisition in phytoplankton. Limnol. Oceanogr. 45, 744-750.

Turley, C.M. and Mackie, P.J. (1994). Biogeochemical significance of attached and free-living bacteria and the flux of particles in the NE Atlantic Ocean. Mar. Ecol. Prog. Ser. 115, 119-203.

Tyrrell, T. and Taylor, A.H. (1995). Latitudinal and seasonal variations in carbon dioxide and oxygen in the northeast Atlantic and the effects on Emiliania huxleyi and other phytoplankton. Global Biogeochem. Cycles 9, 585-604.

Tyrrell, T. and Taylor, A.H. (1996). A modelling study of Emiliania huxleyi in the NE Atlantic. J. Mar. Syst. 9, 83-112.

Vitousek PM and Howarth RW (1991) Nitrogen limitation on land and in the sea: how can it occur? Biogeochemistry 13, 87-115.

Vitousek, P.M. et al. (2002). Towards an ecological understanding of biological nitrogen fixation. Biogeochemistry 57/58, 1-45.

Volk T., and Hoffert, M.I. (1985). Ocean carbon pumps: Analysis of relative strengths and efficiencies in ocean-driven atmospheric CO2 changes. Geophys. Monogr., 32, 99-110.

Watson AJ, Bakker DCE, Ridgwell AJ and Law CS (2000). Effect of iron supply on Southern Ocean CO2 uptake and implications for glacial atmospheric CO2. Nature 407: 730-733.

Wells, M.L., Price, N.M. and Bruland, K.W. (1995). Iron chemistry in seawater and its relationship to phytoplankton: a workshop report. Mar. Chem. 48, 157-182.

Wood, A.W. and Van Valen, L.M. (1990). Paradox lost? On the release of energy-rich compounds by phytoplankton. Mar. Microb. Food Webs 4, 103-116.

Wroblewski, J.S., Sarmiento, J.L. and Flierl, G.R. (1988). An ocean basin scale model of plankton dynamics in the North Atlantic. 1. Solutions for the climatological oceanographic conditions in May. Global Biogeochem. Cycles 2, 199-218.

Yamanaka, Y. and Tajika, E. (1996). The role of vertical fluxes of particulate organic matter and calcite in the oceanic carbon cycle: Studies using an ocean biogeochemical general circulation model. Global Biogeochem. Cycles 10, 361-382.

Ziveri, P., Broerse, A.T.C., van Hinte, J.E., Westbroek, P. and Honjo, S. (2000). The fate of coccoliths at 48°N 21°W, northeastern Atlantic. Deep-Sea Res. II 47, 1853-1875.

Zondervan, I., Zeebe, R.E., Rost, B. and Riebesell, U. (2001). Decreasing marine biogenic calcification: a negative feedback on rising atmospheric pCO2. Global Biogeochem. Cycles 15, 507-516.

Zondervan, I., Rost, B. and Riebesell, U. (2002). Effect of CO2 concentration on the PIC/POC ratio in the coccolithophore Emiliania huxleyi grown under light-limiting conditions and different daylengths. J. Exp. Mar. Biol. Ecol. 272, 55-70.

THE DYNAMICS OF THE MARINE NITROGEN CYCLE AND ITS INFLUENCE ON ATMOSPHERIC CO_2 VARIATIONS

NICOLAS GRUBER

ngruber@igpp.ucla.edu

INTRODUCTION

The bioavailability of nutrients represents one of the most important factors controlling the strength of the biological carbon pump and ultimately the impact of ocean biology on atmospheric CO_2. Among those nutrients, the macro-nutrients nitrate (NO_3^-) and phosphate (PO_4^{3-}) play a particularly important role in limiting biological productivity as evidenced by their often near complete exhaustion in surface waters. As near surface NO_3^- concentrations are generally somewhat lower than those of PO_4^{3-} relative to the demand by phytoplankton, biological oceanographers have argued historically that NO_3^- rather than PO_4^{3-} is the primary macro-nutrient controlling phytoplankton productivity [*Smith*, 1984; *Codispoti*, 1989; *Tyrrell*, 1999]. Geologists, in contrast, regarded PO_4^{3-} as the primary controlling macronutrient [*Codispoti*, 1989]. They argued that while NO_3^- may indeed be the limiting factor at any given location and time, PO_4^{3-} is truly the limiting factor on geological time-scales, because the biologically mediated fixation of the much more abundant dinitrogen gas (N_2) into organic nitrogen is alleviating the scarcity of bioavailable nitrogen (Figure 1). Phosphate on the other hand, does not have such a biologically mediated source (Figure 1). It is therefore the geologically controlled balance between the riverine (and atmospheric) input of PO_4^{3-} and its burial on the sea-floor that ultimately controls marine biological productivity. *Tyrrell* [1999] provided a synthesis of these two views by identifying NO_3^- as the *proximate* nutrient, while giving PO_4^{3-} the role of being the *ultimate* nutrient.

Over the last two decades, the clear distinction between the two views started to disappear. A first reason is the growing evidence that marine N_2 fixation is quantitatively much more important than previously believed [*Capone*, 2001; *Fuhrman and Capone*, 2001; *Karl et al.*, 2002]. Global estimates of N_2 fix-

M. Follows and T. Oguz (eds.), The Ocean Carbon Cycle and Climate, 97–148.

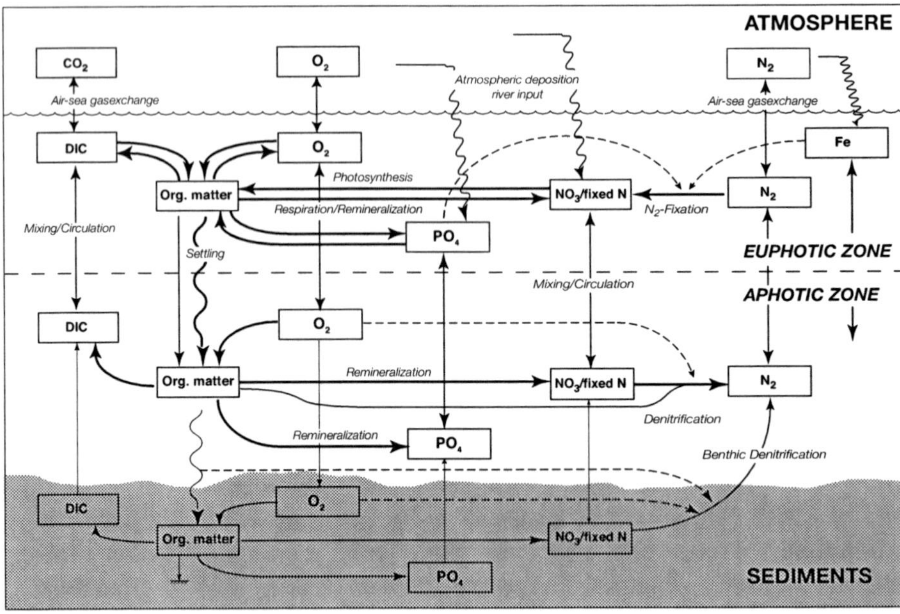

Schematic representation of the marine nitrogen cycle and its coupling to the marine oxygen, phosphorus, and carbon cycles. The marine nitrogen cycle sets itself apart from that of phosphorus by the existence of two biologically mediated processes, N_2 fixation and denitrification, which can alter substantially the amount of fixed nitrogen (mostly NO_3^-) in the ocean on timescales of a few centuries. The solid lines represent the flow and transformation of the various properties, while the dashed lines indicate a controlling role of a property on a particular process.

ation by marine phytoplankton have more than doubled since the 1980s with commonly accepted rates now in the range between 100 and 200 Tg N yr^{-1} (1 Tg = 10^{12}g) [*Falkowski*, 1997; *Gruber and Sarmiento*, 1997; *Capone et al.*, 1997; *Gruber and Sarmiento*, 2002; *Karl et al.*, 2002]. In addition, N_2 fixation appears to represent a significant fraction of total productivity in a number of subtropical and tropical areas (e.g. *Karl et al.* [1997]). Furthermore, the use of molecular techniques has led to the discovery of many new phytoplankton species that are capable of fixing N_2 (diazotrophic species) [*Zehr et al.*, 2000; *Zehr et al.*, 2001]. This expanded dramatically the previously relatively small number of diazotrophs, whose best studied genus is the cyanobacterium *Trichodesmium* [*Capone et al.*, 1997; *Karl et al.*, 2002]. These observations motivated the biological oceanographers to relax their traditional paradigm of a nitrogen limited open ocean environment to adopt one where PO_4^{3-} and possibly iron [*Falkowski*, 1997; *Karl et al.*, 2002] play an equally or even more fundamental role.

The geologists also contributed to the blurring of the distinction of the arguments in the nitrogen vs phosphorus discussion. Over the last two decades, they increasingly recognized that the marine N_2 cycle is more subtle and has numerous feedbacks that permit a behavior that is more complex than simply being slaved to the marine phosphorus content. In particular, they became increasingly aware that the total amount of fixed nitrogen (NO_3^-, ammonium, and all forms of organic nitrogen) in the ocean can be changed substantially by denitrification processes [*Ganeshram et al.,* 1995, 2000; *Altabet et al.,* 1995, 2002; *Devol*, 2002] (see Figure 1). Denitrification refers to a dissimilatory process in which NO_3^- replaces oxygen as the terminal electron acceptor during respiration. As a consequence, NO_3^- is reduced to N_2, which is not bioavailable for most phytoplankton. As the energy yield of this process is substantially lower than that of oxygen based respiration, denitrification usually does not occur in oxygenated waters, and starts becoming quantitatively important only at very low dissolved O_2 concentrations [*Hattori*, 1983]. In stark contrast to N_2 fixation, the majority of marine heterotrophic bacteria can switch to denitrification if oxygen becomes scarce. As the water column of the current ocean is generally well oxygenated, water column denitrification occurs only in a few places, the most important being the Arabian Sea [*Naqvi*, 1987; *Burkill et al.*, 1993; *Howell et al.*, 1997] and the Eastern Tropical Pacific [*Brandhorst*, 1959; *Cline and Richards*, 1972; *Codispoti and Richards*, 1976; *Deutsch et al.*, 2001]. By contrast, benthic denitrification is more uniformly distributed, with highest activities in the organic rich sediments along the continental margins [*Devol*, 1991; *Christensen et al.*, 1987; *Hammond et al.*, 1999; *Jahnke and Jahnke*, 2000].

Recent estimates of global denitrification have also increased dramatically over the last few decades. *Codispoti et al.* [2001] suggested that oceanic denitrification might be as large as 450 Tg N yr^{-1}, with the majority (300 Tg N yr^{-1}) caused by benthic denitrification and the remaining (150 Tg N yr^{-1}) coming from water column denitrification. As evident from Table 1, this would cause a large imbalance (order of 200 Tg N yr^{-1}) in the present-day marine nitrogen budget. However, uncertainties of the terms in Table 1 are large and it is therefore unclear whether such an imbalance exists today, possibly as a consequence of the massive alteration of the global nitrogen cycle by humans [*Codispoti et al.*, 2001; *Falkowski et al.*, 2000] or whether any of the source/sink terms have been seriously under/overestimated [*Brandes and Devol*, 2002]. In this paper, among other issues, I will review this unresolved issue, and propose an updated budget (see Table 1).

Irrespective of whether the current marine nitrogen cycle is balanced or not, it is important to realize that the large biologically mediated fluxes of nitrogen in and out of the marine fixed nitrogen pool lead to a very dynamic marine nitrogen cycle, with residence times of less than 3000 years [*Gruber and*

Present-day Global Marine Nitrogen Budgets of [2001], [2002], and updated in this study

Process	Codispoti et al.[a] 2001 Tg N yr^{-1}	Gruber and Sarmiento[a] 2002 Tg N yr^{-1}	this study Tg N yr^{-1}
Pelagic N_2 fixation	117	117 ± 40	120 ± 50 [b]
Benthic N_2 fixation	15	15 ± 10	15 ± 10
River input (DON)	34	34 ± 10	35 ± 10 [b]
River input (PON)	42	42 ± 10	45 ± 10 [b]
Atmospheric deposition	86	30 ± 5	50 ± 20 [c]
Total sources	294	238 ± 44	265 ± 55
Organic N export	1		1
Benthic denitrification	300	95 ± 20	180 ± 50 [d]
Water column denitrification	150	80 ± 20	65 ± 20 [e]
Sediment Burial	25	25 ± 10	25 ± 10
N_2O loss to atmosphere	6	4 ± 2	4 ± 2
Total sinks	482	204 ± 29	275 ± 55

a: See *Gruber and Sarmiento* [2002] and *Codispoti et al.* [2001] for details.
b: Rounded relative to *Gruber and Sarmiento* [2002].
c: Increased by 20 Tg N yr^{-1} to take into consideration enhanced DON deposition onto the open ocean [*Cornell et al.*, 1995]. Increased error estimate to reflect large uncertainty associated with this estimate.
d: Adjusted to be in agreement with the ^{15}N isotopic constraint proposed by *Brandes and Devol* [2002]. See text for details.
e: Adjusted downward to take the possible contribution of benthic remineralization into account. Increased error bar to reflect this uncertainty.

Sarmiento, 1997; *Brandes and Devol*, 2002]. This stands in stark contrast to the residence time of phosphorus in the ocean, which is presently estimated to be about 30,000 to 50,000 years [*Delaney*, 1998]. As a consequence of this short residence time, small imbalances, if they persist for several hundred years or more, can cause substantial changes in the marine inventory of fixed nitrogen, with potential impacts on marine productivity and atmospheric CO_2 [*Devol*, 2002]. Assuming that fixed nitrogen is the only factor controlling marine productivity, *Gruber and Sarmiento* [1997] estimated that an imbalance of 100 Tg N yr^{-1} can cause a net air-sea flux of about 0.1 Pg C yr^{-1}. If such an imbalance continued for 1000 years, the marine fixed N inventory would change by about 30% and atmospheric CO_2 by about 50 ppm.

McElroy [1983] was first in noticing this large leverage of the marine nitrogen cycle on atmospheric CO_2, and proposed that it contributed substantially to the observed past variations in atmospheric CO_2. In particular, he proposed that the ocean oscillates between a state of excess fixed nitrogen during glacial periods and a state of deficit during interglacial periods. Such imbalances can occur because of changes in both the source and sink terms. *McElroy* [1983] suggested that the nitrogen excess during the cold periods was due to erosion of organic nitrogen rich sediments along the exposed continental margins. On the other hand, *Falkowski* [1997] and *Broecker and Henderson* [1998] argued, for example, that the excess during glacial times is a consequence of increased N_2 fixation, while *Altabet et al.* [2002] and *Ganeshram et al.* [2000] suggested that it is because of reduced denitrification. While these processes were proposed to act mainly in response to climate change, *Michaels et al.* [2001] and *Karl et al.* [2002] went one step further and postulated the existence of fully developed feedback loops, whereby changes in the marine nitrogen cycle would impact climate through changes in atmospheric CO_2, and where a changed climate would impact marine N_2 fixation through changes in the delivery of dust to the ocean. The latter link builds on the hypothesis that iron is one of the main controls on marine N_2 fixation.

Before it is possible to discuss and evaluate these hypotheses, it is necessary to understand the factors that control N_2 fixation and denitrification. Of particular concern is how these two processes are linked to each other. While imbalances might have occurred in the past and contributed to variations in the inventory of fixed nitrogen in the ocean, these variations appear to have remained within relatively well defined limits. This requires feedback mechanisms between N_2 fixation and denitrification that prevent the marine N cycle from running out of control. Accepting the basic tenet that fixed nitrogen is a primary factor controlling marine productivity, these feedback mechanisms must have operated efficiently over the last 10,000 years, as the absence of atmospheric CO_2 variations above $\pm$ 10 ppm [*Indermühle et al.*, 1999] indicates that marine productivity remained relatively constant over this period.

A related puzzle arises by inspecting the global-scale oceanic distribution of NO_3^- versus PO_4^{3-} (Figure 2a), which reveals a global trend that is well described by a line with a slope of about 16:1. This slope corresponds to the N:P ratio of demand and composition of marine phytoplankton and suggests that photosynthesis, respiration, and remineralization are controlling most of the observed variability of NO_3^- and PO_4^{3-}. The existence of a relative constant N:P ratio of these processes and their impact on the chemical composition of seawater has been embodied by the Redfield ratio concept [*Redfield et al.*, 1963]. Although this concept explains why most points lie on a 16:1 line, it does not provide an answer why the intercept of the line is close to zero. A corollary question is why the ratio of the total inventory of NO_3^- and PO_4^{3-} in the ocean is close to the N:P ratio of phytoplankton (see also Figure 2b and 2c).

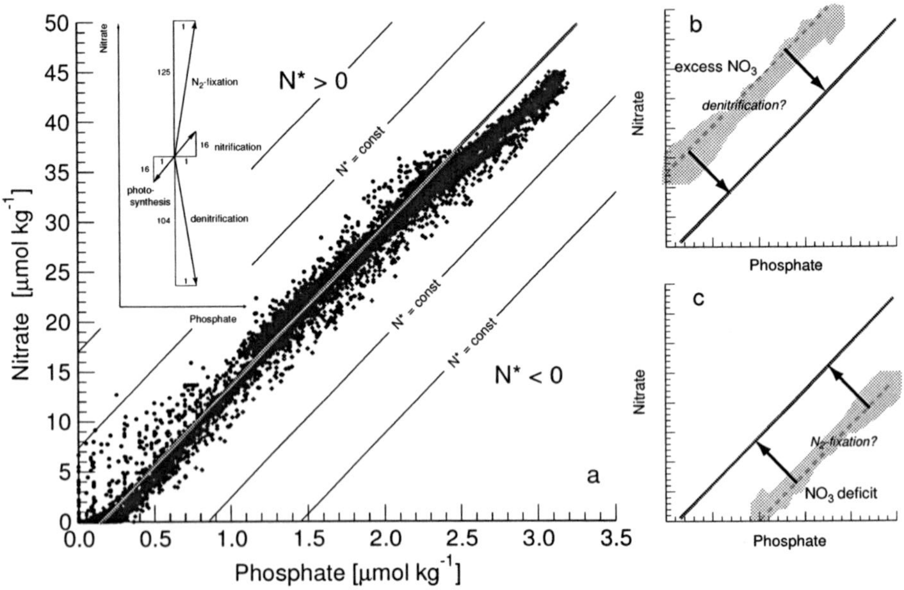

(a) Plot of NO_3^- versus PO_4^{3-} from all depths and from selected WOCE cruises in all ocean basins (A16,P16,I8NI9S). The inset shows how the various processes influence the NO_3^- versus PO_4^{3-} distribution. The solid line represents the mean ocean trend with a slope of 16:1, while the thinner lines show trends of constant N^* (see text for definition). (b) Hypothetical distribution of NO_3^- versus PO_4^{3-} in a situation of excess NO_3^- (c) As (b), except for a situation of a NO_3^- deficit. It is unclear why the intercept of the mean oceanic trendline in (a) is so close to zero.

A first hypothesis is that the mean ocean NO_3^- to PO_4^{3-} ratio reflects the N:P ratio of the main external inputs and outputs. Apart from N_2 fixation and denitrification, the main inputs for NO_3^- and PO_4^{3-} are river flow and atmospheric deposition, while the main loss term is burial in ocean sediments. As it turns out, the ratios of fixed nitrogen and phosphorus in rivers and of atmospheric deposition are typically much higher than 16:1 [*Meybeck*, 1993]. Therefore, if the mean ocean balance were set by these input and loss processes alone, one would expect an intercept with strongly positive NO_3^- values (or similar to Figure 2b). Denitrification appears to be removing this excess fixed nitrogen from the system and brings the intercept down to near zero. But what keeps the system from running into a strong NO_3^- deficit? N_2 fixation is the likely candidate process as it adds nitrogen to the system without adding PO_4^{3-}. The observation of a near zero intercept indicates that these processes are operating in a coupled manner, with limited ability to deviate strongly. But what are the mechanisms that control the two processes and particularly which factors lead to the coupling between them? This is a particularly interesting puzzle, as N_2 fixation and denitrification are generally well separated in both space and time, with N_2 fixation occurring in the surface ocean of the low latitudes, whereas the majority of denitrification occurs in shallow sediments and in the oxygen minimum zones of the Pacific and Indian Oceans.

In summary, two major puzzles of the marine nitrogen cycle and its dynamics emerge. The first refers to the long term mean ocean nitrogen cycle and asks the question why the interactions between the source processes (N_2 fixation) and sink processes (denitrification) cause an ocean inventory of NO_3^- and PO_4^{3-} that is very close to that required by marine phytoplankton. The second puzzle refers to the temporal dynamics of the marine nitrogen cycle and tries to elucidate if and how the marine nitrogen cycle varied in the past and how it contributed to observed atmospheric CO_2 variations. Both puzzles are closely tied to the question of what controls oceanic N_2 fixation and denitrification and how these two processes are linked to each other.

This overview chapter makes an attempt to provide some answers to these questions, starting with a review of what is known about N_2 fixation and denitrification on the basis of oceanic observations of inorganic nutrients and their isotopic composition. I then discuss the implications of these geochemical findings on the present ocean nitrogen budget. I will then review our understanding of the marine nitrogen cycle and its controls. I end the chapter with a critical evaluation of the proposed mechanisms that might have contributed to the observed past variations in atmospheric CO_2. The chapter will not be able to provide answers to all questions, though, and as often, we end up with more new questions than answers. Addressing these challenges will be a topic of great importance in the future.

1. THE WATER COLUMN SIGNATURES OF N_2-FIXATION AND DENITRIFICATION

It is currently difficult to obtain global-scale constraints on N_2 fixation and denitrification from direct rate measurements. This is a result of both methodological challenges and the high spatio-temporal variability of these processes, which makes undersampling a chronic problem [*Karl et al.*, 2002]. While direct measurements should be the method of choice for improving the understanding of the detailed functioning of these processes, indirect geochemical methods that integrate over space and time are providing one of the few ways how the undersampling problem can be alleviated and large-scale constraints be obtained. I discuss here two such geochemical approaches: the first one is based on the tracer N^* [*Gruber and Sarmiento*, 1997] and the second one consists of the $^{14}N/^{15}N$ stable isotopic composition of nitrogen.

1.1 GLOBAL PATTERN OF N_2 FIXATION AND DENITRIFICATION FROM N^*

The quasi-conservative tracer N^* has been defined in such a manner that it eliminates all processes occurring with a N:P ratio of 16:1, i.e. photosynthesis, respiration, and remineralization, so that variations in the distribution of N^* reflect only the net impact of N_2 fixation, denitrification, and any other process that adds or removes NO_3^- with a N:P stoichiometry different than 16:1. The latter category includes processes such as luxury uptake of PO_4^{3-}, preferential remineralization, atmospheric deposition, river input, and PO_4^{3-} scavenging effects.

The elimination of the main processes controlling the variability of NO_3^- and PO_4^{3-} is achieved by defining a linear combination of the two nutrients: $N^* = NO_3^- - 16*PO_4^{3-} + 2.9$ μmol kg^{-1} (see *Gruber and Sarmiento* [2002] for an in depth derivation). As illustrated in Figure 2a, N^* is constant for any line with a slope of 16:1 and represents a measure of the vertical distance for a NO_3^- and PO_4^{3-} pair from the mean oceanic trend line. Points on the right side of the mean trend line have a deficit of NO_3^- relative to PO_4^{3-} and therefore have negative N^* concentrations, while points on the left side have a surplus of NO_3^- relative to PO_4^{3-} and positive N^* values. In order to interpret variations in N^* correctly, it is important to recognize that only the non-conservative component of N^* reflects a net source or sink. The absolute value of N^* is arbitrary. In general, we assume that the contribution of processes that deviate from a 16:1 stoichiometry is small, and interpret therefore all non-conservative variations in N^* as due to denitrification and N_2 fixation. This assumption, however, must be evaluated in each particular application of the method.

The biggest strength of N^* is that it provides a global-scale picture of the sources and sinks for fixed nitrogen. N^* can also be combined with tracers reflecting the ventilation age of waters to compute actual rates. A drawback of the tracer N^* is that it reflects only the net impact of N_2 fixation and denitrification. However, the interpretation is greatly aided by the fact that these two processes are often spatially and temporally separated. In addition, concurrent analyses of the isotopic $^{14}N/^{15}N$ ratio can also help to untangle the various processes that lead to variations in N^* (see below).

1.1.1 N^* distribution. Figure 3 shows mean profiles of N^* in 45° latitude bands within the three ocean basins. A first remarkable observation is the existence of a gradual decrease from the North Atlantic over the Southern Ocean into the Pacific and Indian Oceans. This suggests that, from a global perspective, the North Atlantic ocean acts as a net source of fixed nitrogen to the world ocean, while the Indian and North Pacific Oceans act as net sinks. This implies a net transport of fixed nitrogen from the Atlantic into the other

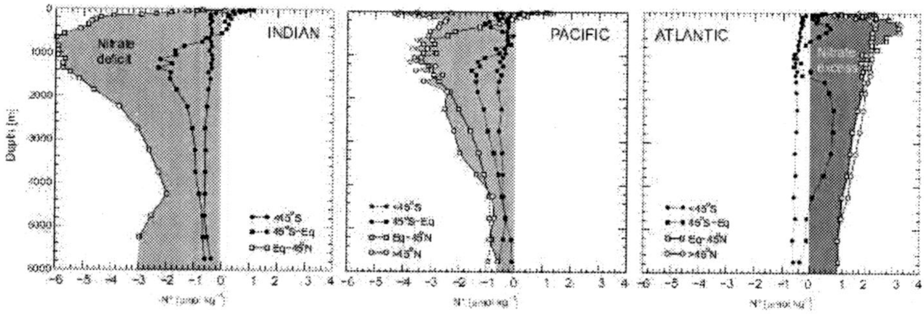

Horizontally averaged profiles of N^* in (a) the Indian, (b) the Pacific, and (c) the Atlantic Oceans. The profiles were computed by averaging all data within the ocean basin and the four latitude bands considered. The NO_3^- and PO_4^{3-} data stem from the World Ocean Circulation Experiment (WOCE) and selected pre-WOCE cruises and were individually adjusted to common deep water trends by the Global Ocean Data Analysis Project (GLODAP).

ocean basins. A closer inspection reveals that these interbasin trends are dominated by a large maximum of N^* in the thermocline of the North Atlantic and deep minima of N^* in the thermoclines of the Pacific and Indian Oceans. A second positive anomaly of N^* can be found in the near surface waters of the equator to 45°N band in the North Pacific, although this feature is substantially less pronounced than the North Atlantic maximum. It's important to note that N^* shows depth trends that are different from those expected as a result of preferential remineralization of PO_4^{3-}, since the latter process would cause a decrease of N^* with depth in the upper ocean followed by a strong increase in the deeper ocean. Although such a pattern can be found in some ocean basins (e.g. tropical Pacific, North Pacific and tropical Indian), we will show below that each occurrence can be traced to a fixed N deficit caused by denitrification. We conclude that preferential remineralization of PO_4^{3-} is relatively unimportant on the basin-scales of interest here.

In order to investigate the N^* distribution in more detail, we turn to global maps of N^* on two isopycnal surfaces (Figures 4). The shallower surface (potential density $\sigma_\theta = 26.50$) represents subtropical mode waters (mean depth of about 300 m) and the deeper surface ($\sigma_\theta = 27.10$) reflects subpolar mode waters and lies at a mean depth of about 800 m. These maps reveal that the N^* maximum in the North Atlantic extends across the entire basin. These maps also show that the thermocline minima of N^* seen in the mean profiles of the Pacific (Figure 3b) are mainly localized in the eastern Pacific and are actually separated into two zones, one north of the equator and the other south of it. Finally, the maps reveal that the minimum of N^* in the northern Indian ocean is mainly caused by the Arabian Sea and only to a lesser degree by the Bay of Bengal. We discuss the N^* maxima in the next section and the N^* minima in a following section.

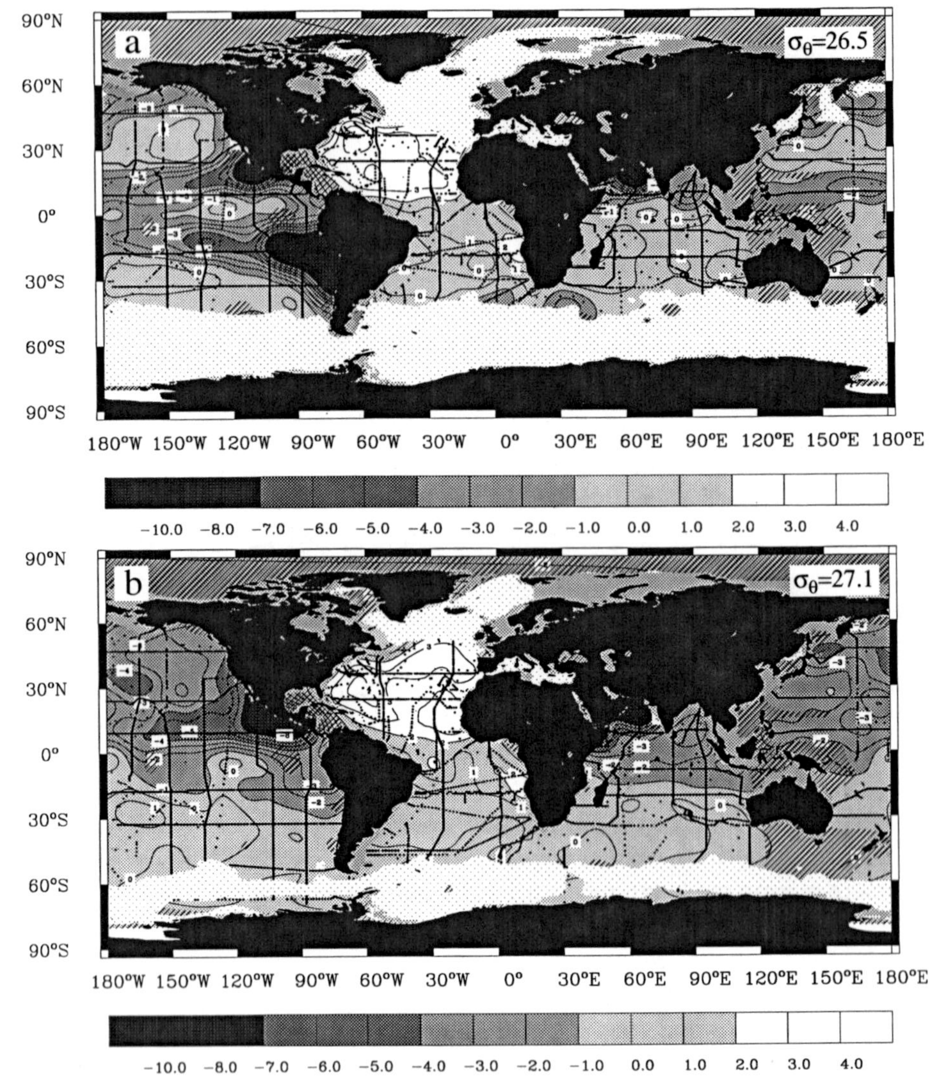

Global plots of objectively analyzed N^* (μmol kg^{-1}) on isopycnal surfaces. (a) N^* on the σ_θ = 26.50 surface. This surface is characteristic of subtropical mode waters, such as the 18° Water in the North Atlantic, (b) N^* on the σ_θ = 27.10 surface representing subpolar mode Waters. Solid circles denote the stations; stippling denotes areas where the waters of this potential density are not present in winter time. Cross-hatched areas are regions where the estimated error in objectively analyzed N^* is greater than 0.7 μmol kg^{-1}. Data are from the WOCE program augmented by selected pre-WOCE cruises and were adjusted by the GLODAP project (see also Figure 3).

1.1.2 Patterns and rates of N_2 fixation. A more detailed inspection of the N^* maximum in the subtropical North Atlantic (Figure 4) reveals that the concentration of N^* increases as waters move from the outcrop region in the northeastern part of the gyre into the center. As most of these isopycnal slabs have single end-members, this indicates that there must be an ocean internal source of N^*, which we assume to be primarily the remineralization of nitrogen rich organic material stemming from diazotrophic organisms. Figure 5 shows that the vertical extent of this maximum is confined to the main thermocline and the waters of Mediterranean origin, which occupy the depth range between about 800 and 1100m in the mid-latitude North Atlantic.

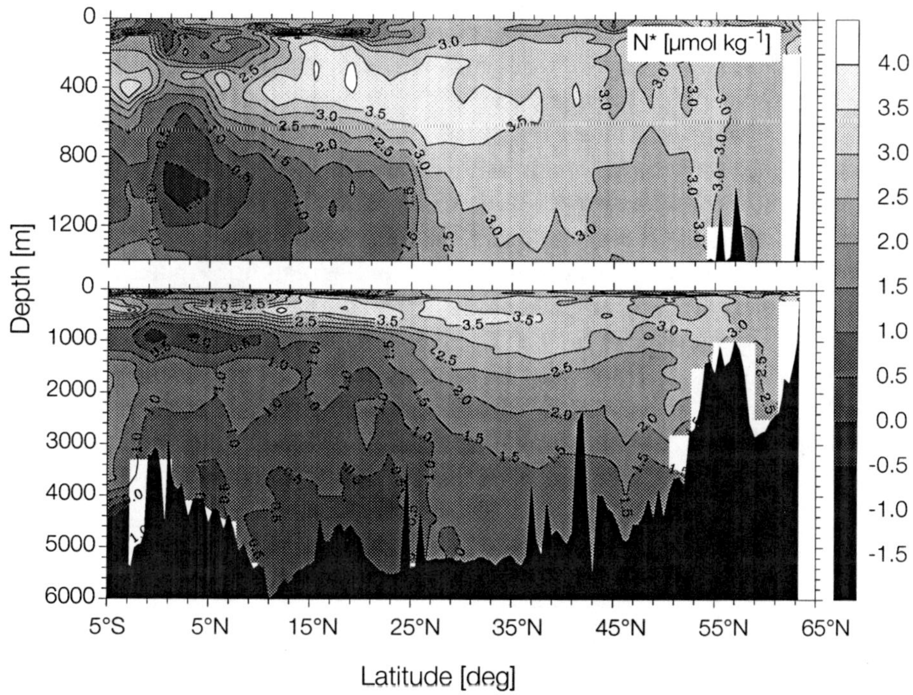

Meridional section of N^* in μmol kg^{-1} in the eastern North Atlantic along approximately 20°W. Data are from the A16N repeat occupation in 2003 by the CO_2/ Climate Variability (CLIVAR) repeat hydrography program.

By combining these non-conservative changes in N^* with tracers that provide an estimate of how the waters age as they move along in the ocean interior, *Gruber and Sarmiento* [1997] estimated a formation rate of N^*. Taking into account a small contribution from atmospheric deposition, *Gruber and Sarmiento* [1997] interpreted this non-conservative increase of N^* as a result of the remineralization of nitrogen rich organic material from diazotrophic organisms. In order to convert a formation rate of N^* into an estimate of N_2 fixation, it is necessary to assume a N:P value for diazotrophic organisms. *Gruber and Sarmiento* [1997] assumed a value of 125:1 based on measurements obtained during a *Trichodesmium* bloom near Hawaii [*Karl et al.*, 1992] and computed a North Atlantic wide N_2 fixation rate of 28 Tg N yr^{-1}. However, recent laboratory measurements and in situ observations indicate substantially lower values during non-bloom conditions [*Karl et al.*, 2002]. With a diazotrophic N:P ratio of 40:1, the estimate by *Gruber and Sarmiento* [1997] would increase to about 40 Tg N yr^{-1}.

Hansell et al. [2003] recently reexamined the North Atlantic N^* distribution using new data collected as part of the World Ocean Circulation Experiment (WOCE). They found an areal fixation rate that is similar to that of *Gruber and Sarmiento* [1997], but concluded that the volume over which the non-conservative increase of N^* is occurring is substantially smaller than was assumed by *Gruber and Sarmiento* [1997]. Employing also a N:P value of 125:1 for the diazotrophs, they suggest that the North Atlantic N_2 fixation rate is about 4 Tg N yr^{-1} only. Their estimate would increase to 6 Tg N yr^{-1} if they had chosen a diazotrophic N:P value of 40:1. The spatial coverage with high quality nutrient and age tracer data is currently insufficient to clearly determine the areal extent over which N^* is accumulating. Figures 4 and 5 suggest, however, that it is larger than the 15°N to 25°and 25°W to 75°W region assumed by *Hansell et al.* [2003]. Support for a higher North Atlantic N_2 fixation rate comes from another geochemical study by *Michaels et al.* [1996]. They used the observed N^* distribution near Bermuda and combined it with a tritium-helium calibrated model of the North Atlantic ventilated thermocline [*Sarmiento*, 1983], and suggested that the North Atlantic N_2 fixation rate might be as high as 50- 90 Tg N yr^{-1}. However, *Gruber and Sarmiento* [1997] showed that this latter estimate is likely an overestimate because *Michaels et al.* [1996] had chosen too low N^* end-members and therefore overestimated the increase in N^*. In summary, while it appears that the exact magnitude of N_2 fixation in the North Atlantic remains elusive, it is likely of the order of ten to several tens of Tg N yr^{-1}.

These geochemical N_2 fixation estimates are an order of magnitude higher than previous estimates based on *Trichodesmium* abundance and direct rate measurements in the same region [*Capone and Carpenter*, 1982]. More recent direct rate experiments of N_2 fixation in *Trichodesmium* colonies from the Bermuda Atlantic Timeseries Study (BATS) site near Bermuda by *Orcutt et al.* [2001] support significantly higher fixation rates than older studies suggested [*Carpenter and Price*, 1977], but still smaller than the geochemical estimates. However, N_2 fixation is known to occur not only in *Trichodesmium* but in a number of additional species [*Villareal and Carpenter*, 1989; *Zehr et al.*,

2000]. Their quantitative contribution to large-scale N_2 fixation is presently unknown, although *Zehr et al.* [2001] speculated that this could be substantial. An alternative possibility is that the majority of the N^* signal in the subtropical gyre of the North Atlantic is not due to local N_2 fixation, but reflects the well characterized high N_2 fixation in the Caribbean [*Carpenter*, 1983; *Carpenter and Romans*, 1991] and western tropical North Atlantic [*Carpenter et al.*, 1999], which export this excess nitrogen as dissolved organic nitrogen (DON) that then gets remineralized within the gyre. This would require substantial gradients in DON between the Caribbean and the North Atlantic subtropical gyre. Presently available DON data are insufficient to investigate this hypothesis (D. Hansell, pers. comm., 2003).

Support for the interpretation of the signals as a local phenomenon come from the observation that the $^{15}N/^{14}N$ ratios of NO_3^- and organic matter in the thermocline of the subtropical gyre of the North Atlantic are near atmospheric values. This suggests that a substantial fraction of the nitrogen comes from the atmosphere by N_2 fixation (see discussion below).

It is also interesting to point out that the North Atlantic areal average N_2 fixation rate determined by *Gruber and Sarmiento* [1997] and *Hansell et al.* [2003] is of about the right magnitude in order to provide the nitrogen source that fuels the biologically driven spring to summer drawdown in dissolved inorganic carbon typically observed near Bermuda [*Keeling*, 1993; *Michaels et al.*, 1994; *Bates et al.*, 1996; *Gruber et al.*, 1998; *Anderson and Pondaven*, 2003]. Although this agreement in magnitude is enticing, the absence of high in situ N_2 fixation rates by *Trichodesmium* [*Orcutt et al.*, 2001] and the unexplained phosphorus source make this suggestion tentative. Vertical migration, large-scale horizontal transport of dissolved organic phosphorus, and PO_4^{3-} sparing effects have been suggested (see discussion in *Karl et al.* [2002]), but none has been demonstrated to be of quantitative importance.

Interpretation of the N^* distribution in the Pacific and Indian oceans is more difficult because of the existence of strong denitrification sites adjacent to the places where one expects to find the N_2 fixation signals. In the subtropical North Pacific, where N_2 fixation is well documented [*Karl et al.*, 1997], the N^* maximum exists just near the surface and does not extend into the thermocline as is the case in the subtropical North Atlantic. This is likely a result of the eroding effect by waters that obtain extremely low N^* values in the eastern Pacific and are then advected westwards in the main thermocline (Figure 4).

Rather than trying to separate the N_2 fixation signal from the denitrification signal, *Deutsch et al.* [2001] decided to estimate N_2 fixation in the Pacific by establishing a nitrogen budget for this basin. Assuming steady state, they estimated a N_2 fixation rate by differencing fixed nitrogen transport across enclosing sections, total rates of denitrification, atmospheric deposition and river fluxes. They arrived at a N_2 fixation estimate of about 59 Tg N yr^{-1}. N_2 fixation is likely not uniformly distributed over the Pacific Ocean. If the near surface distribution of N^* is taken as an indication of where this fixation is most likely taking place, the northwestern and the southwestern Pacific emerge. This distribution is roughly consistent with observations of *Trichodesmium* abundance [*Capone and Carpenter*, 1982; *Carpenter*, 1983; *Carpenter and Capone*, 1992], isotopic values of $^{15}N/^{14}N$ [*Wada and Hattori*, 1991], remote sensing [*Dupoy et al.*, 2000], and dissolved organic nitrogen concentrations [*Hansell and Feely*, 2000].

In the Indian Ocean, near surface values of N^* are slightly elevated [*Gruber and Sarmiento*, 1997]. As these elevated values are located just above some of the lowest N^* values found anywhere in the world oceans, they could indicate rather strong N_2 fixation. Observations of *Trichodesmium* abundance [*Carpenter*, 1983] and a few direct rate measurements (reported in *Capone et al.* [1997]) appear consistent with such a conclusion. However, no basin-scale estimate of N_2 fixation has been attempted yet for the Indian Ocean.

1.1.3 Pattern and rates of water column denitrification. We turn now to the observed N^* minima (Figures 3 and 4), which are indicative of denitrification. We will focus on water-column denitrification, since this process leads to the main N^* signals that we observe in the open ocean. However, N^* is also influenced by benthic denitrification, especially near continental shelf regions [*Christensen et al.*, 1987; *Devol*, 1991]. This can be seen in Figure 4, where N^* exhibits minimum concentrations all along the eastern shelf of North America, a region where benthic denitrification is known to occur [*Seitzinger and Giblin*, 1996].

The strong N^* minima in the eastern tropical North Pacific (ETNP) and eastern tropical South Pacific (ETSP) (Figures 4) coincide with oxygen minimum zones, where water column denitrification has been well studied since *Brandhorst* [1959] noted the existence of a secondary nitrite maximum due to denitrification (see *Hattori* [1983] for a review). The existence of strong N^* minima within these known denitrification regions provides confirmation of our interpretation of N^* as a tracer reflecting N_2 fixation and denitrification. The cores of the two N^* minima have long tails that reach far westward into the center of the Pacific. The shape and position of these tails, together with their high O_2 concentrations, suggest that these features are the result of westward advection of the N^* signals in the North and South Equatorial currents. These waters slowly lose their signature as they mix with eastward moving waters in the equatorial undercurrent system and as they entrain subtropical waters, both having elevated N^* concentrations.

Codispoti and Richards [1976] estimated total denitrification in the ETNP to be about 19 Tg N yr^{-1} on the basis of in situ measurements of denitrification rates and extrapolation of these measurements to the entire denitrification zone. Applying the same technique, *Codispoti and Packard* [1980] arrived at an estimate of about 23 Tg N yr^{-1} for the ETSP. Given the difficulty of extrapolating a few individual rate measurements, the uncertainty of these estimates is large. This problem can be avoided using N^*, which integrates denitrification over space and time and is therefore less susceptible to short-term variability. *Deutsch et al.* [2001] followed this approach and used the disappearance of N^* in these regions together with water age from CFC observations to estimate integrated denitrification in these two regions. They found a water column denitrification of about 22 Tg N yr^{-1} for the ETSP and of about 26 Tg N yr^{-1} for the ETNP, in good agreement with the extrapolation of direct rate measurements. These values were recently challenged by *Codispoti et al.* [2001], who argued on the basis of a few excess N_2 observations that the estimates on the basis of NO_3^- deficits underestimate denitrification by about 50%. He therefore suggested that denitrification in the ETSP and ETNP needs to be increased to about 75 Tg N yr^{-1}.

The third strong minimum in the N^* distribution occurs in the Arabian Sea, another well known oxygen minimum zone (Figure 4) with high rates of water column denitrification [*Naqvi et al.*, 1982; *Naqvi*, 1987; *Mantoura et al.*, 1993]. Low concentrations of N^* are also found in the Bay of Bengal, where denitrification has been less studied [*Naqvi et al.*, 1978]. *Hattori* [1983] speculated that the Bay of Bengal may contribute to oceanic denitrification, as very low oxygen concentrations are found there. However, no active denitrification has been found yet [*Howell et al.*, 1997]. Rates of denitrification in the Arabian Sea have been estimated following various techniques. Recently, *Howell et al.* [1997] used 'NO', a geochemical tracer similar to N^* to arrive at an integrated denitrification of about 21 Tg N yr^{-1}. This compares favorably with an extrapolation of direct rate measurements [*Naqvi and Shailaja*, 1993] and box models [*Naqvi*, 1987]. *Codispoti et al.* [2001] argued, however, that an estimate of 60 Tg N yr^{-1} is more appropriate. They based their argument mainly on the observation that NO_3^- deficits are several times smaller than the observed N_2 excesses and that the reason why methods that rely on NO_3^- deficits such as N^* or NO are biased low is because they fail to take into account denitrification processes that do not use NO_3^-, but other forms of fixed nitrogen.

One such process is the anammox process [*Devol*, 2002], which is the anaerobic oxidation of ammonium with nitrite to N_2. While this process was known to occur in marine sediments, *Kuypers et al.* [2003] and *Dalsgaard et al.* [2003] demonstrated recently that it occurs also in the water column. However, these studies stem from the Black Sea and from a deep coastal Bay in Costa Rica, and there is no evidence yet that this process is quantitatively important in the major water column denitrification zones. Furthermore, the argument that the anammox reaction would not leave a signal in N^* is incorrect. This is because N^* looks only at the end products of remineralization and it is not unreasonable to assume that the ammonium and nitrite that get converted to N_2 were associated with PO_4^{3-} when these nitrogen compounds were still in the form of organic matter. Thus, an organic matter remineralization reaction that involves an anammox reaction would lead to an increase in PO_4^{3-} without a corresponding increase in NO_3^- and therefore would be recorded as a decrease in N^*. This would be interpreted correctly as a loss term for fixed nitrogen. I am therefore reluctant to dismiss the N^* based estimates of water column denitrification and remain doubtful of the N_2 excess based arguments that suggest a doubling of water column denitrification rates.

But the higher than expected N_2 excesses desribed by *Codispoti et al.* [2001] nevertheless need to be explained in order to discount the possibility for much higher water column denitrification rates. One possibility is the influence of N_2 emmanating from the sediments, another is that some of these N_2 excesses are generated at the surface by strong heating and/or injection of bubbles by breaking waves [*Craig and Hayward*, 1987; *Emerson et al.*, 1991].

1.2 CONSTRAINTS FROM THE $^{15}N/^{14}N$ RATIO OF NO_3^-

A complimentary geochemical tracer to N^* exists in the stable isotopic ratio $^{15}N/^{14}N$. As natural deviations in the relative abundance between the abundant isotope ^{14}N (99.634 %) and the rare isotope ^{15}N (0.366 %) are small, the isotopic ratio $^{15}N/^{14}N$ is usually expressed as a reduced isotopic ratio, where $\delta^{15}N = ((^{15}N/^{14}N)_{sample}/(^{15}N/^{14}N)_{standard}-1)$, where the isotopic ratio of atmospheric N_2 has been adopted as $(^{15}N/^{14}N)_{standard}$. While the different isotopes of an element behave essentially identically in chemical reactions and physical transfers, there are nevertheless small differences in their reactivity and transfer speeds, mainly as a result of their differing masses. In most cases, the lighter isotope ^{14}N, reacts faster than the heavier isotope ^{15}N, generating a product pool that tends to be depleted in ^{15}N (low $\delta^{15}N$ values), and an educt pool that tends to be enriched in ^{15}N (high $\delta^{15}N$ values). These isotopic fractionations not only cause variations in isotopic ratios between the various oceanic nitrogen pools, but lead also to spatiotemporal variations of the isotopic ratio within a pool. Nitrogen isotopes act therefore much like "dyes" whose colors are determined by the history of processes acting on the nitrogen pool as well as where the nitrogen in this pool has come from [*Owens*, 1987].

1.2.1 Processes and Fractionations.

The most important process affecting the $\delta^{15}N$ value of NO_3^- is water column denitrification [*Cline and Kaplan*, 1975] (see Figure 6). This process has an isotopic enrichment (ε) of about -20 ‰[*Brandes et al.*, 1998; *Voss et al.*, 2001; *Brandes and Devol*, 2002]. The isotopic enrichment, ε, is very nearly approximated by the difference in the

δ^{15}N value of the product (N_2 in the case of denitrification) and the educt (NO_3^- in this case), i.e. $\varepsilon = \delta^{15}N_{\text{prod}} - \delta^{15}N_{\text{educt}}$. In sharp contrast, benthic denitrification has a small effective isotopic enrichment of -1.5 ‰[*Liu*, 1979; *Brandes and Devol*, 2002; *Sigman et al.*, 2003]. This surprising fact is not due to the enzymatic fractionation of denitrification being different in sediments or the water column, but due to the fact that benthic denitrification tends to use up all available NO_3^-. In a completely closed system, there would be no net fractionation in such a case, as the number of ^{15}N and ^{14}N atoms in the educt pool and the number of these atoms in the product pool is exactly the same. In reality, diffusion of NO_3^- from the water column into the sediments leads to a partial expression of a net fractionation for benthic denitrification (see Figure 6a).

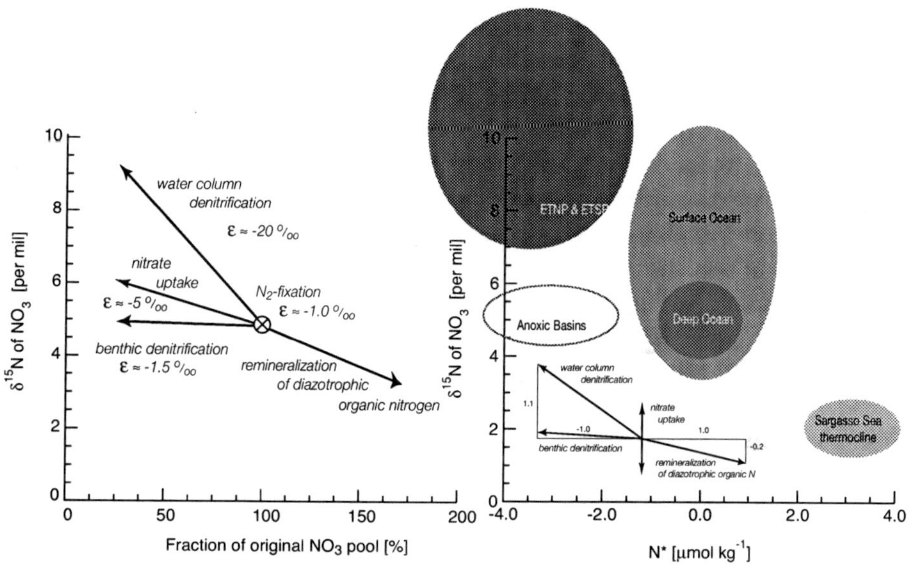

Plots of how the various processes controlling the marine nitrogen cycle affect the δ^{15}N of NO_3^- (a) in relationship to the magnitude of the NO_3^- pool, and (b) in relationship to N^*. The isotopic enrichments, ε, for the different processes are indicated in (a). Although there is little isotopic enrichment associated with N_2 fixation (circle with cross), the addition of NO_3^- from the remineralization of diazotrophic organic nitrogen leads to a reduction of the δ^{15}N of NO_3^-, because the δ^{15}N of oceanic N_2 is lower than that of NO_3^-. The inset in (b) shows how the various processes influence δ^{15}N and N^*. Note that NO_3^- uptake and remineralization has no influence on N^*, but impacts δ^{15}N. The ellipses indicate where existing data from the eastern tropical Pacific [, 1989; , 1998], the Arabian Sea [, 1998], the anoxic Santa Barbara Basin [, 1979; , 2003] and from the North Atlantic thermocline [, 2002] fall on this plot.

During the uptake of NO_3^-, marine phytoplankton prefer the light isotope ^{14}N over ^{15}N with an isotopic enrichment of about -5 ‰[*Sigman et al.*, 1999] (see Figure 6a). During the subsequent cycling of the reduced organic nitrogen within the upper ocean food web, there is a tendency of zooplankton to excrete isotopically lighter ammonium [*Mullin et al.*, 1984]. As a consequence, zooplankton tends to be isotopically enriched relative to their food source. The isotopic value of the exported organic nitrogen reflects then the relative contribution to export from phytoplankton and zooplankton plus their respective N sources [*Altabet*, 1988; *Altabet et al.*, 1991]. This offers the opportunity to study variations in the δ^{15}N of organic nitrogen preserved in marine sediment cores to learn something about the state of the upper ocean nitrogen cycle.

When surface NO_3^- is completely exhausted and steady state can be assumed over the annual cycle, the δ^{15}N of the exported N must reflect the flux weighted mean δ^{15}N of all nitrogen sources [*Altabet*, 1988; *Altabet et al.*, 1991]. If NO_3^- is the dominant source, such as in upwelling systems, variations in the δ^{15}N of the deposited organic N provide therefore a direct record of variations in the δ^{15}N of the upwelled NO_3^- [*Altabet et al.*, 1995, 2002; *Ganeshram et al.*, 1995, 2000, 2002]. When a residual surface NO_3^- remains year-round, such as in the Southern Ocean and in the northern North Pacific, the isotopic fractionation associated with NO_3^- uptake is fully expressed, and the δ^{15}N of the exported organic nitrogen reflects primarily the degree of nutrient utilization [*Altabet and Francois*, 1994; *Sigman et al.*, 1999].

In contrast to denitrification and NO_3^- uptake, N_2 fixation hardly discriminates ^{15}N versus ^{14}N. Direct measurements of *Trichodesmium* [*Capone et al.*, 1997] and laboratory studies [*Minagawa and Wada*, 1986] suggest an isotopic enrichment during N_2 fixation of only about -1‰[*Brandes and Devol*, 2002]. However, as the δ^{15}N of N_2 dissolved in seawater is very close to that of atmospheric N_2 (about 0.6‰[*Emerson et al.*, 1991]), the δ^{15}N of the organic N of diazotrophic organisms is substantially lower than the mean ocean δ^{15}N of NO_3^-, which is about 5 to 6‰[*Liu and Kaplan*, 1989; *Sigman et al.*, 2000; *Brandes and Devol*, 2002]. As a consequence, the addition of isotopically light NO_3^- during the remineralization of diazotrophic organic matter leads to a reduction of the δ^{15}N of the NO_3^- pool (see Figure 6a).

1.2.2 Water column signatures. This isotopic "coloring" of the marine NO_3^- pool provides a wealth of additional information that can be exploited, particularly in conjunction with N^*. For example, the different isotopic signatures associated with benthic denitrification and water column denitrification offer the opportunity to differentiate the relative contribution of these processes. Variations in δ^{15}N of NO_3^- can also provide independent evidence for the interpretation of N^* variations as being driven by denitrification and N_2 fixation. Figure 6b shows how the various processes influence N^* and δ^{15}N of NO_3^-.

Although $\delta^{15}N$ of NO_3^- has not been measured on a global basis, the existing data, summarized schematically in Figure 6b, fall into distinct categories. The high N^* waters from the thermocline of the subtropical North Atlantic are characterized by very low $\delta^{15}N$ of NO_3^- values. As N_2 fixation is the only significant process that adds isotopically light N to the ocean, these $\delta^{15}N$ observations provide strong support for the interpretation of the North Atlantic being a region of intense and widespread N_2 fixation. This interpretation is supported by the observation of generally low $\delta^{15}N$ of organic matter in this region [*Altabet*, 1988; *Mino et al.*, 2002; *Mahaffey et al.*, 2003]. On the opposite extreme, we find data from the eastern tropical Pacific and the Arabian Sea, with very low N^* values and very high $\delta^{15}N$ values. This confirms the interpretation of the low N^* values in these oxygen deficient regions having been generated by water column denitrification.

Data from the anoxic Santa Barbara basin have much lower $\delta^{15}N$, despite very low N^* values. *Sigman et al.* [2003] demonstrated that this is likely due to strong removal of fixed N by benthic denitrification with a very low isotopic enrichment. They used the simultaneous N^* and $\delta^{15}N$ observations to estimate that about 75% of the NO_3^- loss in this basin stems from benthic denitrification.

1.2.3 Mean ocean $\delta^{15}N$ and budget constraints. A remarkable observation in Figure 6b is that $\delta^{15}N$ of NO_3^- is relatively constant in the deep ocean, with a value of about 5‰that is relatively close to the $\delta^{15}N$ of atmospheric N_2. If water column denitrification and N_2 fixation were the only two processes controlling the balance of fixed nitrogen in the ocean, one would expect a substantially higher mean ocean $\delta^{15}N$. This is because the flux weighted $\delta^{15}N$ of the sources and sinks need to be equally far away from the mean $\delta^{15}N$ of the NO_3^- pool in order to achieve an isotopic balance. Since water column denitrification has an isotopic enrichment of -20‰, while N_2 fixation leads to the addition of N with a $\delta^{15}N$ of 0‰, this can only be achieved if the ocean mean $\delta^{15}N$ was 20‰. The observation of an ocean mean $\delta^{15}N$ of 5‰indicates therefore that the flux weighted mean isotopic enrichment of the sink terms in the oceanic N budget has to be substantially closer to 0‰than -20‰. The only process that can reduce the isotopic enrichment of the sinks is benthic denitrification. If the ocean was well mixed and the fixed nitrogen budget was in steady-state, the flux weighted mean isotopic enrichment associated with the sink terms was only about -5‰. Such a low value can only be attained if benthic denitrification is about three times as large as water column denitrification.

Brandes and Devol [2002] investigated the ^{15}N constraint on the marine nitrogen budget in detail, considering also river input, atmospheric deposition, and sediment burial. They confirm the simplified argument above by finding that benthic denitrification has to be about 3 to 4 times larger than water column denitrification in order to achieve an isotopically balanced budget of the current nitrogen budget. However, they also assumed steady-state and a well mixed ocean, which needs to be viewed critically given the strong spatial separation between N_2 fixation and denitrification and their differential impact on $\delta^{15}N$ and NO_3^-. *Deutsch et al.* [2003] investigated the impact of this spatial separation on the ^{15}N constraint for the oceanic N budget, and found that it leads generally to a reduced expression of the isotopic fractionation of water column denitrification on the $\delta^{15}N$ of NO_3^-. This is mainly because denitrification tends to occur (and result) in lower NO_3^- environments, and a mixture of two pools of NO_3^- with differing $\delta^{15}N$ will be biased toward the pool with higher NO_3^- concentrations. *Deutsch et al.* [2003] called this effect a "dilution effect" and found that it reduces the ratio between benthic and water column denitrification to slighly less than 3. Another caveat is the steady-state assumption. If serious imbalances in the current marine fixed nitrogen budget existed, this isotopic balance argument would not hold anymore.

2. IMPLICATIONS FOR THE CURRENT OCEAN BUDGET

Gruber and Sarmiento [1997] tentatively extrapolated their North Atlantic N_2 fixation estimate to the global ocean using the abundance of *Trichodesmium* as an indicator. They arrived at a global N_2 fixation estimate of about 110 Tg N yr^{-1}, with about 50 Tg N yr^{-1} occurring in the Pacific and 30 Tg N yr^{-1} being added in the Indian ocean. *Deutsch et al.* [2001] suggested that N_2 fixation in the Pacific is actually about 59 Tg N yr^{-1}, giving a global estimate of about 117 $\pm$ 40 Tg N yr^{-1} (see Table 1). *Karl et al.* [2002] reviewed the available evidence and suggest that global marine N_2 fixation is likely in the range of 100 to 200 Tg N yr^{-1}. Together with the other sources of fixed nitrogen to the ocean, *Gruber and Sarmiento* [2002] find that the total present-day sources sum up to about 238 $\pm$ 44 Tg N yr^{-1} (188 $\pm$ 44 Tg N yr^{-1} in preindustrial times). Using the N_2 fixation estimate of *Karl et al.* [2002], the total sources would range between 200 and 300 Tg N yr^{-1}. On the other hand, if the conclusion of *Hansell et al.* [2003] from the North Atlantic hold up, N_2 fixation might actually be substantially smaller than 100 Tg N yr^{-1}, giving a minimum estimate for the total sources of the order of 150 Tg N yr^{-1}.

On the basis of the provided evidence, I decided to leave the N_2 fixation estimate of *Gruber and Sarmiento* [2002] unchanged, but increased the associated uncertainty. I also increased the atmospheric deposition estimate, as the previous estimate did not include the possibly enhanced deposition of dissolved organic nitrogen species postulated by *Cornell et al.* [1995]. Their deposition number is uncertain, however, and I therefore adjusted this uncertainty as well.

Table 1 summarizes the global water column denitrification rates that *Gruber and Sarmiento* [2002] inferred from N^* and similar techniques. Additional

10 Tg N yr^{-1} are added to account for water column denitrification in smaller areas, such as the eastern tropical Atlantic [*Christensen et al.*, 1987]. This results in a total estimate for water column denitrification of about 80 Tg N yr^{-1} (Table 1). *Gruber and Sarmiento* [2002] assumed on the basis of the work by *Christensen et al.* [1987] that benthic denitrification is of about equal magnitude, i.e. about 95 Tg N yr^{-1} in the contemporary ocean. In their budget, sediment burial and a small loss of N_2O to the atmosphere bring the total sinks of fixed nitrogen in the ocean to about 204 $\pm$ 29 Tg N yr^{-1} in present times.

Codispoti et al. [2001] challenged the budget of *Gruber and Sarmiento* [2002] by doubling global water column denitrification and tripling benthic denitrification (see Table 1). As discussed above, the evidence provided to support a twofold increase in water column denitrification is not particularly strong, while the N^* based estimates appear relatively robust. In their recent work, *Brandes and Devol* [2002] also shared this view and chose a value of 75 Tg N yr^{-1} for global water column denitrification. However, the global mean oceanic δ^{15}N constraint that benthic denitrification has to exceed water column denitrification several fold requires a substantial increase in benthic denitrification. *Brandes and Devol* [2002] proposed a value of about 280 Tg N yr^{-1}. This estimate would increase the total sinks to about 400 Tg N yr^{-1}, and would generate an imbalance in the current budget of about 100 to 200 Tg N yr^{-1}. As there is little geological evidence that such an imbalance was sustained throughout the Holocene, *Brandes and Devol* [2002] discussed several ways how this imbalance can be reconciled. One way is to increase the source terms, mainly N_2 fixation, another is to decrease the assumed or implied sink terms. The latter can be achieved by increasing the isotopic fractionation for diazotrophic organisms, or by decreasing water column denitrification. Although there is some recent evidence that diazotrophic organisms have a somewhat larger isotopic enrichment than previously assumed (J. Brandes, personal communication, 2003), I consider a smaller water column denitrification as the most likely scenario. One also needs to take into account the isotopic dilution effect identified by *Deutsch et al.* [2003]. If the water column denitrification rate is only 65 Tg N yr^{-1} instead of 80 Tg N yr^{-1}, the isotopic mass balance of ^{15}N in the ocean would require a benthic denitrification rate of about 180 Tg N yr^{-1} [*Brandes and Devol*, 2002; *Deutsch et al.*, 2003]. This value is lower than the 230 to 285 Tg N yr^{-1} range proposed by *Middelburg et al.* [1996] on the basis of a global modeling study, but still within the overall error of their estimate. Adopting these downscaled values, the contemporary marine N budget could be brought back into an approximate, albeit uncertain, balance (see Table 1).

Can such a reduction in the water column denitrification rate be justified on the basis of the N^* distribution? If benthic denitrification is indeed a much more important process than previously thought, then it can also influence the interpretation of the N^* trends within the denitrification zones. Most of the isopycnal surfaces investigated by *Deutsch et al.* [2001], for example, intersect the bottom within the O_2 minimum zone. If sediment denitrification is intense, the outward diffusing N^* signals from the sediments could contribute to the low N^* signals within the water column. These signals would then be interpreted as water column denitrification, while they are, in reality, benthic denitrification signals. As shown by *Sigman et al.* [2003] detailed investigations of the $\delta^{15}N$ of NO_3^- would be able to test this hypothesis. Another reason why the water column denitrification rates might have been biased high is because of an overestimate of the water volume, in which the denitrification occurs.

In summary, the revised marine nitrogen budget presented here differs in two major elements from those previously shown [*Gruber and Sarmiento*, 1997; *Gruber and Sarmiento*, 2002]. The first major change is a shift in the focus of the main sink term in the ocean from water column denitrification to benthic denitrification. Given the strong constraint from the mass balance of ^{15}N, I consider this shift as robust. The second major change is the acknowledgment that the uncertainties in the individual budget terms are larger than previously assumed. Although the revised budget in Table 1 turns out to be approximately balanced, the uncertainties in the individual sink and source terms are so large that the existence of a substantial imbalance cannot be excluded.

Although our ability to quantify the marine nitrogen balance remains limited, the rapid turnover time for fixed nitrogen in the ocean is unquestioned. All budgets in Table 1 suggest a mean ocean residence time of 3,000 years and less. The marine nitrogen cycle can therefore respond quickly to environmental changes. But how does it change and how does it regain its balance after a perturbation? We will investigate these questions in the next section.

3. THE COUPLING BETWEEN N_2-FIXATION AND DENITRIFICATION

On some timescale approaching the turnover time of fixed nitrogen in the ocean (i.e. about 2000 to 3000 years), nitrogen fixation and denitrification in the ocean must be coupled to each other, otherwise the marine fixed nitrogen content of the ocean would undergo massive swings. As discussed above, there is little evidence in the geological record that such massive variations occurred, suggesting that the coupling must be reasonably tight. In order to address the nature of this coupling, we need to review first the factors that control N_2 fixation and denitrification.

3.1 CONTROLS ON N_2 FIXATION

What are the factors controlling the pattern and rates of pelagic N_2 fixation? *Karl et al.* [2002] provide a thorough review, permitting me to summarize the most important factors. These factors are temperature, light, and the ambient concentrations of O_2, NO_3^-, NH_4^+, and essential trace metals. It is well established that *Trichodesmium* colonies grow and fix N_2 successfully only at temperatures well above 20°C [*Capone et al.*, 1997], but it appears that this temperature sensitivity must be caused by factors other than the temperature dependence of the enzyme nitrogenase, which catalyzes the reduction of N_2 [*Karl et al.*, 2002]. Nothing is known about the temperature sensitivity of the newly discovered diazotrophic species [*Zehr et al.*, 2001]. Since the fixation of N_2 is an energy consuming process, there is little advantage for diazotrophic organisms in continuing to fix N_2 if they can obtain their required N by taking up NO_3^- or NH_4^+. As the energy comes from sunlight, *Trichodesmium* dwells best at high insolation near the ocean surface [*Capone et al.*, 1997] and low levels of turbulence.

These environmental factors are successful in explaining the high rates of N_2 fixation in the tropics and in the subtropical gyres in contrast to the low rates found in high-latitudes. However, these factors fail to explain why N_2 fixation appears to be high in the North Atlantic, Arabian Sea and western North Pacific, and low in the South Atlantic and eastern South Pacific, since few differences exist between these regions with regard to the controlling factors mentioned above. Therefore other factors need to be considered, such as the availability of essential trace metals.

A major difference between these regions is the delivery of dust from the atmosphere to the surface ocean. Although there are large uncertainties associated with current estimates of dust delivery, the subtropical North Atlantic receives about one order of magnitude more dust than the South Atlantic, and almost two orders of magnitude more than the eastern South Pacific [*Duce et al.*, 1991; *Tegen and Fung*, 1994; *Mahowald et al.*, 1999]. In these regions, the deposition of dust represents the dominant pathway for the delivery of a number of biologically important trace metals to the surface waters [*Fung et al.*, 2000; *Lefevre and Watson*, 1999]. The most important elements for biology among these trace metals are iron, manganese, cobalt, zinc, copper and molybdenum, since these elements are frequently used as reaction centers of many vital enzymes [*Morel and Price*, 2003]. We restrict our discussion here to iron since it is a critical metal co-factor for nitrogenase [*Rueter et al.*, 1992]. It should be kept in mind that other trace metals, such as molybdenum, which is another co-factor of nitrogenase, have the potential to be limiting as well [*Karl et al.*, 2002]. The essential requirement of iron for the nitrogenase enzyme makes the diazotrophic organisms sensitive to variations in the supply of iron. As the dust to iron ratio appears to be relatively constant [*Duce et al.*, 1991], the strong variations of dust supply over the different regions reflect directly strong variations of the supply of iron to the surface ocean.

The observed pattern of N_2 fixation and the distribution of N^* appear to be consistent with the posit of iron playing an important role for controlling N_2 fixation. If this is correct, rates of N_2 fixation could have varied quite strongly in the past, since it is known that the deposition of dust onto the surface ocean during glacial periods was a least twice the amount of today's deposition [*Mahowald et al.*, 1999]. However, the evidence for large-scale iron limitation of N_2 fixation is largely circumstantial. Results from iron addition experiments are inconclusive [*Hood et al.*, 2000], and recent analyses of the iron content in natural populations of *Trichodesmium* in the central Atlantic reveal neither a significantly higher iron requirement, nor a geographical pattern in cellular iron content consistent with patterns of iron deposition [*Sanudo-Wilhelmy et al.*, 2001]. Rather, these authors assign PO_4^{3-} a key role in explaining the observed variability. As iron levels are generally substantially higher in the central Atlantic than in the iron deficient regions of the southern hemisphere, this finding does not rule out a critical role for iron, but emphasizes the role of phosphorus.

3.2 CONTROLS ON DENITRIFICATION

Since the energy yield for the oxidation of organic matter is smaller when NO_3^- is used as terminal electron acceptor instead of oxygen, marine denitrification is essentially limited to those regions that are oxygen deficient. Whereas oxygen is completely used up nearly everywhere in marine sediments below a certain depth, the water column of the current ocean is generally well oxygenated. Only in the few oxygen depleted zones of the open ocean are oxygen levels low enough to support water column denitrification.

The rate of benthic denitrification is to a first order a function of the organic nitrogen flux to the sediments [*Middelburg et al.*, 1996]. This is because nearly all organic nitrogen that escapes oxic remineralization at the generally oxygenated sediment surface is eventually denitrified. The flux of organic matter to the sediments is itself primarily a function of surface export production and bottom topography, as remineralization in the ocean interior appears to follow relatively constant rules [*Martin et al.*, 1987; *Armstrong et al.*, 2002].

In contrast to benthic denitrification, the main control on water column denitrification is the in situ O_2 concentration. In the interior of the ocean, the condition for the generation of an oxygen depletion zone is given when the supply of O_2 by ocean transport from the surface ocean is smaller than the oxygen demand from the remineralization of organic matter. Therefore both low ventilation of the water column or high rates of remineralization of organic matter can cause oxygen depleted waters and hence water column denitrification. It appears that both processes are important for generating the oxygen depleted conditions in the three regions, where we currently observe such conditions i.e. the ETSP, the ETNP and the Arabian Sea.

The high rates of export in these regions are linked to substantial rates of upwelling of nutrient rich waters into the surface waters, thereby promoting high surface productivity. The low ventilation of subsurface waters in the eastern tropical Pacific can be explained on the basis of theories of the thermocline ventilation (see e.g. *Pedlosky* [1987]). The conservation of potential vorticity (a measure of the angular momentum of a fluid parcel) requires that the interior of a density surface can only be advectively ventilated along streamlines that intersect the outcrop. However, not the entire density surface can be reached by these streamlines, resulting in a "shadow zone", that is only being ventilated by diffusion [*Luyten et al.*, 1983]. As diffusion is much less effective in ventilating than advection, waters in the shadow zones are relatively stagnant. The oxygen minimum zones in the eastern tropical Pacific are such "shadow zones", as are the southeastern part of the North Atlantic subtropical gyre and the northeastern part of the South Atlantic subtropical gyre [*Kawase and Sarmiento*, 1985]. The latter two regions in the Atlantic have also very low oxygen concentrations, but are not anoxic. Since the levels of productivity [*Laws et al.*, 2000] and rates of water renewal in the Atlantic are relatively similar to those in the Pacific, this difference is likely caused by the fact that waters in the Atlantic ocean are generally younger and hence contain, on average, higher O_2 concentrations. This therefore provides a higher supply of O_2 from the ocean interior for the same rate of ventilation.

The conditions that lead to the formation of the oxygen minimum zone in the Arabian Sea are less clear [*Naqvi*, 1987; *Morrison et al.*, 1999], since the Arabian Sea is not a "shadow zone". However, the thermocline structure and circulation of the Arabian Sea is unique [*Wyrtki*, 1973]. The Arabian Sea is characterized by a semi-annual reversal in wind stress associated with the monsoon, water mass intrusions from marginal seas and the absence of a subtropical convergence zone and deep water formation. The core of the oxygen minimum zone is located within the salinity maximum of Persian Gulf Water. Measurements of CFC indicate that these waters are only 2 to 8 years old [*Howell et al.*, 1997]. This is much younger than the CFC ages found in the ETSP and ETNP [*Deutsch et al.*, 2001]. Thus, in order to create the intense oxygen depletion zone found in the Arabian Sea, either the oxygen demand must be substantially higher than that created in the ETSP and ETNP, or the source waters must already have very low O_2 concentrations. *Warren* [1994] suggested that export production in the Arabian Sea is not anomalously large, concluding that the latter mechanism must be the cause of anoxia in the Arabian Sea. A newer study by *Smith et al.* [1998] disputes this conclusion and suggests that low ventilation and high biological production conspire to create suboxic waters in this area.

Denitrification, and particularly that occurring in the water column, can respond strongly and rapidly to changes in physical forcing (changes in ventilation) and changes in the supply of organic material as observationally confirmed in the Arabian Sea [*Naqvi et al.*, 1990] and in the ETSP [*Codispoti et al.*, 1986]. This makes it likely that rates of denitrification could have varied substantially in the past.

3.3 DYNAMIC HOMEOSTASIS?

Given that the rates of N_2 fixation and denitrification are controlled by entirely different processes, it is surprising that the marine N-budget appears to be in an approximate balance (see Table 1) and that the intercept of the NO_3^- and PO_4^{3-} relationship is close to zero (see Figure 2). This approximate balance is particularly difficult to explain when considering how strongly and rapidly each of these two processes can respond to changes in its controlling factors and mechanisms. Evidently, negative feedbacks must exist that couple denitrification and N_2 fixation more strongly than it appears to be the case at first.

Two negative feedback processes have been proposed to stabilize the oceanic nitrogen cycle (see Figure 7). The first feedback process is based on denitrification [*Codispoti*, 1989]. When denitrification increases, the fixed nitrogen content of the ocean decreases, thereby on average reducing the supply of fixed nitrogen to the surface ocean. This results in a smaller export production, leading to a smaller oxygen demand in the thermocline and a smaller organic N deposition onto the marine sediments. As a consequence, both water column and benthic denitrification decrease, closing the negative feedback loop.

The second feedback loop links surface N_2 fixation to the small excess of PO_4^{3-} that is generally observed in low-latitude surface waters [*Tyrrell*, 1999]. This proposed mechanism builds on the assumption that N_2 fixers have a lower growth rate in comparison with "normal" phytoplankton [*Capone et al.*, 1997]. They can thrive, however, in the case when NO_3^- gets depleted in the surface ocean before PO_4^{3-}, since they can circumvent nitrogen limitation. Therefore, if fixed nitrogen becomes scarce relative to PO_4^{3-}, N_2 fixation would go up and resupply the fixed nitrogen to the ocean. This process continues until the diazotrophic organisms effectively reduce their habitat until N_2 fixation is back in balance with denitrification. *Haug et al.* [1998] provided evidence that this feedback might have operated across glacial-interglacial cycles in the Cariaco Basin. During interglacial cycles, such as the present Holocene, surface ocean N:P ratios are depleted because of intense denitrification in the anoxic water column of this deep basin. This fuels high levels of N_2 fixation [*Walsh*, 1996], as evidenced by the low $\delta^{15}N$ of the organic N deposited onto the seafloor.

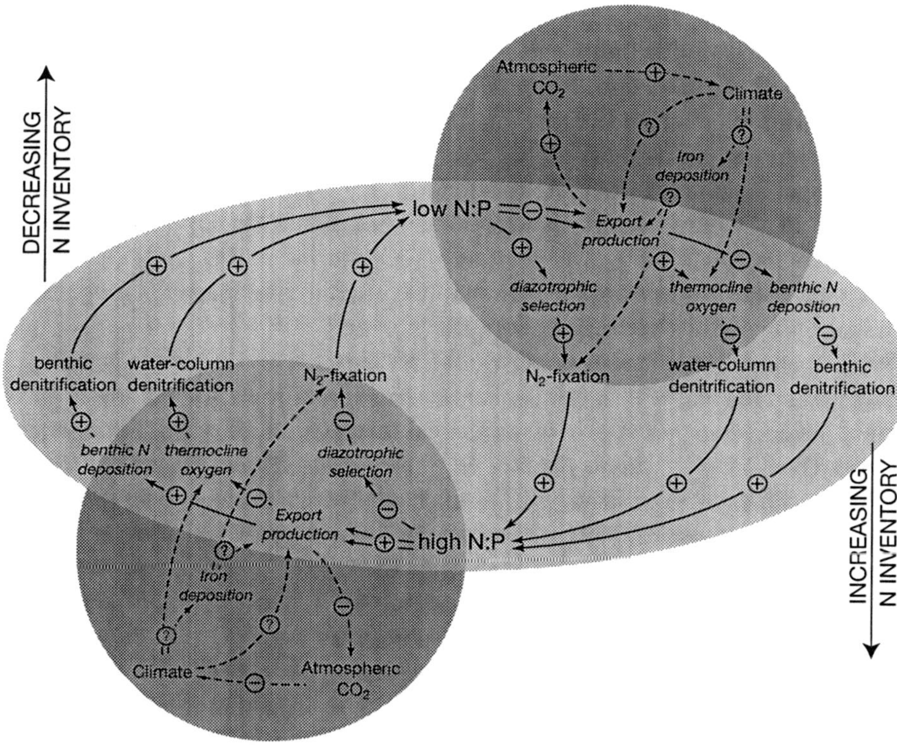

Schematic representation of feedbacks within the marine nitrogen cycle. The inner light grey ellipse shows the two nitrogen cycle "internal" feedback loops that tend to be stabilizing. These two feedback loops are proposed to be mainly controlled by the N:P ratio in the surface waters. If the surface N:P ratio is low, denitrification tends to get reduced because of lower surface productivity, while N_2 fixation tends to increase because of favorable conditions for diazotrophic organisms. Both processes lead to an increase in the oceanic fixed nitrogen inventory. In contrast, when the surface N:P ratio is high, N_2 fixation tends to decrease, while denitrification increases, leading to a reduction of the oceanic fixed nitrogen inventory. These nitrogen internal feedback loops can be altered significantly by their interaction with the carbon cycle and climate (dark grey ellipses). The sign of these feedback loops are currently not understood and can be either positive or negative, leading to stabilizing or destabilizing feedbacks.

During glacial times, the water column was relatively well oxygenated, and therefore litte denitrification occurred. The resulting near Redfield N:P ratios gave little habitat for N_2 fixers, and the organic N deposited onto the seafloor was close to mean ocean $\delta^{15}N$.

These two stabilizing feedback loops not only provide tight control on the oceanic nitrogen inventory relative to that of PO_4^{3-}, but also explain why the intercept of the marine NO_3^- to PO_4^{3-} relationship in Figure 2 is close to the origin. If the system deviated to the left (Figure 2b), it would be in a situation of excess NO_3^-, corresponding to the lower state of the feedback loop in

Figure 7. This would tend to cause an increase in marine denitrification fueled by higher marine productivity and a decrease in N_2 fixation, both of which would reduce this excess until the system runs into a situation of a NO_3^- deficit (Figure 2c or upper state in Figure 7). In this situation, a decrease in marine denitrification and an increase in N_2 fixation would tend to bring the marine fixed nitrogen back toward Redfield stoichiometry with PO_4^{3-}. Therefore, the fixed N to PO_4^{3-} ratio of surface waters emerges as the key controlling variable for this feedback loop. The near zero intercept in the NO_3^- to PO_4^{3-} plot can then be viewed as direct evidence that these stabilizing feedback loops are successfully operating in the present ocean.

While these proposed loops provide possible answers to our questions, many details remain unclear. For instance, the observation of high N_2 fixation in the North Atlantic is inconsistent with the second feedback mechanism, because in this ocean basin, PO_4^{3-} runs out before NO_3^- [*Wu et al.*, 2000]. Therefore, there should be no habitat for diazotrophs in this basin, which is clearly not the case [*Carpenter and Romans*, 1991]. Conversely, over most of the low latitudes of the Pacific and Indian, NO_3^- is depleted before PO_4^{3-}, yet diazotrophs are not as abundant as they should be [*Carpenter*, 1983]. Clearly, other factors in addition to PO_4^{3-}, such as iron, must be important as well.

Lenton and Watson [2000] recently investigated these two feedbacks in a simple model of ocean biogeochemical cycles. They find that these two feedbacks are indeed very effective in keeping NO_3^- close to the Redfield ratio with PO_4^{3-}, as long as other factors, such as iron, do not limit N_2 fixation to any substantial degree. They find that, in the case of strong iron limitation of N_2 fixation, increasing the PO_4^{3-} input into the ocean can cause PO_4^{3-} to deviate above Redfield ratio to NO_3^-. Thus, under certain circumstances, NO_3^- can become the ultimate limiting nutrient even over geological time-scales. But in all other cases, PO_4^{3-} is controlling the amount of fixed nitrogen in the ocean.

It needs to be pointed out that both feedbacks operate on the time-scale of whole ocean overturning, i.e. decades to a few hundred years. This implies that, on time-scales shorter than this, imbalances between N_2 fixation and denitrification can exist and significantly alter the N content of the ocean. A second point to make is that these proposed feedback mechanisms do not necessarily imply that the N content of the ocean has to be exactly 16 times the PO_4^{3-} content, they only imply a rough Redfield stoichiometry. Therefore the existence of negative feedbacks that couple the fixed nitrogen content of the ocean to that of PO_4^{3-}, are not at odds with the possibility that variations in the N-content of the ocean contributed to the glacial inter-glacial CO_2 variations [*Falkowski*, 1997; *Broecker*, 1998]. These feedbacks just confine the marine N inventory to within relative narrow boundaries set by the ocean inventory of PO_4^{3-}.

Figure 7 show how these nitrogen cycle feedbacks internal to the ocean are connected to atmospheric CO_2 and climate, and how these connections can give raise to new feedbacks that can generate more complex behaviors. This includes the climate - dust deposition - N_2 fixation feedback proposed by *Michaels et al.* [2001] as well numerous other linkages between climate and ocean biogeochemical cycles. The impact of climate change on ocean circulation and hence nutrient supply to the surface with subsequent changes in ocean interior oxygen concentration and water column denitrification should be mentioned in particular. *Altabet et al.* [2002] provides an example how millennial timescale variations in monsoonal circulation can lead to dramatic alteration of water column denitrification in the Arabian Sea with possible influences back on atmospheric CO_2.

The emerging view is a dynamic homeostasis, where an internally stabilized system (the nitrogen cycle "homeostat") can be significantly modified and altered by its interactions with other systems, such as the global carbon cycle or the physical climate system. With this picture in mind, we are now ready to re-evaluate the hypotheses that have been proposed to explain the large atmospheric CO_2 changes between glacial and interglacial periods.

4. THE ROLE OF THE OCEANIC N-CYCLE FOR GLACIAL-INTERGLACIAL CO_2 VARIATIONS.

The quest for the causes of the large changes in atmospheric CO_2 between glacial and interglacial periods (see Figure 8) represents one of the biggest challenges of current global carbon cycle research. More than 20 years after these changes have been documented [*Neftel et al.*, 1982], no commonly accepted scenario of mechanisms exists that explains how the global carbon cycle changed in order to generate such a different atmospheric CO_2 concentration (see reviews by *Sigman and Boyle* [2000] and *Archer et al.* [2000]). The only well established fact is that alterations of the ocean carbon cycle must be at the root of the atmospheric CO_2 changes, since the ocean controls atmospheric CO_2 on timescales longer than a few hundred years as a result of it being the largest reservoir of carbon in the global carbon cycle [*Sarmiento and Gruber*, 2002; *Sabine et al.*, 2003]. Given the profound impact of the marine nitrogen cycle on the marine carbon cycle, it is therefore not surprising that alterations of the marine nitrogen cycle have been proposed from the very beginning as a possible mechanism for the glacial-interglacial changes in atmospheric CO_2 [*McElroy*, 1983]. Many additional nitrogen cycle related hypotheses have been proposed since (e.g. [*Altabet et al.*, 1995; *Ganeshram et al.*, 1995; *Falkowski*, 1997; *Broecker and Henderson*, 1998; *Michaels et al.*, 2001]. I will review here the main proposed mechanisms and discuss how paleoclimatic constraints can be used to test them. The conclusion that emerges from this overview is that variations in the marine fixed nitrogen inventory likely contributed to atmospheric CO_2 variations, but were unlikely the major cause.

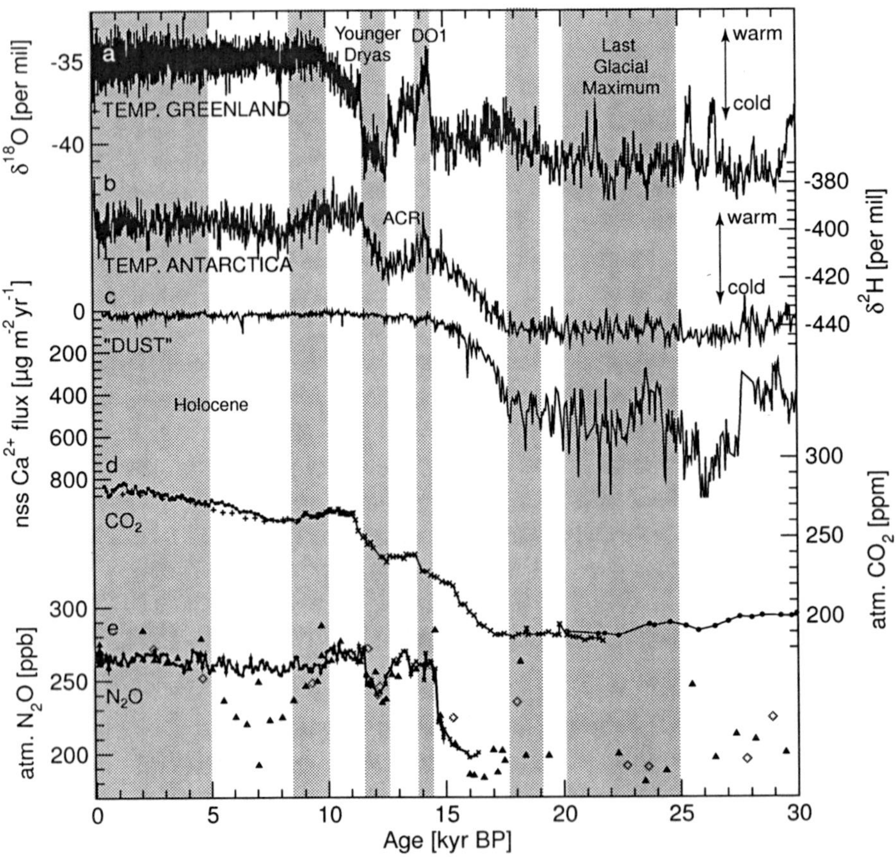

Timeseries of important climatic properties over the last 30 kyrs. (a) $\delta^{18}O$ of ice as a proxy for temperature from GRIP Greenland) [, 1997]. (b) δ^2H of ice as a proxy for temperature from Dome C (Antarctica) [, 2001]. (c) Dust flux (inverted scale) at Dome C site as reconstructed from the sea-salt corrected Ca^{2+} and the accumulation rate [, 2002]. (d) Atmospheric CO_2 as recorded in air bubbles from various Antarctic ice cores [*Indermühle* , 1999, 2000; *Flückiger* , 2002; , 2002]. (e) Atmospheric N_2O as recorded in air bubbles from various Antarctic and Greenland ice cores. The filled triangles denote data from the GISP icecore drilled in Greenland [, 2003], the open circles are from the Taylor Dome [, 2003], the solid line with the crosses denote data from the GRIP icecore [*Flückiger* , 1999], and the solid line with the circles are from Dome C [*Flückiger* , 2002]. The data are plotted versus age before the year 1950, using age-scales of the respective records. ACR denotes Antarctic Cold Rerversal, DO1 stands for Dansgaard-Oeschger event 1.

4.1 EVALUATION OF OCEANIC NITROGEN-CYCLE MECHANISMS

Table 2 shows a summary of the ocean-nitrogen cycle based mechanisms together with arguments that support or challenge them. The first hypothesis is the continental margin erosion mechanism of *McElroy* [1983]. He pointed out that the low sea-level stand during the last glacial maximum (about 120 m lower than today [*Cutler et al.*, 2003]) resulted in the exposure of many continental margin areas, where most of the organic matter preservation occurs during interglacial periods [*Wollast*, 1998]. According to this hypothesis, the nutrients that had accumulated in these margin sediments during interglacial periods would be eroded, transported to the ocean, and would increase the inventory of fixed nitrogen (and phosphorus) during the glacial periods.

This mechanism has the advantage that it could increase the oceanic fixed nitrogen inventory without causing any substantial deviation from a Redfield relationship with PO_4^{3-}. However, the timing of the CO_2 rise versus the sea-level change questions whether this mechanism can play a major role. Detailed reconstructions reveal that sea level started to rise significantly several 1000 years after atmospheric CO_2 [*Bard et al.*, 1996; *Broecker and Henderson*, 1998]. In particular, the first large increase in sea level associated with melt water pulses occurred at around 14,600 BP [*Weaver et al.*, 2003], i.e. at a time when atmospheric CO_2 had already completed about half of its glacial-interglacial transition (see Figure 8). With the continental margin mechanism unlikely having played a substantial role, we are left with the hypotheses centering around changes in marine nitrogen fixation [*Falkowski*, 1997; *Broecker and Henderson*, 1998; *Michaels et al.*, 2001; *Karl et al.*, 2002] and denitrification [*Ganeshram et al.*, 1995; *Altabet et al.*, 1995; *Ganeshram et al.*, 2000].

The N_2 fixation based mechanism builds mainly on the observation that atmospheric iron deposition during the last glacial was substantially larger than it is today. Globally, atmospheric iron deposition on the surface ocean is estimated to be about twice of that today [*Mahowald et al.*, 1999], while the increase over the Southern Ocean and other low iron deposition regions was likely manifold [*Martin*, 1990] (see also Figure 8c). This increase in iron availability not only alleviated the iron stress of phytoplankton in general, but is proposed to also have enhanced N_2 fixation substantially [*Falkowski*, 1997; *Broecker and Henderson*, 1998; *Michaels et al.*, 2001; *Karl et al.*, 2002]. This increase in the input of fixed nitrogen to the ocean, while the other terms of the marine nitrogen balance remained more or less unchanged, would then lead to an increase in the fixed nitrogen inventory. This would stimulate surface production and export of organic carbon in the nitrogen limited regions and finally draw down atmospheric CO_2. The denitrification based arguments also argue for a larger fixed nitrogen inventory during glacial periods, but propose that this was a result of lower denitrification [*Ganeshram et al.*, 1995; *Altabet et al.*, 1995; *Ganeshram et al.*, 2000]. One of the strongest argument in support of these two hypotheses is their agreement with the timing of the atmospheric CO_2 increase, as their driving mechanisms start to set in at about 18,000 to 20,000 yrs BP (see Figure 8).

Summary of marine nitrogen-cycle related mechanisms proposed to explain the glacial-interglacial atmospheric CO_2 variations.

Mechanism	Advantages	Problems	Authors[a]
N erosion on margins	reservoir	timing, denitr. feedback	(1)
Iron deposition and N_2 fixation	timing	low-lat. mechanism, P, lysocline	(2,3)
Water column denitrification	timing	low-lat. mechanism, lysocline	(4-6)

a: 1: *McElroy* [1983]; 2: *Falkowski* [1997]; 3: *Broecker and Henderson* [1998]; 4: *Altabet et al.* [1995]; 5: *Ganeshram et al.* [1995]; 6: *Ganeshram et al.* [2000];

In the case of the proposed connection between dust deposition and increased nitrogen fixation, dust deposition at Dome C in central eastern Antarctica started to decrease at around 18,000 yrs BP, more or less synchronously with the beginning of the atmospheric CO_2 increase. A similar agreement in timing between changes in forcing and atmospheric CO_2 can be found for the proposed link between climate change and oceanic denitrification. The temperature proxies recorded in Antarctica indicate that the southern hemisphere climate started to warm 800±600 years before the onset of the atmospheric CO_2 rise [*Monnin et al.*, 2002] (Figure 8). *Caillon et al.* [2003] specified this lag for termination III ($\tilde{2}$40,000 yrs BP) to about 800±600 years. The denitrification hypothesis postulates that this warming caused a reorganization of ocean circulation and ocean biology that resulted in an increase in oceanic denitrification. The consequent decrease of the ocean fixed nitrogen inventory would begin to reduce global scale oceanic export production and release CO_2 into the atmosphere. The observed lag between the CO_2 increase and Antarctic temperatures is consistent with such a hypothesis, as one would expect that it takes several hundred years for oceanic export production to respond to a change in the loss rate of fixed nitrogen.

It is interesting to point out that this particular scenario asks for an increase in denitrification while global-scale biological productivity is supposed to go down. This is not necessarily inconsistent, since the reduced oxygen demand from remineralization in the thermocline can easily be compensated by a decrease in the ventilation of the subsurface waters, creating anoxic conditions with lowered productivity. Furthermore, it is known that biological productivity did not change uniformly [*Kohfeld et al.*, 2003] and while global ocean productivity likely decreased from the glacial to the interglacial period, certain regions have experienced an increase. This appears to be the case in major coastal upwelling regions such as in the Arabian Sea [*Altabet et al.*, 2002] and along the west coast of the Americas [*Ganeshram et al.*, 2000; *Kienast et al.*, 2002], thus exactly in today's major water column denitrification regions.

While the timing arguments provide support for the N_2 fixation and denitrification based hypotheses, a number of potential caveats exist. One issue is that low and mid-latitude centered hypotheses appear to have a relatively small leverage on atmospheric CO_2 [*Broecker and Peng*, 1998; *Broecker et al.*, 1999]. Only very large changes in surface CO_2 chemistry in these regions would be able to cause a substantial change in atmospheric CO_2, while the leverage on atmospheric CO_2 of changes in the surface ocean of the high latitudes appears to be much larger [*Knox and McElroy*, 1984; *Sarmiento and Toggweiler*, 1984; *Siegenthaler and Wenk*, 1984]. This is mainly a result of the small oceanic reservoir of carbon that communicates with the atmosphere via the low and mid-latitude surface ocean, while the high latitudes represent the conduit between the atmosphere and the deep ocean, where most of the oceanic carbon is stored. The N_2 fixation and denitrification based hypotheses need to be considered as low and mid-latitude processes, as an increase in the fixed nitrogen inventory is likely causing increased productivity only there. This is because of strong evidence that other factors, such as iron bioavailability, appear to limit biological productivity in many high latitude oceans, particularly in the Southern Ocean [*Martin et al.*, 1991; *Boyd et al.*, 2000].

Although early research showed that the low leverage of the low latitude hypotheses appeared to eliminate all such mechanisms [*Broecker and Peng*, 1998], more recent studies pointed out that the dominance of the high-latitudes is likely smaller than previously assumed [*Broecker et al.*, 1999; *Archer et al.*, 2000; *Follows et al.*, 2002]. This opens the door for low-latitude mechanisms again. It needs to be pointed out, however, that even with this revision the leverage per area of the low latitudes is still maximally only about 30 to 40% of that of the high latitudes. This is compensated, however, with the much larger areas covered by the low and mid latitudes. In summary, the high-latitude dominance constraint eliminates neither of the two nitrogen-based mechanisms, but requires the oceanic changes in fixed nitrogen inventory to be large. *Archer et al.* [2000] and *Sigman and Boyle* [2000] showed that the fixed nitrogen inventory needs to be increased by substantially more than 50% in order to cause a 80 ppm change in atmospheric CO_2.

The required change depends, however, strongly on the response of the oceanic $CaCO_3$ system. If the export of $CaCO_3$ from the surface ocean remains proportional to the export of organic matter (called rain ratio and is today about 7% [*Sarmiento et al.*, 2002]), the requirement for increased dissolution at the seafloor to compensate for the increased $CaCO_3$ deposition leads to a whole ocean carbonate compensation that works against the biologically induced lowering of the surface ocean partial pressure of CO_2. However, if oceanic carbonate production was lower during glacial periods, the same increase in the oceanic fixed nitrogen inventory would be able to cause a larger atmospheric CO_2 increase. *Sigman and Boyle* [2000] showed that a 30 to 40%

increase in the marine fixed nitrogen inventory would be sufficient to lower atmospheric CO_2 by 80 ppm if marine $CaCO_3$ production was reduced coincidentally.

There exist, however, strong constraints on the magnitude of such changes in the rain ratio [*Sigman et al.*, 1998] by the observed changes in the mean depth of the lysocline (i.e. the depth horizon above which $CaCO_3$ is preserved and below which it is dissolved [*Berger*, 1970]). On millennial timescales, the burial of $CaCO_3$ needs to balance the input of CO_3^{2-} ions by rivers. The ocean carbon cycle accomplishes this by shifting the position of the lysocline up and down, so that only the $CaCO_3$ that is deposited above the lysocline gets preserved. The position of the lysocline is closely controlled by the saturation level for $CaCO_3$ in the seawater overlying the sediments, which is itself governed by the interplay between organic matter remineralization and $CaCO_3$ dissolution. As a consequence, the lysocline responds very sensitively to changes in the relative and absolute rates of delivery of organic matter and $CaCO_3$ to the deep ocean.

Changes in organic matter delivery to the deep ocean lead to transient responses of the depth of the lysocline, whereas changes in $CaCO_3$ delivery lead to steady-state changes. Available paleoceanographic evidence suggest that the mean depth of the lysocline during the last glacial maximum was similar to its mean depth today [*Catubig et al.*, 1998], while the possible occurrence of transient changes has not been established yet reliably. The absence of a steady-state change in the lysocline depth puts limits on the magnitude of the changes in the export of organic matter and $CaCO_3$. If the magnitude of marine $CaCO_3$ export is directly coupled to the magnitude of organic matter export, i.e. if the rain ratio is constant, the lysocline constraint eliminates all scenarios that call for a substantial increase in organic matter export.

Only if marine $CaCO_3$ export is permitted to change independently of organic matter export can the steady-state lysocline constraint be circumvented. The mechanisms controlling the formation and export of $CaCO_3$ are currently not well understood, but it is not unreasonable to assume that, to zeroth order, $CaCO_3$ production varies in concert with organic matter production. Possible exceptions include the recently proposed idea that low latitude production of $CaCO_3$ was lower during glacial times as a result of a replacement of organisms that produce $CaCO_3$ shells (coccolithophorids) by those that produce opal (SiO_2) shells (diatoms) [*Brzezinski et al.*, 2002; *Matsumoto et al.*, 2002]. These authors proposed that such a phytoplankton community shift was caused by iron fertilization in the Southern Ocean, which led to a reduced Si:N uptake ratio of diatoms, permitting more Si to escape into the low latitude and fueling increased diatom production there.

In summary, the lysocline constraint requires that unless marine $CaCO_3$ production was equal or lower during the glacial period than today, biological export production could not have been substantially larger during this period compared to today. A somewhat lower $CaCO_3$ production would also increase the leverage of an ocean productivity change on atmospheric CO_2 because of its associated $CaCO_3$ compensation.

In addition to the low leverage of low-latitude mechanisms and the lysocline constraint, the N_2 fixation mechanism faces a number of additional challenges. First, it builds directly on the assumption that the availability of iron is the main factor limiting N_2 fixation. As reviewed in section 1.4 above, there exist conflicting evidence with regard to the role of iron. While the global-scale distribution of N^* and direct observations of N_2 fixing organisms support the importance of iron [*Karl et al.*, 2002], direct measurements of diazotrophic organisms point more toward a limiting role for phosphorus [*Sanudo-Wilhelmy et al.*, 2001]. A further problem exists in the requirement of the N_2 fixation hypothesis that marine phytoplankton are able to grow under a wide range of N:P ratios. Although laboratory incubation experiments support such a flexibility [*Goldman et al.*, 1979; *Goldman*, 1980], field analyses generally do not show a dependence of the N:P ratios of marine phytoplankton on the NO_3^- to PO_4^{3-} ratio of their environment. It is therefore questionable that an increase in the fixed nitrogen inventory without a corresponding increase in the marine PO_4^{3-} inventory will lead to a substantial increase in marine export production [*Ganeshram et al.*, 2002]. There exists, nevertheless, some leverage, as there are many oceanic regions today, where NO_3^- is exhausted before PO_4^{3-} (as evidence by the observation that there is a positive PO_4^{3-} intersect in Figure 2). This global-scale NO_3^- deficit amounts to about 3 μmol kg^{-1}, which is about 10% of the global mean NO_3^- concentration. Assuming a flexibility in the N:P ratio of 16±2:1, and filling in this deficit, marine productivity could have been maximally about 20 to 30% higher during the last glacial maximum in comparison to today.

The same constraint applies to the maximum marine export production effect arising from reduced denitrification, therefore also limiting its maximum effect on atmospheric CO_2. While there is strong evidence (see below) that water column denitrification was lower during glacial times, this does not imply automatically that the marine fixed nitrogen inventory was larger. This requires that the N_2 fixation and benthic denitrification feedbacks were not operating at full strength.

We know virtually nothing about the strength of the benthic denitrification feedback, although it may be argued that this feedback was weaker during the last glacial period because of lowered sea level, which exposed many of the margin sediments that are responsible for a substantial fraction of today's benthic remineralization [*Christensen et al.*, 1987; *Devol*, 1991; *Walsh*, 1991]. It may also be possible that the N_2 fixation feedback was weaker during the last glacial because of the higher input of iron from the atmosphere. The higher bioavailability of iron might have compensated for the competitive disadvantage for diazotrophs due to increased availability of fixed nitrogen relative to phosphorus in the surface ocean. It is therefore conceivable that reduced denitrification and increased (or at least constant) N_2 fixation worked in concert to generate a higher fixed nitrogen inventory in the ocean during the glacial. However, as discussed above, there is little evidence that this potentially larger inventory was a major cause for the reduced atmospheric CO_2 concentration at that time.

4.2 PALEOCLIMATIC CONSTRAINTS

All marine nitrogen cycle based mechanisms ask for higher productivity during the last glacial period. The paleoclimatic records of marine productivity provide, however, a mixed picture of the differences between the last glacial maximum and today [*Kohfeld et al.*, 2003]. The only region where all proxy records indicate higher export production is the Southern Ocean region north of the present-day Polar Front. A few regions indicate a general reduction (e.g. the regions poleward of the Polar Front in the Southern Ocean), while the majority show inconclusive changes. Reconstructions of global productivity from measurements of the triple isotopic composition of atmospheric O_2, suggest that global productivity was only about 80% of the present level during the glacial, and that ocean productivity was between 80% and 140% of the present [*Blunier et al.*, 2002]. Therefore, the available proxies for ocean productivity show that oceanic export production was different during the last glacial, but that its global integral was unlikely dramatically different from today. Therefore, the biological productivity proxies also do not support a major role for the marine nitrogen cycle based mechanisms to explain the glacial-interglacial atmospheric CO_2 changes.

Even if variations in the marine nitrogen cycle might not be involved strongly in changes of the productivity of the ocean and atmospheric CO_2, it is nevertheless important to understand how the marine nitrogen cycle changed over time. For that purpose, we can look at the $\delta^{15}N$ of organic nitrogen in marine sediment cores [*Altabet et al.*, 1995; *Ganeshram et al.*, 1995] and at atmospheric N_2O extracted from ice cores [*Flückiger et al.*, 1999].

The $\delta^{15}N$ of organic nitrogen in marine sediments can provide a record of past variations of $\delta^{15}N$ of NO_3^- in regions where surface NO_3^- is entirely used up. In this case, the fractionation of ^{15}N over ^{14}N during the uptake of NO_3^- by phytoplankton is not expressed, and the organic nitrogen formed has, on net, the same isotopic value as the NO_3^- that was supplied to the surface ocean. Such conditions are generally met in the major upwelling regions of the low- and mid latitudes, where all upwelled NO_3^- is eventually consumed. Figure 9 shows sedimentary records of $\delta^{15}N$ of organic nitrogen from three different regions of the world. In the case of the Arabian Sea and the west coast of the Americas [*Altabet et al.*, 2002; *Ganeshram et al.*, 2000; *Kienast et al.*, 2002], the NO_3^- that upwells to the surface comes to a substantial degree from a depth zone where water column denitrification occurs today. As a result, the $\delta^{15}N$ of the organic matter tends to acquire the high $\delta^{15}N$ values of the NO_3^- in these areas (see also Figure 6). Variations in $\delta^{15}N$ of organic matter in these areas are therefore interpreted mainly as a record of past variations in water column denitrification [*Altabet et al.*, 1995, 2002; *Ganeshram et al.*, 1995, 2000, 2002; *Kienast et al.*, 2002] .

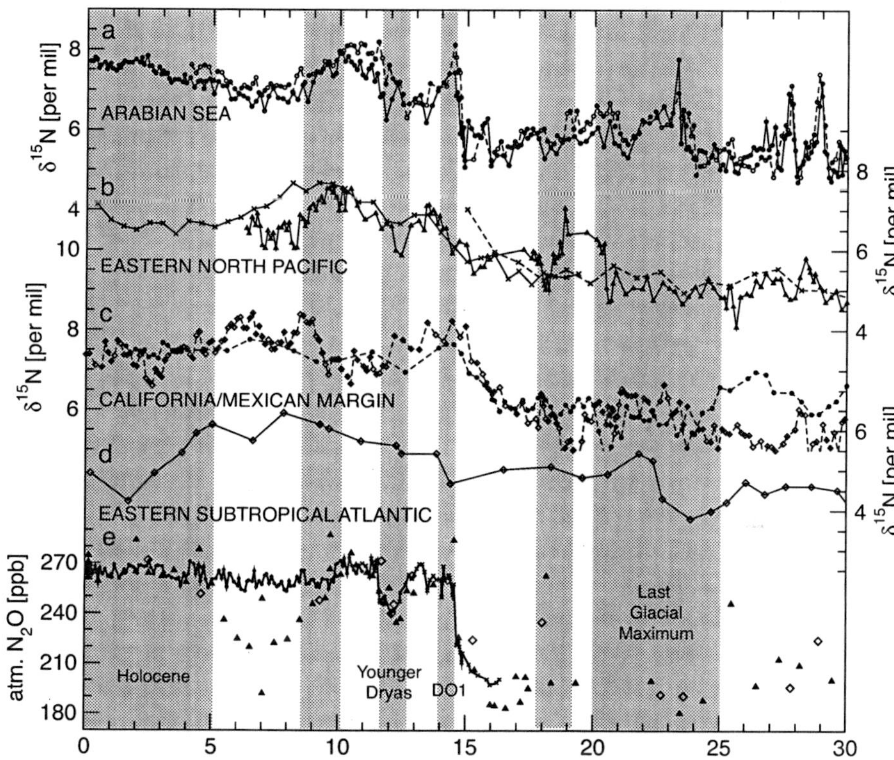

Timeseries of oceanic and atmospheric nitrogen cycle properties over the last 30'000 years. (a) $\delta^{15}N$ of organic nitrogen from the marine sediment cores RC27-14 and RC27-23 from the Oman margin in the Arabian Sea [, 2002]. (b) as (a), but from the cores W8709-13 and W8709-8 (crosses) taken near the Oregon coast [, 2002]. The data from core W8709-8 have been shifted down by 1.2 per mil in order to highlight the similar time evolution. (c) as (a) but from the core NH8P taken off Mexico and from the core ODP1017E taken off Point Conception, California [, 2000]. (d) as (a) but from the core 11K taken off the Mauretanian coast [, 2000], (e) Atmospheric N_2O as recorded in air bubbles from various Antarctic and Greenland ice cores (see Figure 8 for details). The data are plotted on their respective timescales without adjustments.

Betrand et al. [2000] and *Giraud et al.* [2003] pointed out that changes in sea-level can also impact the δ^{15}N of the organic nitrogen deposited on the seafloor. They argue that the change in the sealevel alters the distance between the core site where the signals are recorded and the near-shore area, where upwelling occurs. As the degree of NO_3^- utilization and hence the isotopic value of the remaining NO_3^- depends on the time and distance since upwelling, a sea-level change would result in a change in the δ^{15}N of organic matter independently of any other changes. *Betrand et al.* [2000] and *Giraud et al.* [2003] argue that this mechanism generated a substantial fraction of the glacial-interglacial changes observed in the upwelling area off the northwestern coast of Africa.

These essentially local interpretations of variations in δ^{15}N have to be contrasted with a global-scale interpretation, wherein some of the changes observed at the various sites are simply caused by whole ocean changes in the isotopic value of δ^{15}N. As the global mean δ^{15}N is mainly determined by the relative contribution of water column denitrification to total denitrification [*Deutsch et al.*, 2003], any increase in water column denitrification relative to benthic denitrification would cause a change of the mean ocean δ^{15}N of NO_3^-, which then would be recorded at all sites simultaneously.

Inspection of Figure 9 reveals that the δ^{15}N of organic nitrogen was everywhere substantially lower during the last glacial period, and that it started to shift toward higher values between 17,000 and 12,000 years ago, i.e. in the middle of the transition from the last glacial maximum to the present Holocene. All records tend to show a maximum in δ^{15}N at around 10,000 years BP, followed by a decrease toward the middle Holocene and a light increase toward the present. Most of the glacial-interglacial increase in δ^{15}N in the Arabian Sea and the west coast of the Americas have been interpreted as a local or regional signal, i.e. as an increase in water column denitrification in the source area of the waters that upwell into the surface ocean [*Altabet et al.,* 1995, 2002; *Ganeshram et al.,* 1995, 2000, 2002; *Kienast et al.,* 2002] . The remarkably simultaneous changes in the δ^{15}N in the Atlantic Ocean, far away from the major water column denitrification zones, suggest, however, that either the records in the Arabian Sea and along the west coast of the Americas also contain a strong sea-level change component or that part of the signal recorded in all records represents a whole ocean change in the δ^{15}N of NO_3^-. I would argue for the second scenario, since the sea-level argument would produce different temporal evolutions of δ^{15}N at the various locations. This is because the temporal evolution of δ^{15}N depends on the interaction of the global sea-level change with the local topography, which is different at the various sites.

If we accept that sea-level change had a small effect, the δ^{15}N data indicate a substantial shift toward a greater relative importance of water column denitrification as the ocean moved out of the glacial period into the present interglacial. As the isotopic signatures of benthic denitrification and N_2 fixation are nearly the same, the whole ocean shift provides only weak constraint on the question whether the glacial ocean contained more fixed nitrogen [*Deutsch et al.*, 2003]. The maximum of $\delta^{15}N$ toward the end of the glacial-interglacial transition has, however, been interpreted as evidence for a period of net losses of fixed nitrogen by water column denitrification, which would indeed suggest that the fixed nitrogen inventory was higher during the glacial period. *Deutsch et al.* [2003] constrained this increase to maximally 30%.

It is important to point out that the $\delta^{15}N$ of organic nitrogen shows also a substantial amount of millennial scale variability, some of which have been shown to correspond to major climate events, such as the Younger Dryas that occurred around 11,500 years ago [*Pride et al.*, 1999; *Emmer and Thunell*, 2000; *Altabet et al.*, 2002]. Sediment records in the Arabian Sea and along the west coast of the Americas show a clear decrease in $\delta^{15}N$ during this period, which suggests that water column denitrification became smaller as the northern hemisphere climate slided into this major cold phase. Similar correspondence between northern hemisphere climate variability and water column denitrification has been shown for a seried of millennial scale climate variations during the last glacial in the Arabian Sea [*Altabet et al.*, 2002]. This supports our previous assertion that water column denitrification is very sensitive to alterations in climate.

A second paleoclimatic constraint on the marine nitrogen cycle consists in atmospheric N_2O. N_2O is produced in the ocean via two mechanisms. The first one is associated with the oxidation of ammonium to NO_3^- (nitrification), wherein a relatively fixed fraction of the original ammonium is converted to N_2O [*Cohen and Gordon*, 1979]. As nitrification occurs when organic matter is remineralized back to its inorganic constituents, this mechanism results in a distribution of N_2O production that follows oceanic export production. A second suite of mechanisms to produce N_2O exists at oxygen concentrations below about 50 μmol kg^{-1}. Here, an oxygen concentration dependent fraction of the initial nitrogen is converted to N_2O. At oxygen concentrations of a few μmol kg^{-1}, nearly all organic nitrogen is converted to N_2O [*Goreau et al.*, 1980; *Law and Owens*, 1990]. The exact mechanisms of N_2O production in these low O_2 environments is not well known, but likely includes the interaction of nitrification and denitrification reactions [*Codispoti et al.*, 1992]. This second mechanism results in N_2O production that is localized in the low oxygen regions of the world ocean, i.e. around the areas of water column denitrification. At oxygen concentrations below about 1-2 μmol kg^{-1}, N_2O is consumed and denitrified to N_2 [*Elkins et al.*, 1978; *Cohen and Gordon*, 1978]. This is also the reason that sediments are generally not a source of N_2O, as all N_2O is consumed there. On a global scale, it is currently estimated that about half of the global oceanic emissions of N_2O stems from the nitrification pathway and the other half from the low oxygen pathway [*Jin and Gruber*, 2003; *Nevison et al.*, 2003]. The global oceanic emissions of N_2O are therefore an integral signal of changes in global export production and changes in the oceanic volume with low oxygen concentration. Perturbation simulations have shown that the response of oceanic N_2O production to changes in export production are very sensitive to where these productivity changes occur [*Jin and Gruber*, 2003; *Goldstein et al.*, 2003].

Because of the non-linearity induced by the low oxygen pathway, an increase in oceanic export production in regions underlain by low oxygen concentrations results in a large change in oceanic N_2O production, while the same change in export production in a region whose thermocline waters are well oxygenated, results in a much smaller N_2O response. Overall, variations in oceanic N_2O emissions are therefore more likely caused by variations in the low oxygen pathway than by changes in the nitrification pathway. As there is a correspondence between the volume of waters at low oxygen concentrations with the amount of water column denitrification, variations in the oceanic emissions of N_2O tend to be proportional to the amount of water column denitrification.

The interpretation of the atmospheric N_2O record is hampered by the fact that only about 30 to 50% of the natural emissions of N_2O stem from the ocean [*Ehhalt et al.*, 2001]. Water logged soils are the other main source. A further complicating effect is the limited lifetime of N_2O in the atmosphere, as this compound is destroyed in the upper atmosphere by photodissociation and reactions with singlet atoms of oxygen [*Ehhalt et al.*, 2001]. Model simulations showed, however, that hanges in the oxidizing capacity of the atmosphere over the last few 10,000 years were too small to have caused substantial changes in atmospheric N_2O [*Crutzen and Bruhl,* 1993]. Therefore, variations in atmospheric N_2O can be interpreted to be caused mainly by changes in the source strengths.

Figure 9 reveals that N_2O was about 80 ppb (parts per billion) lower during the last glacial period than during the Holocene. The record also shows that the transition occurred relatively late (around 16,000 to 15,000 years ago), but very rapidly [*Flückiger et al.*, 1999]. Based on isotopic analyses of N_2O, *Sowers et al.* [2003] argued that this glacial-interglacial change is caused in about equal parts by the ocean and the land, albeit with large uncertainties.

It is tempting to compare the atmospheric N_2O record with the marine $\delta^{15}N$ records (Figure 9), as there exists a remarkable degree of correspondence between these different records. As both tracers can be viewed as proxies for water column denitrification, this correspondence might lead to the conclusion that variations in this process are the primary reason for the observed variability in atmospheric N_2O. However, as changes in the physical climate system affect simultaneously land temperature and the water cycle over land (which represents a major forcing mechanism for N_2O production in soils) and ocean circulation and upwelling (which represents a major forcing mechanism for oceanic N_2O production), the simultaneous changes in N_2O and $\delta^{15}N$ cannot be interpreted conclusively as a result of an oceanic dominance on atmospheric N_2O variations.

How can we use now these records to constrain the role of the marine nitrogen cycle on atmospheric CO_2 variations? The glacial-interglacial variations in both $\delta^{15}N$ of organic nitrogen and atmospheric N_2O clearly point toward lower water column denitrification during glacial periods. The generally lower N_2O levels during the last glacial period also suggests that marine productivity was unlikely much larger than today, as this would have caused an increase in the N_2O production by the nitrification pathway that would have had to be compensated by a decrease in any of the other sources beyond the already observed difference. The late beginning of the transition of many oceanic $\delta^{15}N$ records and of atmospheric N_2O indicates that changes in the marine nitrogen cycle were unlikely one of the initial causes of the CO_2 rise.

Combining this evidence with the arguments discussed previously, I conclude that changes in marine N_2 fixation and denitrification likely did not play a central role in explaining the observed glacial-interglacial changes in atmospheric CO_2. The low leverage of any low-latitude mechanism, the lysocline constraint, the existence of internal negative feedbacks, and the limited N:P flexibility of marine phytoplankton appear to conspire against these two hypotheses. However, there is ample evidence suggesting that N_2 fixation and denitrification varied substantially between glacial and interglacial periods and likely contributed to marine productivity changes and even atmospheric CO_2 variability. But their contribution was likely secondary in timing and magnitude.

5. SUMMARY

I began with the quest to understand two puzzles of the marine nitrogen cycle and its connection with atmospheric CO_2. Why are the marine inventories of nitrogen and phosphorus so close to the requirement of phytoplankton (or a corollary question, why is the intercept of the global NO_3^- to PO_4^{3-} relationship near zero), and what is the role of the marine nitrogen cycle for the atmospheric CO_2 variations over glacial-interglacial cycles? No definitive answers can be given to either puzzle yet. I argue, however, that both puzzles are closely related to the question of how N_2 fixation and denitrification in the ocean are coupled to each other. I interpret the existence of a near zero intercept in the global NO_3^- to PO_4^{3-} relationship as evidence for a relatively tight coupling between N_2 fixation and denitrification, as the fixed nitrogen content of surface waters and its ratio to phosphorus can be viewed as the most important control variable in a suite of negative feedbacks. These negative feedbacks tend to contain the fixed nitrogen content of the ocean within relative narrow boundaries of the oceanic reservoir of phosphorus, which is primarily controlled by the balance between river input and sediment burial. These negative feedbacks internal to the marine nitrogen cycle are intertwined with external feedback loops involving atmospheric CO_2 and climate, which could lead to deviations from the tight internal control. I find, however, little evidence that these external feedbacks are strongly developed, despite the fact that the oceanic nitrogen cycle seems to respond rather sensitively to changes in climate. It turns out that the changes in the marine nitrogen inventory are too small to cause substantial alterations of atmospheric CO_2, particularly over glacial-interglacial cycles.

ACKNOWLEDGMENTS

This article was written based on a lecture given at the NATO Advanced Study Institute on Carbon and Climate in Ankara, Turkey in August 2002. I thank Mick Follows and Temel Oguz for being wonderful hosts and for having invited me with financial support from NATO. This work profited from numerous discussions, most importantly those with Curtis Deutsch. I thank Bob Key for sharing his oceanic data base, the NOAA World Data Center for Paleoclimatology, Boulder for providing ice core data to the community, Raja Ganeshram for making his marine $\delta^{15}N$ data available to me, and Regine Roethlisberger, Jacqueline Flueckiger and Thomas Blunier for giving me access to their ice core data and helpful discussion on their interpretation. This work was supported by the U.S. National Science Foundation (OCE-0097337) and the Office of Science (BER), U.S. Department of Energy, Grant No. DE-FG03-00ER63010.

REFERENCES

Altabet, M. A., Variations in nitrogen isotopic composition between sinking and suspended particles: implications for nitrogen cycling and particle transformation in the open ocean, (4), 535–554, 1988.

Altabet, M. A., W. G. Deuser, S. Honjo, and C. Stienen, Seasonal and depth-related changes in the source of sinking particles in the North Atlantic, , 136–139, 1991.

Altabet, M. A., and R. Francois, Sedimentary nitrogen isotopic ratio as a recorder for surface nitrate utilization, (1), 103–116, 1994.

Altabet, M. A., R. Francois, D. W. Murray, and W. L. Prell, Climate-related variations in denitrification in the Arabian Sea from sediment $^{15}N/^{14}N$ ratios, , 506–509, 1995.

Altabet, M. A., M. J. Higginson, and D. W. Murray, The effect of millenial-scale changes in Arabian Sea denitrification on atmospheric CO_2, , 159–162, 2002.

Anderson, T. R., and P. Pondaven, Non-redfield carbon and nitrogen cycling in the Sargasso Sea: pelagic imbalances and export flux, , 573–591, 2003.

Archer, D. E., G. Eshel, A. Winguth, and W. Broecker, Atmospheric CO_2 sensitivity to the biological pump in the ocean, , 1219–1230, 2000.

Archer, D. E., A. Winguth, D. Lea, and N. Mahowald, What caused the glacial/interglacial atmospheric pCO_2 cycles?, (2), 159–189, 2000.

Armstrong, R. A., C. Lee, J. I. Hedges, S. Honjo, and S. G. Wakeham, A new, mechanistic model for organic carbon fluxes in the ocean based on the quantitative association of poc with ballast minerals, , 219–236, 2002.

Bard, E., B. Hamelin, M. Arnold, L. Motaggioni, G. Cabioch, G. Faure, and F. Rougerie, Deglacial sea-level record from tahiti corals and the timing of global meltwater discharge, , 241–244, 1996.

Bates, N. R., A. F. Michaels, and A. H. Knap, Seasonal and interannual variability of oceanic carbon dioxide species at the U.S. JGOFS Bermuda Atlantic Time-series Study (BATS) site, (2-3), 347–383, 1996.

Berger, W. H., Planktonic foraminifera: Selective solution and the lysocline, , 111–138, 1970.

Betrand, P., T. F. Pedersen, P. Martinez, S. Calvert, and G. Shimmicld, Sea level impact on nutrient cycling in coastal upwelling areas during deglaciation: Evidence from nitrogen isotopes, (1), 341–355, 2000.

Blunier, T., B. Barnett, M. L. Bender, and M. B. Hendricks, Biological oxygen productivity during the last 60,000 years from triple oxygen isotope measurements, (3), 3.1–3.15, doi:10.1029/GB001460, 2002.

Boyd, P. W., A. J. Watson, C. S. Law, E. R. Abraham, T. Trull, R. Murdoch, D. C. E. Bakker, A. R. Bowie, K. O. Buesseler, H. Chang, M. Charette, P. Croot, K. Downing, R. Frew, M. Gall, M. Hadfield, J. Hall, M. Harvey, G. Jameson, J. Laroche, M. Liddicoat, R. Ling, M. T. Maldonado, R. M. McKay, S. Nodder, S. Pickmere, R. Pridmore, S. Rintoul, K. Safi, P. Sutton, R. Strzepek, K. Tanneberger, S. Turner, A. Waite, and J. Zeldis, A mesoscale phytoplankton bloom in the polar Southern Ocean stimulated by iron fertilization, , 695–702, 2000.

Brandes, J. A., and A. H. Devol, A global marine-fixed nitrogen isotopic budget: Implications for Holocene nitrogen cycling, (4), doi: 10.1029/2001GB001856, 2002.

Brandes, J. A., A. H. Devol, T. Yoshinari, D. A. Jayakumar, and S. W. A. Naqvi, Isotopic composition of nitrate in the central Arabian sea and eastern tropical pacific: A tracer of mixing and nitrogen cycles, (7), 1680–1689, 1998.

Brandhorst, W., Nitrification and denitrification in the eastern tropical North Pacific, , 3–20, 1959.

Broecker, W. S., Paleocean circulation during the last deglaciation: a bipolar seesaw?, (2), 119–121, 1998.

Broecker, W. S., and G. M. Henderson, The sequence of events surrounding Termination II and their implications for the cause of glacial-interglacial CO_2 changes, (4), 352–364, 1998.

Broecker, W. S., J. Lynch-Stieglitz, D. Archer, M. Hofmann, E. Maier-Reimer, O. Marchal, T. F. Stocker, and N. Gruber, How strong is the Harvardton-Bear constraint?, (4), 817–820, 1999.

Broecker, W. S., and T. H. Peng, , Eldigio Press, Lamont Doherty Earth Observatory of Columbia University, Palisades, NY, 1998.

Brzezinski, M. A., J. L. Sarmiento, K. Matsumoto, C. J. Pride, D. M. Sigman, N. Gruber, G. H. Rau, and K. H. Coale, A switch from $Si(OH)_4$ to NO_3^- depletion in the glacial Southern Ocean, (12), 5.1 – 5.4, doi: 10.1029/2001GL014349, 2002.

Burkill, P. H., R. F. C. Mantoura, and N. J. P. Owens, Biogeochemical cycling in the northwestern Indian Ocean: a brief overview, (3), 643–649, 1993.

Caillon, N., J. P. Severinghaus, J. Jouzel, J.-M. Barnola, J. Kang, and V. Y. Lipenkov, Timing of atmospheric CO_2 and Antarctic temperature changes across termination iii, , 1728–1731, 2003.

Capone, D. G., Marine nitrogen fixation: what's the fuss?, , 341–348, 2001.

Capone, D. G., and E. J. Carpenter, Nitrogen fixation in the marine environment, , 1140–1142, 1982.

Capone, D. G., J. P. Zehr, H. W. Paerl, B. Bergman, and E. J. Carpenter, Trichodesmium, a globally significant marine cyanobacterium, , 1221–1229, 1997.

Carpenter, E., and C. Price, Nitrogen fixation, distribution, and production of ( ) spp. in the western Sargasso and Caribbean Seas, , 60–72, 1977.

Carpenter, E. J., Nitrogen fixation by marine Oscillatoria ( ) in the world's oceans, in , edited by E. J. Carpenter and D. G. Capone, pp. 65–103, Academic Press, San Diego, Calif., 1983.

Carpenter, E. J., and D. G. Capone, Nitrogen fixation in Trichodesmium blooms, in , edited by E. J. Carpenter, pp. 211–217, Kluwer Academic Publishers, Dordrecht, 1992.

Carpenter, E. J., J. P. Montoya, J. Burns, M. R. Mulholland, A. Subramaniam, and D. G. Capone, Extensive bloom of a n_2-fixing diatom/cyanobacterial association in the tropical Atlantic Ocean, , 273–283, 1999.

Carpenter, E. J., and K. Romans, Major role of the cyanobacterium in nutrient cycling in the North Atlantic Ocean, , 1356–1358, 1991.

Catubig, N. R., D. E. Archer, R. Francois, P. deMenocal, W. Howard, and E.-F. Yu, Global deep-sea burial rate of calcium carbonate during the last glacial maximum, (3), 298–310, 1998.

Christensen, J. P., J. W. Murray, A. H. Devol, and L. A. Codispoti, Denitrification in continental shelf sediments has major impact on the oceanic nitrogen budget, (2), 97–116, 1987.

Cline, J. D., and I. R. Kaplan, Isotopic fractionation of dissolved nitrate during denitrification in the Eastern Tropical North Pacific Ocean, , 271–299, 1975.

Cline, J. D., and F. A. Richards, Oxygen deficient conditions and nitrate reduction in the eastern tropical North Pacific ocean, (6), 885–900, 1972.

Codispoti, L., J. Wlkins, T. Yoshinari, G. Friederich, C. Sakamoto, and T. Packard, On the nitrous oxide flux from productive regions that contain low oxygen waters, in , edited by B. Desai, pp. 271–284, Oxford and IBH Publishing Co.,New Delhi, India, 1992.

Codispoti, L. A., Phosphorus vs. nitrogen limitation of new and export production, in , edited by W. H. Berger, V. S. Smetacek, and G. Wefer, pp. 377–394, John Wiley & Sons, New York, 1989.

Codispoti, L. A., J. A. Brandes, J. P. Christensen, A. H. Devol, S. W. A. Naqvi, H. W. Paerl, and T. Yoshinari, The oceanic fixed nitrogen and nitrous oxide budgets: Moving targets as we enter the anthropocene?, (Suppl. 2), 85–105, 2001.

Codispoti, L. A., G. E. Friederich, T. T. Packared, H. E. Glover, P. J. Kelly, R. W. Spinrad, R. T. Barber, J. W. Elkins, B. B. Ward, F. Lipschultz, and N. Lostaunau, High nitrite levels off Northern Peru: A signal of instability in the marine denitrification rate, , 1200–1202, 1986.

Codispoti, L. A., and T. T. Packard, Denitrification rates in the eastern tropical South Pacific, (3), 453–477, 1980.

Codispoti, L. A., and F. A. Richards, An analysis of the horizontal regime of denitrification in the eastern tropical North Pacific, (3), 379–388, 1976.

Cohen, Y., and L. Gordon, Nitrous oxide in the oxygen minimum of the eastern tropical North Pacific: evidence for its consumption during denitrification and possible mechanisms for its production, , 509–524, 1978.

Cohen, Y., and L. Gordon, Nitrous oxide production in the ocean, , 347–353, 1979.

Cornell, S., A. Rendell, and T. Jickells, Atmospheric inputs of dissolved organic nitrogen to the oceans, , 243–246, 1995.

Craig, H., and T. Hayward, Oxygen supersaturation in the ocean : Biological versus physical contributions, , 199–202, 1987.

Crutzen, P. J., and C. Bruhl, A model study of atmospheric temperatures and the concentration of ozone, hydroxyl, and some other photochemical active gases during the glacial, the preindustrial holocene and the present, (11), 1047–1050, 1993.

Cutler, K. B., R. L. Edwards, R. W. Taylor, H. Cheng, J. Adkins, C. D. Gallup, P. M. Cutler, G. S. Burr, and A. L. Bloom, Rapid sea-level fall and deep-ocean temperature change since the last interglacial period, , 253–271, 2003.

Dalsgaard, T., D. E. Canfield, J. Petersen, B. Thamdrup, and J. A. Gonzalez, N_2 production by the anammox reaction in the anoxic water column of Golfo Dulce, Costa Rica, , 606–608, 2003.

Delaney, M. L., Phosphorus accumulation in marine sediments and the oceanic phosphorus cycle, (4), 563–572, 1998.

Deutsch, C., D. M. Sigman, R. C. Thunell, A. N. Meckler, and G. H. Haug, Isotopic constraints on glacial/interglacial changes in the oceanic nitrogen budget, , submitted, 2003.

Deutsch, C. D., N. Gruber, R. M. Key, A. Ganachaud, and J. L. Sarmiento, Denitrification and N_2 fixation in the Pacific Ocean, (2), 483–506, 2001.

Devol, A. H., Direct measurements of nitrogen gas fluxes from continental shelf sediments, , 319–321, 1991.

Devol, A. H., Getting cool with nitrogen, , 131–132, 2002.

Duce, R. A., P. S. Liss, J. T. Merrill, E. L. Atlas, P. Buat-Menard, B. B. Hicks, J. M. Miller, J. M. Prospero, R. Arimoto, T. M. Church, W. Ellis, J. N. Galloway, L. Hansen, T. D. Jickells, A. H. Knap, K. H. Reinhardt, B. Schneider, A. Soudine, J. J. Tokos, S. Tsunogai, R. Wollast, and M. Zhou, The atmospheric input of trace species to the world ocean, (3), 193–259, 1991.

Dupoy, C., J. Neveux, A. Subramaniam, M. R. Mullholland, J. P. Montoya, L. Campbell, E. J. Carpenter, and D. G. Capone, Satellite captures blooms in the southwestern tropical Pacific, (1), 15–16, 2000.

Ehhalt, D., M. Prather, F. Dentener, R. Derwent, E. Dlugokencky, E. Holland, I. Isaksen, J. Katima, V. Kirchhoff, P. Matson, P. Midgley, and M. Wang, Atmospheric chemistry and greenhouse gases, in , Chapter 4, pp. 241–287, Cambridge University Press, Cambridge, 2001.

Elkins, J., S. C. Wofsy, M. B. McElroy, C. E. Kolb, and W. A. Kaplan, Aquatic sources and sinks for nitrous oxide, , 602–606, 1978.

Emerson, S., P. Quay, C. Stump, D. Wilbur, and M. Knox, O_2, Ar, N_2 and ^{222}Rn in surface waters of the subarctic ocean: Net biological O_2 production, (1), 49–69, 1991.

Emmer, E., and R. C. Thunell, Nitrogen isotope variations in Santa Barbara Basin sediments: Implications for denitrification in the eastern tropical North Pacific during the last 50,000 years, (4), 377–387, 2000.

Falkowski, P. G., Evolution of the nitrogen cycle and its influence on the biological sequestration of CO_2 in the ocean, , 272–275, 1997.

Falkowski, P. G., R. J. Scholes, E. Boyle, J. Canadell, D. Canfield, J. Elser, N. Gruber, K. Hibbard, P. Hogberg, S. Linder, F. T. Mackenzie, B. Moore III, T. Pedersen, Y. Rosenthal, S. Seitzinger, V. Smetacek, and W. Steffen, Integrated understanding of the global carbon cycle: A test of our knowledge, , 291–296, 2000.

Flückiger, J., A. Dallenbach, T. Blunier, B. Stauffer, T. F. Stocker, D. Raynaud, and J.-M. Barnola, Variations in atmospheric N_2O concentration during abrupt climate changes, , 227–230, 1999.

Flückiger, J., E. Monnin, B. Stauffer, J. Schwander, T. F. Stocker, J. C. D. Raynaud, and J.-M. Barnola, High-resolution holocene N_2O ice core record and its relationship with CH_4 and CO_2, (1), doi: 10.29/2001GB001417, 2002.

Follows, M. J., T. Ito, and J. Marotzke, The wind-driven, subtropical gyres and the solubility pump of CO_2, (4), 1113, doi:10.1029/2001GB001786, 2002.

Fuhrman, J. A., and D. G. Capone, Nifty nanoplankton, , 593–594, 2001.

Fung, I. Y., S. K. Meyn, I. Tegen, S. C. Doney, J. G. John, and J. K. B. Bishop, Iron supply and demand in the upper ocean, , 281–295, 2000.

Ganeshram, R. S., T. F. Pedersen, S. E. Calvert, and R. Francois, Reduced nitrogen fixation in the glacial ocean inferred from changes in marine nitrogen and phosphorus inventories, , 156–159, 2002.

Ganeshram, R. S., T. F. Pedersen, S. E. Calvert, G. W. McNeil, and M. R. Fontugne, Glacial-interglacial variability in denitrification in the world's ocean: Causes and consequences, (4), 301–376, 2000.

Ganeshram, R. S., T. F. Pedersen, S. E. Calvert, and J. W. Murray, Large changes in oceanic nutrient inventories from glacial to interglacial periods, , 755–758, 1995.

Giraud, X., P. Bertrand, V. Garcon, and I. Dadou, Interpretation of the nitrogen isotopic signal variations in the Mauretanian upwelling with a 2D physical-biogeochemical model, (2), 28.1 – 18.19, 1059: doi:10.1029/2002GB001951, 2003.

Goldman, J. C., Physiological processes, nutrient availability, and the concept of relative growth rate in marine phytoplankton ecology, in , edited by P. G. Falkowksi, pp. 179–194, Plenum Press, New York, 1980.

Goldman, J. C., J. J. McCarthy, and D. G. Peavey, Growth rate influence on the chemical composition of phytoplankton in oceanic waters, , 210–215, 1979.

Goldstein, B., F. Joos, and T. F. Stocker, A modeling study of oceannic nitrous oxide during the Younger Dryas, (2), 64.1 – 64.4, 1092, doi:10.1029/2002GL016418, 2003.

Goreau, T., W. Kaplan, S. Wofsy, M. McElroy, F. Valois, and S. Watson, Production of no_2^- and n_2o by nitrifying bacteria at reduced concentrations of oxygen, , 526–532, 1980.

Gruber, N., C. D. Keeling, and T. F. Stocker, Carbon-13 constraints on the seasonal inorganic carbon budget at the BATS site in the northwestern Sargasso Sea, , 673–717, 1998.

Gruber, N., and J. L. Sarmiento, Global patterns of marine nitrogen fixation and denitrification, (2), 235–266, 1997.

Gruber, N., and J. L. Sarmiento, Biogeochemical/physical interactions in elemental cycles, in , edited by A. R. Robinson, J. J. McCarthy, and B. J. Rothschild, vol. 12, pp. 337–399, John Wiley and Sons,, New York, 2002.

Hammond, D. E., P. Giordani, W. M. Berelson, and R. Poletti, Diagenesis of carbon and nutrients and benthic exchange in sediments of the Northern Adriatic Sea, (1-2), 53–79, 1999.

Hansell, D. A., N. R. Bates, and D. B. Olson, Excess nitrate and nitrogen fixation in the Subtropical North Atlantic, , submitted, 2003.

Hansell, D. A., and R. A. Feely, Atmospheric intertropical convergence impacts surface ocean carbon and nitrogen biogeochemistry in the western tropical Pacific, (7), 1013–1016, 2000.

Hattori, A., Denitrification and dissimilatory nitrate reduction, in , edited by E. J. Carpenter and D. G. Capone, pp. 191–232, Academic Press, San Diego, Calif., 1983.

Haug, G. H., T. F. Pedersen, D. Sigman, S. Calvert, and B. Nielsen, Glacial/interglacial variations in productivity and nitrogen fixation in the Cariaco basin during the last 550 ka, (5), 427–432, 1998.

Hood, R. H., A. F. Michaels, and D. G. Capone, The enigma of marine nitrogen fixation: answers on the horizon, (13), 133,138,139, 2000.

Howell, E. A., S. C. Doney, R. A. Fine, and D. B. Olson, Geochemical estimates of denitrifications in the Arabian Sea and the Bay of Bengal during WOCE, (21), 2539–2552, 1997.

Indermühle, A., E. Monnin, B. Stauffer, T. F. Stocker, and M. Wahlen, Atmospheric CO_2 concentration from 60 to 20 kyr BP from the the Taylor Dome ice core, Antarctica, (5), 735–738, 2000.

Indermühle, A., T. F. Stocker, F. Joos, H. Fischer, H. J. Smith, M. Wahlen, B. Deck, D. Mastroianni, J. Tschumi, T. Blunier, R. Meyer, and B. Stauffer, Holocene carbon-cycle dynamics based on CO_2 trapped in ice at Taylor Dome, Antarctica, , 121–126, 1999.

Jahnke, R. A., and D. B. Jahnke, Rates of C, N, P, and Si recycling and denitrification at the U.S. Mid-Atlantic continental slope depocenter, (8), 1405–1428, 2000.

Jin, X., and N. Gruber, Offsetting the radiative benefit of ocean iron fertilization by enhancing N_2O emissions, , in press, 2003.

Johnsen, S. J., H. Clausen, W. Dansgaard, N. Gundestrup, C. Hammer, U. Andersen, K. Andersen, C. Hvidberg, D. Dahl-Jensen, J. Steffensen, H. Shoji, A. Sveinbjörnsdóttir, J. White, J. Jouzel, , and D. Fisher, The $\delta^{18}O$ record along the Greenland Ice Core Project deep ice core and the problem of possible Eemian climatic instability, , 26397–26410, 1997.

Jouzel, J., V. Masson, O. Cattani, S. Falourd, M. Stievenard, B. Stenni, A. Longinelli, S. Johnsen, J. Steffensen, J. Petit, J. Schwander, R. Souchez, , and N. Barkov, A new 27 ky high resolution east antarctic climate record, (16), 3199–3202, 2001.

Karl, D., R. Letelier, L. Tupas, J. Dore, J. Christian, and D. Hebel, The role of nitrogen fixation in the biogeochemical cycling in the subtropical North Pacific Ocean, , 533–538, 1997.

Karl, D. M., R. Letelier, D. V. Hebel, D. F. Bird, and C. D. Winn, Trichodesmium blooms and new production in the North Pacific gyre, in

, edited by E. J. Carpenter, pp. 219–237, Kluwer Academic Publishers, Dordrecht, 1992.

Karl, D. M., A. F. Michaels, B. Bergman, D. Capone, E. Carpenter, R. Letelier, F. Lipschultz, H. Paerl, D. Sigman, and L. Stal, Dinitrogen fixation in the world's ocean, , 47–98, 2002.

Kawase, M., and J. L. Sarmiento, Nutrients in the Atlantic thermocline, (C5), 8961–8979, 1985.

Keeling, C. D., NATO lecture 2: Surface ocean CO_2, in , edited by M. Heimann, pp. 413–430, Springer-Verlag, New York, 1993.

Kienast, S. S., S. E. Calvert, and T. F. Pedersen, Nitrogen isotope and productivity variations along the northeast Pacific margin over the last 120kyr: surface and subsurface paleoceanography, (4), 7.1 – 7.17, 1055, doi:10.1029/2001PA000650, 2002.

Knox, F., and M. McElroy, Changes in atmospheric CO_2: influence of marine biota at high latitudes, , 4629–4637, 1984.

Kohfeld, K., C. Le Quéré, and S. P. Harrison, Limited role of marine export production for glacial-interglacial CO_2 variations, , submitted, 2003.

Kuypers, M. M. M., A. O. Sliekers, G. Lavik, M. Schmid, B. B. Jorgensen, J. G. Kuenen, J. S. S. Damste, M. Strous, and M. S. M. Jetten, Anaerobic ammonium oxidation by anammox bacteria in the Black Sea, , 608–610, 2003.

Law, C. S., and N. J. P. Owens, Significant flux of atmospheric nitrous oxide from the northwest Indian ocean, , 826–828, 1990.

Laws, E. A., P. Falkowski, E. Carpenter, and H. Ducklow, Temperature effects on export production in the open ocean, (4), 1231–1246, 2000.

Lefevre, N., and A. J. Watson, Modeling the geochemical cycle of iron in the oceans and its impact on atmospheric CO_2 concentrations, (3), 727–736, 1999.

Lenton, T. M., and A. J. Watson, Redfield revisited 1. regulation of nitrate, phosphate, and oxygen in the ocean, (1), 225–248, 2000.

Liu, K.-K., 1979, , Ph.d. thesis, Univ. of California, Los Angeles.

Liu, K. K., and I. R. Kaplan, The eastern tropical Pacific as a source of N-15 enriched nitrate in seawater off southern California, (5), 820–830, 1989.

Luyten, J. L., J. Pedlosky, and H. Stommel, The ventilated thermocline, , 292–309, 1983.

Mahaffey, C., R. G. Williams, G. A. Wolff, N. M. Mahowald, W. Anderson, and M. Woodward, Biogeochemical signatures of nitrogen fixation in the eastern North Atlantic, (6), 33.1 – 33.4, 1300, doi:10.1029/2002GL016542, 2003.

Mahowald, N., K. E. Kohfeld, M. Hansson, Y. Balkanski, S. P. Harrison, I. C. Prentice, M. Schulz, and H. Rodhe, Dust sources and deposition during the Last Glacial Maximum and current climate: A comparison of model results with paleodata from ice cores and marine sediments, , 15,895–15,916, 1999.

Mantoura, R. F. C., C. S. Law, N. J. P. Owens, P. H. Burkill, E. M. S. Woodward, R. J. M. Howland, and C. A. Lewellyn, Nitrogen biogeochemical cycling in the northwestern Indian ocean, (3), 651–671, 1993.

Martin, J. H., Glacial-interglacial CO_2 change: the iron hypothesis, (1), 1–13, 1990.

Martin, J. H., S. E. Fitzwater, and R. M. Gordon, Iron deficiency limits phytoplancton growth in antarctic waters, (1), 5–12, 1991.

Martin, J. H., G. A. Knauer, D. M. Karl, and W. W. Broenkow, Vertex: carbon cycling in the northeast Pacific, (2), 267–285, 1987.

Matsumoto, K., J. L. Sarmiento, and M. A. Brzezinski, Silicic acid leakage from the Southern Ocean as possible mechanism for explaining glacial atmospheric pCO_2, (10), doi:10.1029/2001GB001442, 2002.

McElroy, M. B., Marine biological controls on atmospheric CO_2 and climate, , 328–329, 1983.

Meybeck, M., C, N, P and S in rivers: from sources to global inputs, in , edited by R. Wollast, F. T. Mackenzie, and L. Chou, pp. 163–193, Springer-Verlag, Berlin, 1993.

Michaels, A. F., N. R. Bates, K. O. Buesseler, C. A. Carlson, and A. H. Knap, Carbon-cycle imbalances in the Sargasso Sea, , 537–540, 1994.

Michaels, A. F., D. M. Karl, and D. G. Capone, Element stoichiometry, new production nitrogen fixation, (4), 68–77, 2001.

Michaels, A. F., D. Olson, J. L. Sarmiento, J. Ammerman, K. Fanning, R. Jahnke, A. H. Knap, R. Lipschultz, and J. Prospero, Inputs, losses and transformations of nitrogen and phosphorus in the pelagic North Atlantic Ocean, , 181–226, 1996.

Middelburg, J. J., K. Soetaert, P. M. J. Herman, and C. H. R. Heip, Denitrification in marine sediments: A model study, (4), 661–673, 1996.

Minagawa, M., and E. Wada, Nitrogen isotope ratios of red tide organisms in the East China Sea - a characterization of biological nitrogen fixation, (3), 245–259, 1986.

Mino, Y., T. Saino, K. Suzuki, and E. M. Nón, Isotopic composition of suspended nitrogen ($\delta^{15}n_{sus}$) in surface waters of the Atlantic from 50°n to 50°s, (4), 7–1 – 7–9, 1059, doi:10.1029/2002GB001635, 2002.

Monnin, E., A. Indermühle, A. Dallenbach, J. Ëkr , B. Stauffer, T. F. Stocker, D. Raynaud, and J.-M. Barnola, Atmospheric CO_2 concentrations over the last glacial termination, , 112–114, electronic copy, 2002.

Morel, F. M. M., and N. M. Price, The biogeochemical cycles of trace metals in the ocean, , 944–947, 2003.

Morrison, J. M., L. A. Codispoti, S. L. Smith, K. Wishner, C. Flagg, W. D. Gardner, S. Gaurin, S. W. A. Naqvi, V. Manghnani, L. Prosperie, and J. S. Gundersen, The oxygen minimum zone in the Arabian Sea during 1995, , 1903–1931, 1999.

Mullin, M. M., G. H. Rau, and R. W. Eppley, Stable nitrogen isotopes in zooplancton: Some geographic and temporal variations in the North Pacific, (6), 1267–1273, 1984.

Naqvi, S., S. D. Souza, and C. Reddy, Relationship between nutrients and dissolved oxygen with special reference to water masses in western Bay of Bengal, , 15–17, 1978.

Naqvi, S. W. A., Some aspects of the oxygen-deficient conditions and denitrification in the Arabian sea, , 1049–1072, 1987.

Naqvi, S. W. A., R. J. Noronha, and C. V. G. Reddy, Denitrification in the Arabian Sea, (4A), 459–469, 1982.

Naqvi, S. W. A., R. J. Noronha, K. Somasundrar, and R. Sen Gupta, Seasonal changes in the denitrification regime of the Arabian Sea, (4), 593–611, 1990.

Naqvi, S. W. A., and M. S. Shailaja, Activity of the respiratory electron transport system and respiration rates within the oxygen minimum layer of the Arabian Sea, (3), 687–695, 1993.

Neftel, A., H. Oeschger, J. Schwander, B. Stauffer, and R. Zumbrunn, Ice core sample measurements give atmospheric CO_2 content during the past 40,000 yr, , 220–223, 1982.

Nevison, C. D., J. Butler, and J. Elkins, Global distribution of N_2O and the δN_2O/AOU yield in the subsurface ocean, , submitted, 2003.

Orcutt, K. M., F. J. Lipschultz, K. Gundersen, R. Arimoto, J. R. Gallon, A. F. Michaels, and A. H. Knap, A seasonal study of the significance of N_2 fixation by spp. at the Bermuda Time-Series Study (BATS) site, (8-9), 1583–1608, 2001.

Owens, N. J. P., Natural variations in ^{15}n in the marine environment, , 389–451, 1987.

Pedlosky, J., , Springer Verlag, New York, 1987.

Pride, C., R. Thunell, D. Sigman, L. Keigwin, M. Altabet, and E. Tappa, Nitrogen isotope variations in the Gulf of California since the last glaciation: response to global climate change, (3), 397–409, 1999.

Redfield, A. C., B. H. Ketchum, and F. A. Richards, The influence of organisms on the composition of sea-water, in , edited by M. N. Hill, vol. 2, pp. 26–77, Wiley-Interscience, New York, 1963.

Roethlisberger, R., R. Mulvaney, E. W. Wolff, M. A. Hutterli, M. Bigler, S. Sommer, and J. Jouzel, Dust and sea salt variability in central East Antarctica (Dome C) over the last 45 kyrs and its implications for southern high-latitude climate, (20), 24.1 – 24.4, 1963, doi:10.1029/2002GL015186, 2002.

Rueter, J. G., D. A. Hutchins, R. W. Smith, and N. L. Unsworth, Iron nutrition of Trichodesmium, in , edited by E. J. Carpenter, pp. 289–306, Kluwer Academic Publishers, Dordrecht, 1992.

Sabine, C. S., M. Heimann, P. Artaxo, D. Bakker, C. Chen, C. Field, N. Gruber, C. Le Quéré, R. G. Prinn, J. E. Richey, P. Romero-Lankao, J. Sathaye, and R. Valentini, Current status and past trends of the carbon cycle, in $_2$, edited by C. B. Field and M. R. Raupach, Chapter 2, Island Press, Vienna, 2003.

Sanudo-Wilhelmy, S. A., A. B. Kustka, C. J. Gobler, D. A. Hutchins, M. Yang, K. Lwiza, J. Burns, D. G. Capone, J. A. Raven, and E. J. Carpenter, Phosphorus limitation of nitrogen fixation by in the central Atlantic ocean, , 66–69, 2001.

Sarmiento, J., A tritium box model of the North Atlantic thermocline, , 1269–1274, 1983.

Sarmiento, J. L., J. P. Dunne, A. Gnanadesikan, R. M. Key, K. Matsumoto, and R. Slater, A new estimate of the $CaCO_3$:C_{org} ratio, (4), 1107, doi:1029/2002GB001919, 2002.

Sarmiento, J. L., and N. Gruber, Anthropogenic carbon sinks, , 30–36, 2002.

Sarmiento, J. L., and J. R. Toggweiler, A new model for the role of the oceans in determining atmospheric pCO_2, , 621–624, 1984.

Seitzinger, S., and A. Giblin, Estimating denitrification in North Atlantic continental shelf sediments, (1), 235–260, 1996.

Siegenthaler, U., and T. Wenk, Rapid atmospheric CO_2 variations and ocean circulation, , 624–626, 1984.

Sigman, D., D. C. McCorkle, and W. R. Martin, The calcite lysocline as a constraint on glacial/interglacial low-latitude production changes, (3), 409–427, 1998.

Sigman, D. M., M. A. Altabet, D. D. McCorkle, R. Francois, and G. Fischer, The $\delta^{15}N$ of nitrate in the Southern Ocean: Consumption of nitrate in surface waters, (4), 1149–1166, 1999.

Sigman, D. M., M. A. Altabet, D. D. McCorkle, R. Francois, and G. Fischer, The $\delta^{15}N$ of nitrate in the Southern Ocean: Nitrogen cycling and circulation in the ocean interior, (C8), 19599–19614, 2000.

Sigman, D. M., and E. A. Boyle, Glacial/interglacial variations in carbon dioxide: searching for a cause, , 859–869, 2000.

Sigman, D. M., R. Robinson, A. N. Knapp, A. van Geen, D. C. McCorkle, J. A. Brandes, and R. C. Thunell, Distinguishing between water column and sedimentary denitrification in the Santa Barbara Basin using the stable isotopes of nitrate, (5), 1–20, doi:10.1029/2002GC000384, 2003.

Smith, S., M. Roman, K. Wishner, M. Gowing, L. Codispoti, R. Barber, J. Marra, I. Prusova, and C. Flagg, Seasonal response of zooplankton to monsoonal reverals in the Arabian Sea, , 2369–2404, 1998.

Smith, S. V., Phosphorus versus nitrogen limitation in the marine environment, (6), 1149–1160, 1984.

Sowers, T., R. B. Alley, and J. Jubenville, Ice core records of atmospheric N_2O covering the last 106,000 years, , 945–948, 2003.

Tegen, I., and I. Fung, Modeling of mineral dust in the atmosphere: Sources, transport and optical thickness, , 22897–22914, 1994.

Tyrrell, T., The relative influences of nitrogen and phosphorus on oceanic primary production, , 525–531, 1999.

Villareal, T. A., and E. J. Carpenter, Nitrogen fixation, suspension characteristics, and chemical composition of rhizosolenia mats in the central Pacific gyre, , 327–345, 1989.

Voss, M., J. W. Dippner, and J. P. Montoya, Nitrogen isotope patterns in the oxygen deficient waters of the Eastern Tropical North Pacific Ocean, (8), 35–49, 2001.

Wada, E., and A. Hattori, , CRC Press, Boca Raton, FL, 1991.

Walsh, J. J., Importance of continental margins in the marine biogeochemical cycling of carbon and nitrogen, , 53–55, 1991.

Walsh, J. J., Nitrogen fixation within a tropical upwelling ecosystem: Evidence for a redfield budget of carbon/nitrogen cycling by the total phytoplankton community, (C9), 20607–20616, 1996.

Warren, B. A., Context of the suboxic layer in the Arabian Sea, (2), 301–314, 1994.

Weaver, A. J., O. A. Saenko, P. U. Clark, and J. X. Mitrovica, Meltwater pulse 1A from Antarctica as a trigger of the Bolling-Allerod warm interval, , 1709–1713, 2003.

Wollast, R., Evaluation and comparison of the global carbon cycle in the coastal zone and in the open ocean, in , edited by K. H. Brink and A. R. Robinson, vol. 10, Chapter 9, pp. 213–252, John Wiley & Sons, New York, 1998.

Wu, J., W. Sunda, E. A. Boyle, and D. M. Karl, Phosphate depletion in the Western North Atlantic Ocean, , 759–762, 2000.

Wyrtki, K., The physical oceanography of the Indian Ocean, in ,
pp. 18–36, Springer-Verlag, 1973.

Zehr, J. P., E. Carpenter, and T. A. Villareal, New perspectives on nitrogen-fixing microrganisms in tropical and subtropical oceans, (2), 68–73, 2000.

Zehr, J. P., J. B. Waterbury, P. J. Turner, J. P. Montoya, E. Omoregie, G. F. Steward, A. Hansen, and D. M. Karl, Unicellular cyanobacteria fix N_2 in the subtropical North Pacific Ocean, , 635–638, 2001.

EXPLORATION OF BIOGEOCHEMICAL TEMPORAL VARIABILITY

TOMMY D. DICKEY
Ocean Physics Laboratory, University of California, Santa Barbara,
tommy.dickey@opl.ucsb.edu

INTRODUCTION

There is growing interest in better understanding the biogeochemistry of the world oceans, largely because of the rapid increase of atmospheric carbon dioxide, CO_2, and the anticipated consequences for the oceans as well as the atmosphere. Interestingly, measurements are presently showing increasing levels of CO_2 in the surface layers of the ocean in relatively pristine areas, namely at the Joint Global Ocean Flux Study (JGOFS) time series sites off Hawaii and Bermuda (e.g., Karl et al., 2001). Suggested effects of rising levels of CO_2 include increases in global atmospheric and oceanic temperature, shifts in weather patterns causing regional droughts and flooding, melting of ice caps, sea level rise, and changes in ocean ecology and population dynamics that may affect fisheries. A fundamental scientific, societal, and political issue concerns the partitioning of the causes of these phenomena between natural variability versus anthropogenic influences.

Several important biogeochemical- and carbon-related questions stimulated the development and execution of the ambitious international JGOFS program and several related research efforts on a global scale (e.g., Brewer et al., 1986; Buesseler, 2001; Murray and other authors, this volume). Simply put, JGOFS and other recent studies were designed to study oceanic biogeochemical cycles and their roles in affecting global change and *vice versa* (see other chapters in this volume).

M. Follows and T. Oguz (eds.), The Ocean Carbon Cycle and Climate, 149–188.

It is now generally thought that the world oceans primarily control the atmospheric concentration of CO_2 on very long geological time scales (see other chapters in this volume). Higher concentrations of dissolved inorganic carbon (DIC) are found below the surface mixed layer of the ocean. Yet, the upper portion of the water column is in equilibrium with the atmosphere to first order. This gradient is maintained by two carbon "pumps," the "solubility pump" and the "biological pump." The solubility pump involves the thermal contrast between the upper ocean and the ocean interior with the solubilities of cold waters of the deep ocean being on the order of twice as great as some of the warmer near surface waters. Thus, the net effect of sinking high latitude surface waters, via deep ocean thermohaline circulation, is to enrich deeper waters in carbon.

The present chapter focuses primarily on processes, observations, and recent studies related to the second type of carbon pump, the biological pump (see reviews by Bishop, 1989; Ducklow et al., 2001; and other chapters in this volume). Quantification and understanding of this process drive much carbon research. We emphasize *in situ* methodologies in this contribution as remote sensing aspects are covered by Korotaev in this volume). A brief description of the biological pump concept follows for completeness (see Figure 1; more in-depth discussions are presented in this volume by Anderson and Toterdell. Phytoplankton living in the euphotic layer of the ocean use carbon dioxide to form organic matter. A high percentage of the organic matter is metabolized (used for maintaining life functions); however, a significant portion (roughly 20%, but highly variable in terms of location and time) sinks to the deep sea. There it is converted back to carbon dioxide (through a process called remineralization; e.g., via bacteria). Physical transport processes later bring the carbon dioxide back to the surface (e.g., through upwelling in various forms or advection of water masses and mixing from depth, Anderson and Toterdell, this volume). The biological pump also includes pathways for getting carbon to the deeper layers through dissolved organic carbon (DOC) molecules (e.g., Kirchman, this volume; Hansell and Carlson, 2001) and particulate organic carbon (POC) matter. Importantly, the overall effect is to transport carbon to the deep ocean. The determination of primary productivity, the rate of plant tissue build up through photosynthesis, in the upper ocean is a vital step in quantifying the carbon flux associated with the biological pump. Primary producers, predominantly phytoplankton, convert inorganic materials like nitrate, phosphate, and silicate into organic compounds such as lipids and proteins through the process of photosynthesis (e.g., Lalli and Parsons, 1993). Secondary producers, mainly zooplankton (small generally drifting, but some mobile, animal organisms), are important players also as they excrete ammonia (another nutrient), graze phytoplankton, and contribute to the downward flux of materials (e.g., detrital or waste and decomposition (after death) components). Together, the solubility and biological pumps are now believed to affect atmospheric carbon dioxide levels significantly.

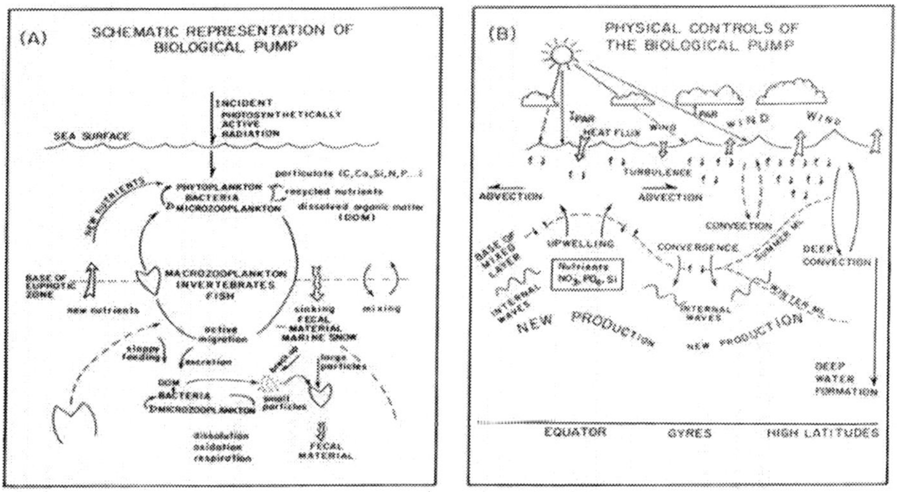

*Figure 1.(*a) Schematic representation of the biological pump and the various contributing biological processes. (b) Illustration of some of the physical mechanisms affecting the biological pump from equatorial regions to mid to high latitudes. Based on figure by Bishop (1989).

Here we introduce a few of the key issues regarding primary productivity in relation to the biological pump. This line of research offers several opportunities to cross traditional disciplinary boundaries. For our interests, we highlight several connections between biogeochemistry, bio-optics (the study of biological effects on the subsurface light field and *vice versa*), upper ocean physics, and the biological pump (see Dickey and Falkowski, 2002). Phytoplankton biomass depends on photosynthetic processes, which involve the availability of light (e.g., photosynthetically available radiation or PAR, simply light in the visible wavelengths) and plant nutrients. Two classes of plant nutrients, macro-nutrients (e.g., nitrate, silicate, and phosphate) and micro-nutrients (e.g., iron), are important. The spectral quality and intensity of light vary with depth and are also important for specific phytoplankton species with special pigmentation or photoadaptive characteristics and niches (e.g., Bidigare et al., 1990). Light exposure for individual phytoplankton is affected by variation in physical conditions such as mixed layer depth, turbulent mixing, and currents and of course the incident light field, which varies spectrally (e.g., blue light penetrates more deeply than other wavelengths in the open ocean) and in time and space (e.g., daily or diel cycle, seasonal cycle, cloud variability, etc.). An interesting feedback concerns the modulation of the spectral light field and radiant heating at depth as phytoplankton concentrations and communities wax and wane (Dickey and Falkowski, 2002). Some of the driving mechanisms, processes, and feedbacks for phytoplankton production are illustrated in a conceptual physical/biogeochemical model (Figure 2), which includes a basic nutrient-phytoplankton-zooplankton-detrital model component.

A major biological research effort has involved the estimation and modeling of primary production. This has been driven in part by a host of new technologies involving *in situ*-, satellite- and aircraft-based optical (color) sensors capable of measuring key parameters (with varying spectral resolution; often 7 and recently up to 100 wavelength bands in the visible) on unprecedented time and space scales. Primary productivity, P(z), as a function of depth, z, has also been estimated with relatively simple models using chlorophyll *a*, Chl(z), and PAR(z) data. Best estimates or measurements of the chlorophyll *a* specific light absorption coefficient for phytoplankton, a*, and the quantum yield for carbon fixation, $\phi(z)$, have been utilized as in the following empirical formulation

$$P(z) = a^* \, \phi(z) \, Chl(z) \, PAR(z) \qquad (1)$$

Some of examples presented in this chapter utilize this type of relationship (see (e.g., see Bidigare et al., 1990, and Behrenfeld and Falkowski, 1997a,b, for more detailed information). Importantly, neither a* nor ϕ is constant because of community structure changes and varying light and nutrient stresses.

Another of our major remaining research questions concerns the relationship of upper ocean primary production and the flux of biogeochemical materials, especially carbon, to the deep sea. This is a complicated problem that again involves physical, chemical, and biological interactions and transformations. Some of the examples presented in this chapter outline recent work in this important study area.

The present chapter is intended to provide an introduction to our current understanding of biogeochemical processes, their observation, and their relationships with ocean physics. Our focus is primarily on the time domain, with consideration of processes that have time scales from minutes to the interannual (see Levy, this volume, concerning small time and space scale processes). Long-term, high frequency time series studies are ideal, and likely crucial, if we are to begin to understand subtle changes in the ocean habitat which are likely greatly affected by periodic forcing as well as episodic events that perturb the ocean's ecosystem and thus biogeochemical variability (e.g., Karl et al., 2001). Likewise, such studies are needed to help us to begin to unravel vexing biogeochemical questions related to presently unpredictable, large-scale climate variations such as El Niño-Southern Oscillation (ENSO), Pacific Decadal Oscillation (PDO), the North Atlantic

Oscillation (NAO) and the Arctic Oscillation (AO) and their likely interrelationships.

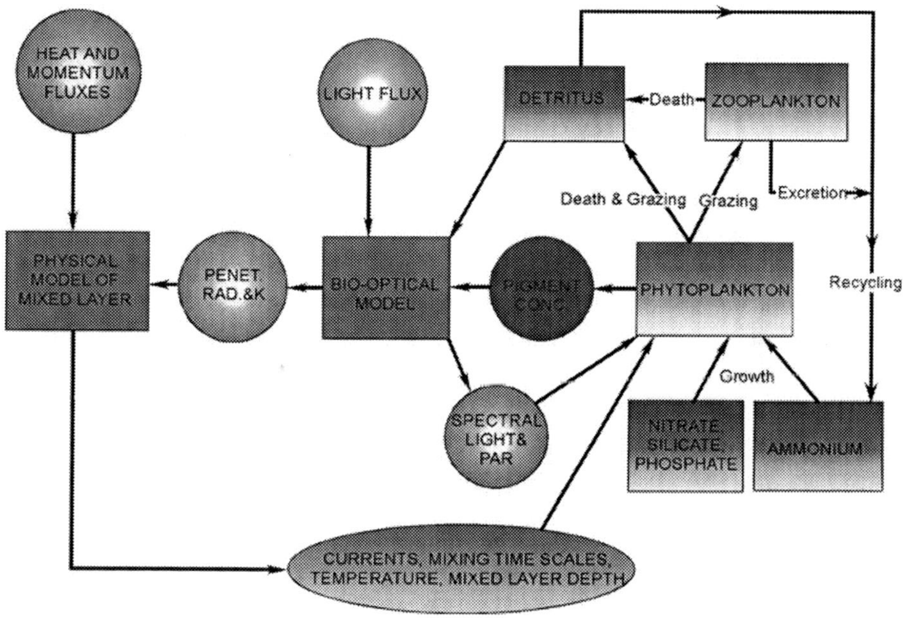

Figure 2. A conceptual model illustrating coupling of physical and biological processes through optical processes. PENET. RAD. and K represent the penetrative component of solar radiation and the diffuse attenuation coefficient of solar radiation. The schematic portrayal of the biological state variables (phytoplankton, zooplankton, nitrate, silicate, phosphate, ammonium, and detritus) basically follows Friedrichs and Hofmann (2001).

The present chapter is organized as follows. First, a brief introduction to relevant processes and their time-space domains is presented. Next, some of the emergent methodologies involving sampling strategies, platforms, sensors, and systems are reviewed. Then, examples of biogeochemical studies in different oceanic environments are used to highlight some new research advances and paradigm shifts. Finally, we consider some emerging research approaches that involve merging observations and models.

1. PROCESSES AND SCALES

The oceans are naturally dynamic, with large amplitude periodic and episodic variability that confounds attempts to quantify long-term trends and changes. Oceanographers are confronted with challenges unlike those faced by laboratory scientists, as data must be obtained from an uncontrolled and often harsh, if not violent, environment. In addition, most oceanographic problems are interdisciplinary in nature and thus require a very large number of observational variables with relevant processes spanning over 10 orders of magnitude (Figure 3). Further, the processes are often nonlinear, turbulent, or chaotic, and biological and chemical aspects involve behavior and reactions that force the use of a large number of coupled nonlinear equations and empirical relations for models (see Matear, this volume). In addition, recent studies suggest that only a few strong episodic events can often be of far greater importance than more slowly varying relatively small amplitude variations. Despite these daunting challenges, recent biogeochemical studies have begun to characterize, quantify, and improve our understanding of ocean processes that cause biogeochemical variability in the form of carbon inventories and carbon fluxes However, it must be recognized at the outset that insufficient diversity of data and undersampling limit our ability to make further advances. Hence, a major challenge remains to greatly increase the variety and quantity of biogeochemical and other needed interdisciplinary data. Ideally, the relevant data should be collected simultaneously (concept of synopticity) and span broad time and space scales to observe the processes of interest (Figure 3). For global problems, this means sampling variability that extends well over ten orders of magnitude in space and much longer in time for climate problems. Next we review some sampling strategies, platforms, sensors, and systems that are presently being used or are planned for future biogeochemical research. Already, many measurements of several biogeochemical variables can be done on virtually the same time and space scales as physical variables. The optimal utilization of these platforms and measurement systems for regional and global studies is considered in the final section.

2. OBSERVATIONAL METHODS

2.1 PLATFORMS

Several major interdisciplinary oceanographic programs, including JGOFS, have adopted multi-platform approaches. The schematics shown in Figure 4 illustrate a variety of platforms, several of which can utilize physical, chemical, bio-optical, acoustical, and geophysical sensors or systems and the types of data sets that can be obtained (examples come from the Sargasso Sea off Bermuda). The time-space diagram shown in Figure 3 provides a rough means of estimating the utility of different platforms in space (horizontal aspect depicted in this figure) and time. It also re-emphasizes the need for deploying sensors from both *in situ* and remote platforms. Nesting of platforms and various modeling approaches can be used to optimize utilization of these observational assets as discussed in the final section. Below, brief summaries of capabilities and future directions for a few platforms are presented.

2.2 SHIPS

Ships have provided access to ocean observations since early expeditionary voyages. Ships are important for manned on-site observations and data collection and for deployment of other sampling platforms such as remotely operated vehicles (ROVs), autonomous underwater vehicles (AUVs), moorings, drifters, floats, and others described below. Common modes of ship sampling include: 1) on-station profiling of instruments, 2) underway sampling of meteorological and optical variables and surface waters using flow-through systems, 3) underway sampling to provide data in the horizontal and vertical dimensions using towed undulating or fixed depth bodies with sampling chains, which act as platforms for sensor suites, and 4) underway acoustical measurements (e.g. for acoustic Doppler current profilers (ADCPs), hydrophone arrays, and sidescan sonar). These various modes of sampling are useful for regional process-oriented studies and for long transect sampling programs designed to provide important spatial maps. Research ships can be used to deploy advanced analytical instrumentation, which presently cannot be easily deployed from other platforms, can be utilized, often with real-time data analysis. In addition, large volumes of water and net towing are still required for some research efforts. Commercially operated voluntary observing ship (VOS) or ships-of-opportunity observational programs are especially valuable, particularly for obtaining data in remote oceanic regions where few dedicated research sampling programs can be routinely executed or where ferries or ships transit regularly (essentially enably spatial time series). Dedicated research ships as well as VOSs have also played important roles in calibrating and validating ocean color satellite data.

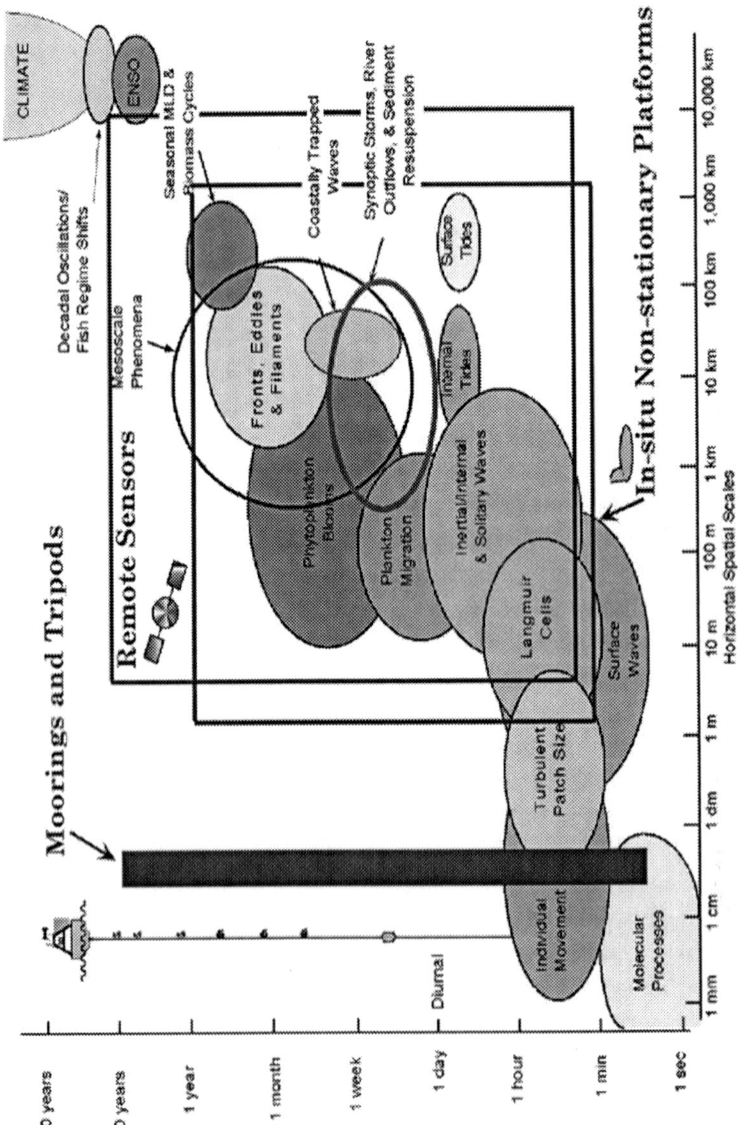

Figure 3. (a) Time-horizontal space scale diagram illustrating several physical and biological processes in ovals. (b) Rectangles indicate the approximate sampling domain capabilities of several platforms at present.

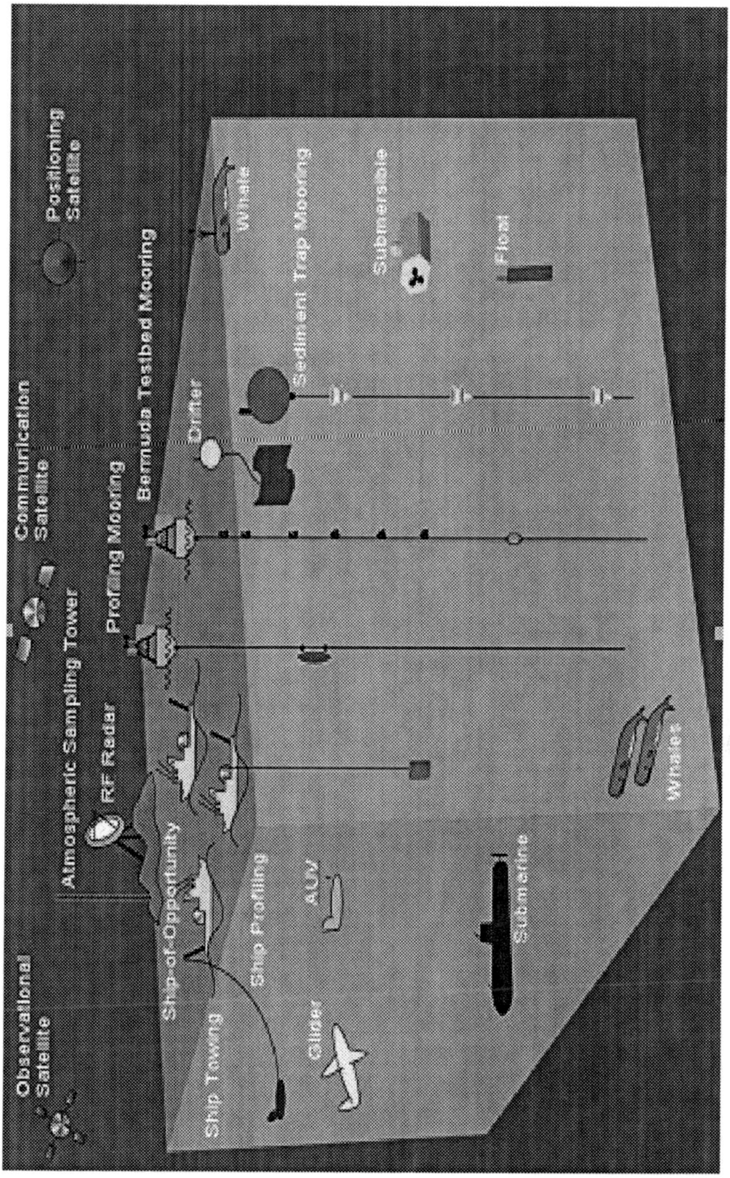

Figure 4a. Conceptual illustration of several sampling platforms for the open ocean. These platforms all exist and most have been used off the island of Bermuda

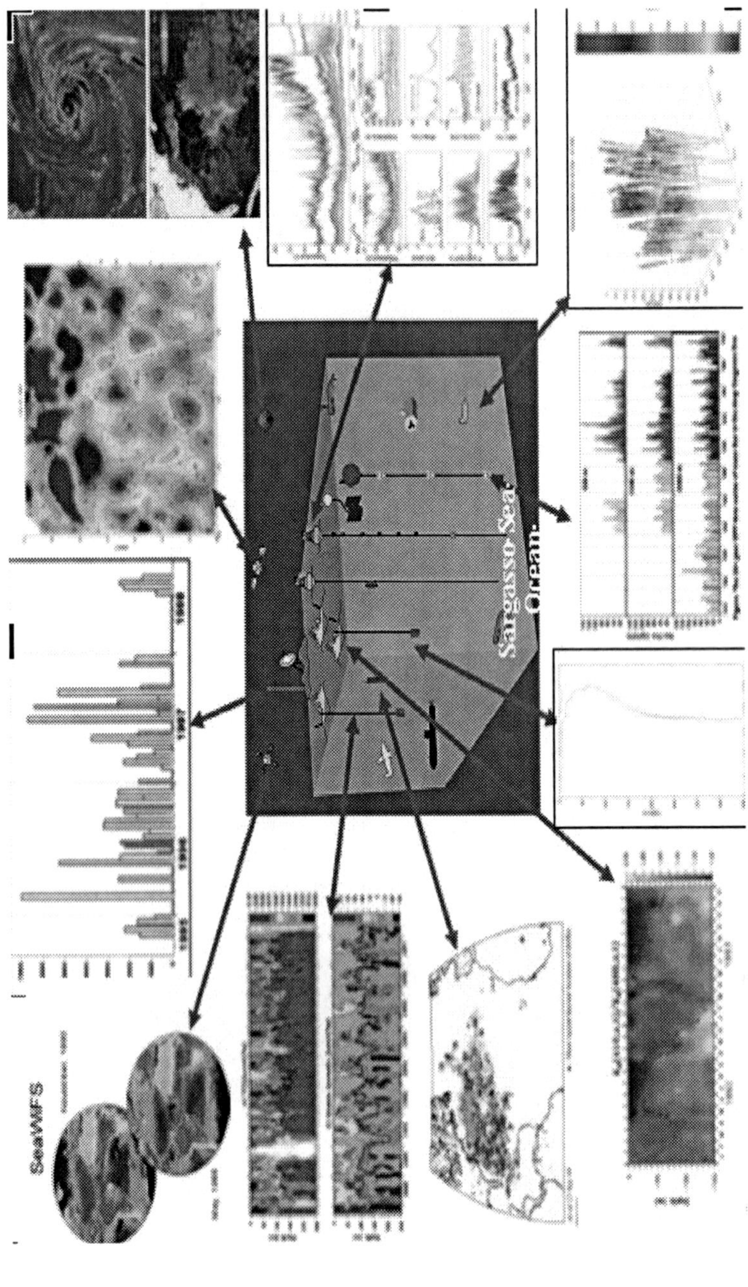

Figure 4b. . Panels indicate the types of data sets that have been collected with various platforms off Bermuda.

Ships have served the oceanographic community well, but their limitations in terms of cost, availability, limited synoptic sampling capability, and sample degradation and contamination have necessitated development of other complementary platforms as described below.

2.3 MOORINGS

Moorings have been used for physical and meteorological measurements for several decades. However the use of moorings, some with data telemetry capabilities, for interdisciplinary studies including bio-optics and biogeochemistry began relatively recently. Moorings are especially well suited for observing high frequency, episodic, and slowly varying phenomena (see Dickey et al., 1998a,b, 2000, 2001; Dickey and Falkowski, 2002). They are also valuable for testing of new sensors and systems that can later be used on other platforms that are either difficult or impossible to retrieve (e.g., drifters, profiling floats, and autonomous underwater vehicles). Some of the more relevant time series measurements that can now be made from moorings include: nitrate, trace elements, partial pressure of carbon dioxide (pCO_2), dissolved oxygen (DO), primary production (via time series samplers and the ^{14}C method), particle fluxes (time series sampling sediment traps; Berelson, 2001), and a broad range of relevant optical variables for spectral light fields for estimating primary production and other carbon parameters (e.g., POC; Bishop, 1999). These have enabled the oceanographic community to gain new insights into a variety of biogeochemical processes (examples are given in the next section). Several diverse, and often adverse, oceanic regions have been studied using interdisciplinary moored systems. These range from the equatorial Pacific to high latitude areas south of Iceland and in the Southern Ocean (see Dickey and Falkowski, 2002). Because of biofouling of sensors, useful data from moorings have often been limited to 3-6 months in the open ocean; however, work is advancing to extend useful deployment periods for biogeochemical and bio-optical sensors and systems (e.g., Dickey et al., 2000). High temporal resolution will continue to be needed to minimize sampling induced uncertainties (via undersampling and aliasing).

Time series observations at selected sites of expected high environmental consequence (e.g., equatorial Pacific, high latitude sites of water formation and/or CO_2 uptake; Follows and Williams, this volume) and special long-term monitoring value (e.g., oligotrophic areas of the gyres of the Pacific and Atlantic Oceans, the Arctic region, the Southern Ocean, and near Antarctica; Follows and Williams, this volume) will require mooring platforms (Send et al., 2001; Dickey et al., 2002). Optimal selections of locations will be essential because of costs and return on investment. There is also need for essentially expendable moorings, which can be deployed in remote areas, which require excessive shiptime for recovery, and for special observational programs. Novel uses of mooring platforms will evolve. For example, moored profilers can be used to good advantage for situations requiring high vertical as well as temporal resolution data. A variant of moored profilers is the "pop-up" system, which could be deployed as an expendable system with a telemetry module. The package would be dropped from a ship or airplane and rest on the ocean bottom until it rose to the surface either on-command or at a pre-specified time. Advanced communication satellites and fiber-optic cables will be needed for transmittal of data types requiring high bandwidth; examples include multi-frequency acoustical and multi-wavelength optical as well as video data.

2.4 DRIFTERS AND FLOATS

Drifters and floats provide spatial data by effectively following water parcels (e.g., see Bishop, this volume, Griffiths et al., 2001; Dickey 2002a). Physical oceanographers have utilized these methodologies for several decades. Drifters and floats can collect data in regions of the world oceans that are rarely visited by oceanographic cruises. These collective platforms now utilize the Global Positioning System (GPS), giving very accurate position data. Surface drifters provide important near surface data whereas isobaric and isopycnal following subsurface floats give information concerning circulation in the ocean's interior. Recently developed profiling floats that use buoyancy changes to move vertically have been used to provide physical data during rise and descent through the water column as part of their function to telemeter data. Present projections suggest that roughly 1000 surface drifters and 3000 profiling floats (near surface to 2000 m) will be in operation annually within the next few years (Argo Science Team, 2001). Emerging drifter and float systems will be simplified and ruggedized to enable easy deployment from ships-of-opportunity and aircraft. Within the past decade, a few oceanographers have begun to deploy optical and/or chemical sensors from drifters and floats (e.g., see Bishop this volume and Dickey, 2002a). It is anticipated that an increasing number of interdisciplinary variables will be sampled from these platforms in the future as size, weight, and power become less limiting. Biofouling will require special measures as mentioned earlier, but even more so for non-recoverable platforms that cannot be post-calibrated.

2.5 AUVs AND GLIDERS

Autonomous underwater vehicles (AUVs) and gliders are rapidly emerging as valuable sampling platforms (e.g., Griffiths et al., 2001; Dickey, 2002a). AUVs are essentially robotic underwater submarines (varying in size and capabilities) that take advantage of breakthroughs in microelectronics, navigation, computer control systems, and in-water and in-air data telemetry. Gliders are essentially unpowered AUVs, with adaptive sampling control, that are designed to observe the ocean interior over long ranges. They use wings and hydrodynamic shape to induce horizontal travel from buoyancy control. The key elements enabling long-range glider missions are efficient buoyancy control, modest hydrodynamic performance, low-power electronics, on-board computing, satellite navigation, and two-way communication for near real-time data transmission and remote control. Gliders are limited by energy considerations to move through the ocean at speeds comparable to ocean currents and their transect sections are thus prone to aliasing. However, one creative way to use gliders is to control them so as to maintain their geographic position by stemming the ambient currents, and thus operating them as "virtual moorings."

Several recent activities have begun to exploit AUVs and gliders for scientific studies (e.g., Griffiths et al., 2000; Yu et al., 2002). This has become possible because of the development of new sensors and systems, which are relatively small in size, consume moderate power, and can be interfaced to the vehicles. Some of the key advantages of AUVs and gliders include cost per deployment, capability to sample in environments generally inaccessible to ships (e.g., in hurricane or typhoon conditions and under ice), good spatial coverage and sampling over repeated sections, ability to do feature-based or adaptive sampling, and potential deployment of several vehicles from moorings, mother ships, offshore platforms, and coastal stations. Creative uses of AUVs and gliders will involve networking and informational feedback loops to guide sampling programs (in some areas involving predictive models), and responses to extreme natural and anthropogenic driven events.

2.6 SENSORS AND SYSTEMS

Next we consider a few of the newer chemical and bio-optical sensors and systems that are relevant to biogeochemical studies.

2.6.1 Chemical Sensors and Systems Until quite recently, most ocean chemistry was done using water samples collected by ships with bottles at-sea. Laboratory methods using wet chemical techniques require that samples be stored, preserved, and transported from the field into the laboratory before chemical analysis can occur. This approach is problematic in that biological, chemical and photochemical, and physical processes can cause changes in the chemical properties of interest.

Moored water samplers are now being used to obtain discrete preserved samples periodically over the course of a few months to create chemical time series, and thus provide important complements for *in situ* sensors. Similar systems have also been developed for drifters. Trace metals, as well as macronutrients, can be analyzed using the water samples, because of improved laboratory analytical systems (work by Ed Boyle; see Dickey et al., 1998a). Underway surface water sampling can now be done using shipboard water intake systems and automated laboratory chemical analyzers (e.g., for dissolved oxygen, partial pressure of carbon dioxide (pCO_2), and plant nutrients such as nitrate, phosphate, silicate, and ammonia).

Several new sensors and systems are capable of making *in situ* time series measurements autonomously with sampling intervals of a few minutes and durations of months (Dickey et al., 1998a, 2000, 2001). Measurements include nitrate and other nutrients, dissolved oxygen, pCO_2, pH, and alkalinity. The various measurements utilize a variety of methods including polarigraphic electrodes, colorimetry (multiple reagents used for different analyses), and spectrophotometry. The latter method has taken advantage of long pathlength absorbance spectrometry enabled by using fiber optics liquid core waveguides. It should be noted that some chemical sensors and analyzers have been successfully deployed from moorings, drifters, and AUVs. Real-time and near real-time telemetry of chemical variables has also been demonstrated. The importance of atmospheric input of dust and aerosols into the ocean has been recognized (e.g., role of iron fertilization). Samplers for dust, aerosols, and particular chemical species are being developed for deployment from surface buoys to avoid land contamination (Ed Sholkovitz, personal communication). Verification of *in situ* chemical sensor data using discrete water samples (e.g., from ships and moored serial samplers) will likely be necessary during the developmental phases. There remains an urgent need for increasing capabilities for sampling broader suites of chemicals autonomously. It is expected that chemical sensors rather than analyzers will be preferable, if not required, for some platforms such as towed bodies, floats, drifters, and AUVs because of payload and power constraints.

Looking toward the future, nother promising types of sensors include very small microelectromechanical systems (MEMS) and nanotechnology devices (Kaku, 1997; Tokar and Dickey, 2000). MEMS are based on a relatively new technology, which is used for making, and combining miniaturized mechanical and electronic components out of silicon wafers using micro-machining. MEMS have shown encouraging results for sensing physical parameters, but work is needed to realize their full potential for chemical sensing. Most work with MEMS has been done in laboratories thus far; however, transitioning to *in situ* applications seems feasible. Potential advantages of MEMS and nanotechnologies include: auto-calibration, self-testing, digital compensation, small size, and economical production.

2.6.2 Bio-optical Sensors and Systems Bio-optics have gained increasing attention in part because of new technologies and growing realization of their central importance to several ocean problems involving biological-chemical-optical-physical interactions including primary productivity, upper ocean ecology, biogeochemical cycling and the biological pump, and bio-optically modulated variability in upper ocean heating rates (see Figure 2 and reviews by Dickey, 2001a,b, 2002a, Chang and Dickey, 2001, and Dickey and Falkowski, 2002). Optical properties are typically biologically modulated in the open ocean whereas terrigenous input and resuspended sediment also play major roles in the coastal oceans.

Two operational classifications of bulk optical properties are most useful for the following discussion: inherent optical properties (IOPs) and apparent optical properties (AOPs). IOPs depend only on the ocean water medium (pure seawater and its various constituents) and are independent of the ambient light field. On the other hand, AOPs are those properties depending on both the IOPs and the geometric structure of the subsurface light field. Instruments designed for underwater light observations are usually described as measuring either IOPs (e.g., spectral beam attenuation, absorption, and scattering coefficients) or AOPs (e.g., spectral diffuse light attenuation coefficients). Recently developed instruments are capable of measuring light absorption, scattering, and attenuation at multiple wavelengths (from 9 to about 100 different wavelengths). Likewise measurements of diffuse light attenuation have increased spectral resolution to a few nanometers in the visible. The power of these types of instruments lies in their ability to distinguish phytoplankton from detritus and dissolved materials, and potentially to identify phytoplankton (perhaps including harmful algae) at least by community groups. See Dickey (2001a) for more details.

Estimation of primary productivity is an important goal as mentioned earlier and several different optical measurements have been used (e.g., Dickey and Falkowski, 2002). Again, examples include the use of chlorophyll fluorescence and photosynthetically available radiation (PAR) measurements in empirical models and more sophisticated measurements using "pump and probe" fluorometers (Kolber et al., 1998). The latter instruments have the important advantage of providing information about biophysical parameters related to photosynthesis (e.g., quantum yield as affected by nutrient and light conditions). The pump and probe technique is being used in the development of an aircraft instrument as well (Yoder et al., 2001). Many of these optical instruments have been deployed from ships, moorings, and drifters. Thus, sampling with both high spatial and temporal resolution, comparable to those of physical parameters, is now possible. One of the common uses of *in situ* optical measurements has also been for groundtruthing and algorithm development of ocean color imagers (e.g., Dickey et al, 2002).

Other optical instruments, using the Fraunhofer light diffraction effect, have been used successfully to obtain particle size distributions. The Fraunhofer effect occurs when planar light passes through a small aperature or slit and changes direction (bent) or is spread out into finite lines (for slit aperture) or circles (for circular aperture). A different and extremely powerful optical technology, flow cytometry, has been successfully used onshore and onboard ships for counting and distinguishing particles and phytoplankton as well as for characterizing their optical properties. Work is advancing to miniaturize and ruggedize flow cytometers for deployment at-sea from buoys and AUVs. The time series obtained from many of the aforementioned *in situ* optical systems show remarkable variability associated with high frequency and episodic events as well as longer term processes. The interpretation of these time series is often difficult, because of the complex nature of the observed medium and organismal physiology. Thus, more effort in understanding these types of data will be essential. Nonetheless, some unexpected correlations have been noted (e.g., very high correlations between beam attenuation (660nm) and particulate organic carbon and between optical backscatter and particulate organic carbon; Bishop, 1999).

3. EXAMPLES OF BIOGEOCHEMICAL TIME SERIES STUDIES

Several times series experiments and programs have been devoted to studies of biochemical variability, particularly as it relates to physical forcing and processes, within approximately the past decade and a half (e.g., see reviews by Dickey, 2001a; Dickey and Falkowski, 2002). Several of these have utilized a multi-platform approach with interdisciplinary moorings serving as the primary means of obtaining high temporal resolution (sampling intervals on scales of minutes) data with durations of at least one year. Several of these have been done as part of the JGOFS program and earlier and subsequent Office of Naval Research programs (e.g., Biowatt, Marine Light in the Mixed Layer, Coastal Mixing and Optics, HyCODE; see references on www.opl.ucsb.edu). Here we highlight results from three distinct oceanographic domains: subtropical gyres of the North Atlantic and North Pacific Oceans, the Arabian Sea, and the equatorial Pacific Ocean. We have selected these as case studies because they portray several different and important physical and biogeochemical processes as well as sharing some in common. It should be emphasized that the results described below were enabled by major breakthroughs in platform and interdisciplinary sensor and system technologies including those described earlier.

3.1 BERMUDA TESTBED MOORING AND HALE-ALOHA MOORING PROGRAMS

The JGOFS Bermuda Atlantic Time-series Study (BATS) and the Hawaii Ocean Time-series (HOT) programs, established in 1988, have been used extensively to observe, quantify, and model temporal variability of the biogeochemistry and ecology of the oligotrophic oceans off Bermuda in the North Atlantic and off Hawaii in the North Pacific, respectively (e.g., Karl et al., 2001; Deep-Sea Research II, 43(2-3), 1996; 48(8-9), 2001). The ship-based studies along with complementary mooring observations document seasonal effects quite well. The BATS site is characterized by quite strong seasonal variability in physics (Figure 5) and biogeochemistry whereas the seasonal effect is subtler at the HOT site (located at a lower latitude than BATS), which is perhaps, influenced relatively more by interannual variability. Both sites will likely serve as bellwethers for decadal oscillations as their records continue; already hints of influences of interannual and decadal variability have been reported (Karl et al., 2001). Traditional shipboard sampling at two or, more commonly, four week intervals has been used as the primary data collection methodology for these two programs. Since this observational mode cannot capture important phenomena with characteristic time scales from minutes to a month, high frequency, long-term, autonomous mooring observations were established first using the Bermuda Testbed Mooring (BTM; Dickey et al., 1998a, 2001) near the BATS and Ocean Flux Study (OFP; e.g., Conte et al., 2002) sampling sites in 1994 and in 1996 using the HALE-ALOHA mooring near the HOT site (Letelier et al., 2000). These programs have clearly shown the importance of several processes, which cannot be resolved by bi-weekly or monthly shipboard sampling (see Figures 5 and 6).

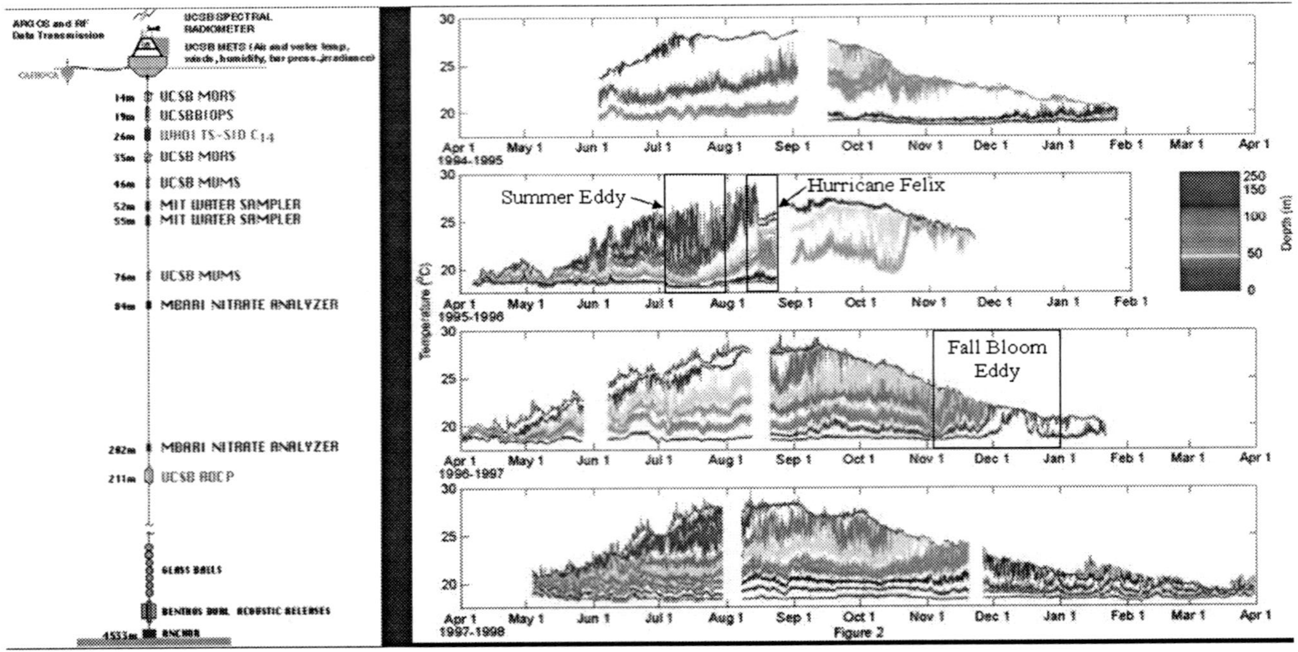

Figure 5. Left. schematic of the Bermuda Testbed Mooring (BTM). Right. Time series of temperature for several depths obtained from the Bermuda Testbed Mooring (BTM; see Dickey et al., 2001). Seasonal cycles are evident in the upper and lower panels. High frequency variability and episodic events were also observed. Examples of eddy and mesoscale feature passages are highlighted and are discussed in the text.

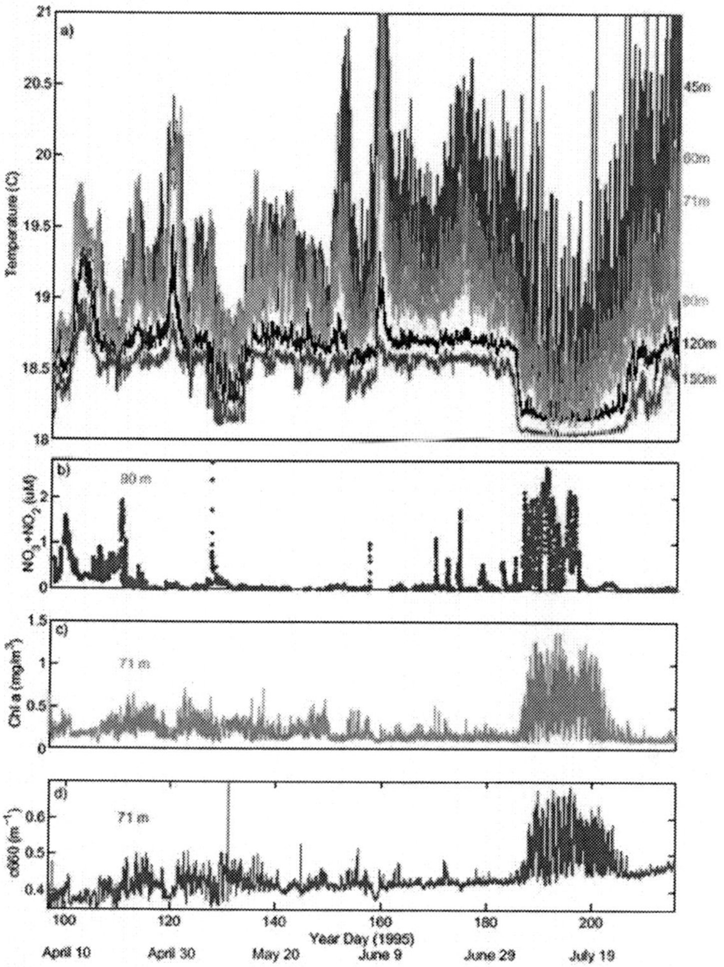

Figure 6. Time series from BTM showing passage of a second baroclinic mode eddy in July 1995 showing nutrient injection into the euphotic layer and extremely high levels of chlorophyll (and primary production). Inertial oscillations are also evident.

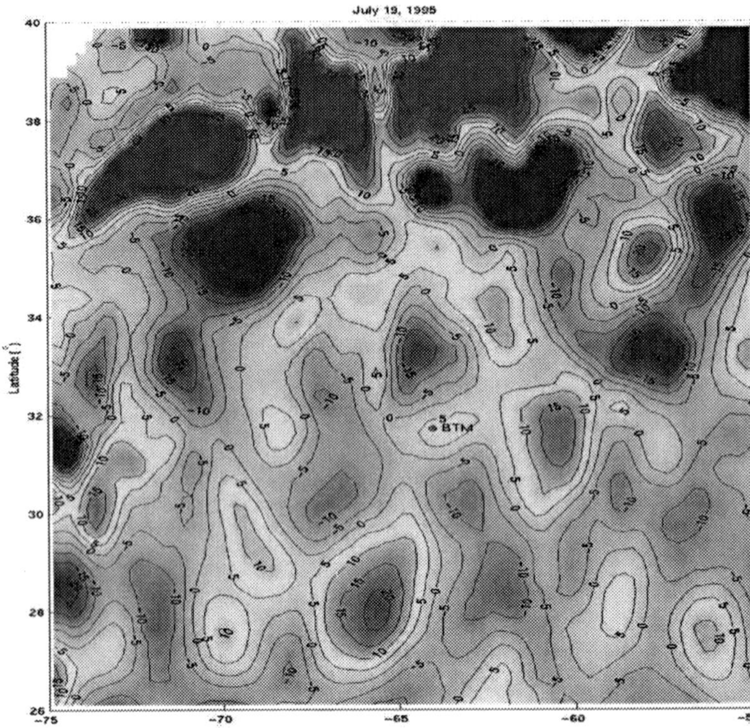

Figure 7. Sea surface height anomaly map derived from satellite altimetry data sets in the region of Bermuda. Location of the BTM is indicated.

Key processes at the two sites with time scales on the order of or less than one month include: mesoscale eddies (described below), storms, hurricanes, dust deposition events, episodic shoaling of the mixed layer and transient phytoplankton blooms, inertial oscillations, diel and shorter time scale variability in phytoplankton and bio-optical properties, and internal gravity waves (Dickey et al., 1998a, 2001; Letelier et al., 2000).

Several recent studies have suggested that mesoscale eddies may play major roles in supplying new nutrients to the upper ocean (i.e., importantly, the euphotic layer), and thus enhance primary production and carbon export in otherwise nutrient-deficient systems as well as shift phytoplankton community structure (e.g. McGillicuddy, et al. 1998; Oschlies and Garcon, 1998; McNeil et al., 1999; Letelier et al. 2000; Dickey et al., 2001; McGillicuddy, 2001; Conte et al., 2002; see recent reviews by McGillicuddy, 2001; Dickey and Falkowski, 2002; Lewis 2002). One depiction of this process is schematicized for a first baroclinic mode cyclonic, cold core eddy in Figure 8a.

Mesoscale eddies are now known to be ubiquitous oceanographic features (see BTM time series in Figures 5 and 6 and altimetry data set shown in Figure 7).

Several recent studies have suggested that mesoscale eddies may play major roles in supplying new nutrients to the upper ocean (i.e., importantly, the euphotic layer), and thus enhance primary production and carbon export in otherwise nutrient-deficient systems as well as shift phytoplankton community structure (e.g. McGillicuddy, et al. 1998; Oschlies and Garcon, 1998; McNeil et al., 1999; Letelier et al. 2000; Dickey et al., 2001; McGillicuddy, 2001; Conte et al., 2002; see recent reviews by McGillicuddy, 2001; Dickey and Falkowski, 2002; Lewis 2002). One depiction of this process is schematicized for a first baroclinic mode cyclonic, cold core eddy in Figure 8a.

Interestingly, current estimates suggest a broad range of 10 to 50% of global new primary production to be due to eddy-induced nutrient fluxes (e.g., McGillicuddy et al. 1998, Oschlies and Garcon 1998; Siegel et al, 1999; Letelier et al. 2000). Clearly, lack of sufficient relevant data sets has inhibited our ability to understand and quantify the biogeochemical significance of eddies leading to the ongoing controversies (e.g., Oschlies and Garcon, 1998; McGillicuddy et al., 1998).

Mesoscale features have been observed with remote sensing tools including sea surface temperature and color sensors such as the Sea-viewing Wide Field-of-view Sensor, SeaWiFS (e.g., Yoder et al., 2001; McGillicuddy, 2001); however not all eddies are shown with these methods and cloud obscuration remains problematic for developing continuous time series of features. Also, analyses of satellite altimetry data sets in conjunction with biogeochemical models have suggested the importance of first baroclinic mode cyclonic eddies (surface depression at center of the eddy) for input of nutrients into the euphotic layer and the enhancement of primary productivity and thus carbon flux to depth (McGillicuddy et al., 1998; Siegel et al., 1999).

Other types of eddies and mesoscale features (see Figure 8b) have been shown to exhibit high nutrient injection events and elevated primary productivity as well, thus confounding relatively simple models of new production and nitrate fluxes based on altimetry data. Further, the evolution of eddies (spin-up or formation versus spin-down and decay phases; Figure 8b) must be important as well because of the vertical motion of the thermocline and nutricline into or out of the euphotic layer (e.g., Flierl and McGillicuddy, 2002).

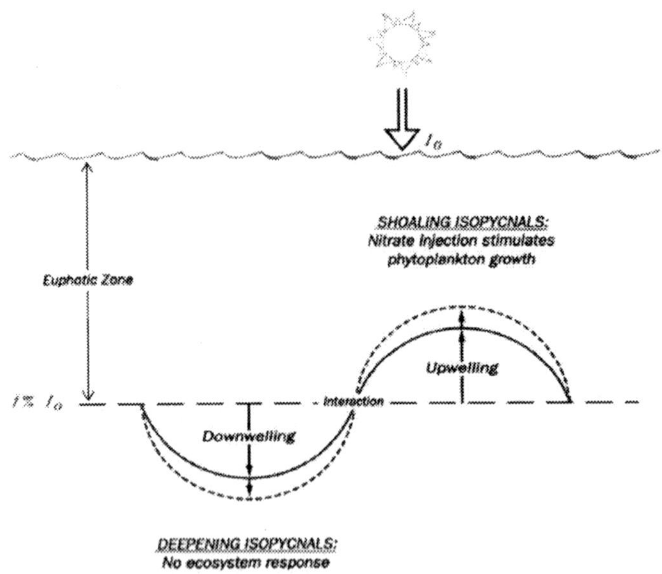

Figure 8a. Conceptual model illustrating nutrient injection into euphotic layer by eddy pumping for a first baroclinic mode eddy (from Mcillicuddy et al., 1998).

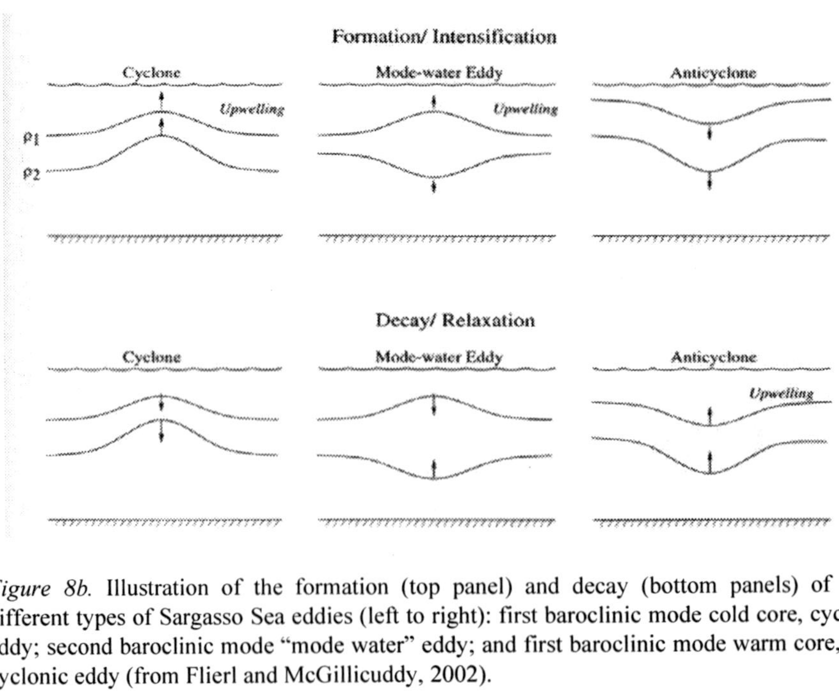

Figure 8b. Illustration of the formation (top panel) and decay (bottom panels) of three different types of Sargasso Sea eddies (left to right): first baroclinic mode cold core, cyclonic eddy; second baroclinic mode "mode water" eddy; and first baroclinic mode warm core, anticyclonic eddy (from Flierl and McGillicuddy, 2002).

Both the BTM and HALE-ALOHA mooring time series programs have documented passages of mesoscale features with high nutrient and phytoplankton concentrations. For example, BTM time series (Figures 5 and 6) indicate 1) the passage of a major eddy and its impact on nutrients and phytoplankton biomass in July 1995 and 2) upper ocean response to Hurricane Felix which passed over the BTM in August 1995 (Dickey et al., 1998c). Interestingly, the July 1995 eddy was a "mode water" or second baroclinic mode eddy whose surface expression in satellite altimetry was an elevation (see McGillicuddy, 2001) implying an anti-cyclonic feature (belying the subsurface cyclonic feature). However, isotherms and isopleths of nutrients were uplifted at the eddy's center. The direct BTM measurements of nitrate and chlorophyll (Figure 6) indicate that this was perhaps the most productive eddy, which has passed the BATS, BTM, and OFP sites since initiation of the programs. A species succession was noted as diatoms dominated the phytoplankton community of the interior of the eddy. Recently, OFP data sets have been analyzed in conjunction with BTM data sets for the late fall to early winter period of 1996 (Conte et al., 2002). These records (indicated as Fall Bloom in 1996 in Figure 5) indicate the passage of a warm mesoscale feature that was accompanied by elevated chlorophyll and high and rapid mass flux to depth as inferred from BTM and OFP sediment trap data. Species succession was also recorded in the deep sediment trap. Similar time series collected by the HALE-ALOHA mooring in March 1997 have documented an important eddy bloom event in the subtropical North Pacific off Hawaii at the HOT site (Letelier et al., 2000). The examples presented by the BTM and HALE-ALOHA time series are not atypical and clearly emphasize the danger in over-reliance upon satellite derived products and the need for comprehensive *in situ* and satellite-based biogeochemical and physical data sets on a multiplicity of time and space scales.

3.2 ARABIAN SEA

JGOFS process studies were conducted in the Arabian Sea during the period 1994-1996 (Deep-Sea Research II, vols. 45(10-11), 46(3-4, 8-9), 47(7-8)). The Arabian Sea is an especially interesting study region. Prior to the JGOFS field programs, it was unknown whether the Arabian Sea was a source or sink of CO_2, and even the upper ocean physical and biogeochemical responses to monsoonal atmospheric forcing were undocumented. Interestingly, mesoscale processes turned out to be very important for this system, like the oligotrophic regions described in the previous subsection. The various Arabian Sea studies utilized ships (underway sampling (with ADCP current and acoustic backscatter data), profile, and tow-yo (SeaSoar) modes), multiple moorings, and satellites (for sea surface temperature and altimetry). A novel aspect of the observational program was the deployment of an array of five moorings with meteorological and physical instruments covering a square (roughly 7 km X 7 km; e.g., see Dickey et al., 1998). The location was selected to be near the axis of the atmospheric Findlater Jet (see Figure 5.11 in Colling et al., 2001). The Findlater Jet occurs during the southwest monsoon with maximum wind speeds occurring in July. Its central axis moves over time. A mooring was placed at the center of the square (15° 30'N, 61° 30'E; Figure 9) and included multi-variable moored systems (MVMSs; also used on BTM) at four depths in the upper 80m. Important attributes of the MVMS are that it concurrently measures bio-optical and chemical as well as physical variables at sampling intervals of a few minutes for periods of up to 6 months. The array was designed to enable quantification of variability associated with the passage of mesoscale and other spatially varying features (see altimetry data in Figure 9).

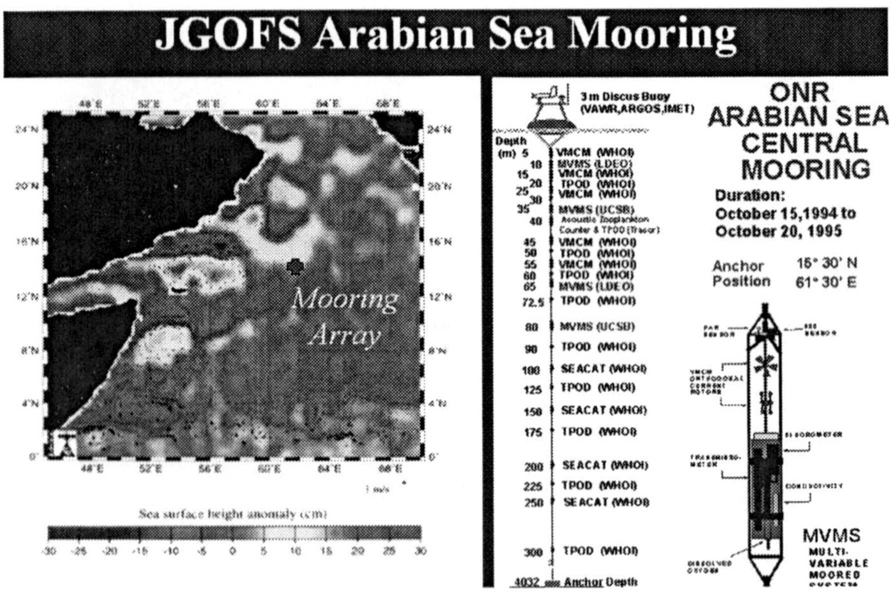

*Figure 9.(*Left). Sea surface height anomaly map derived from satellite altimetry data sets in the Arabian Sea. Location of a mooring array is indicated. (Right). The central mooring configuration and a multi-variable moored system (MVMS).

The physical mooring data featured two mixed layer deepening and shoaling cycles per year (Figure 10; Dickey et al., 1998b). The northeast (NE) monsoon was characterized by steady northeasterly winds of moderate intensity (~6 m/sec), surface cooling, and convection whereas the southwest

(SW) monsoon featured strong, persistent southwesterly winds with greater intensity (up to 15 m/sec). The NE monsoon forced deeper mixed layers (~110 m depth) than the SW monsoon (~80 m depth) because of the convective forcing. The MVMS data set showed that chlorophyll concentrations and primary production correlate with the seasonal physical cycle associated with the NE and SW monsoons (Figure 10). A half-yearly cycle in chlorophyll *a* is an important feature with seasonal blooms occurring late in each monsoon season and into the respective intermonsoon periods. One of the more interesting results was the finding that mesoscale eddies play such an important role in the evolution of chlorophyll *a* at the observational site (Figure 10; Dickey et al., 1998b). A sediment trap mooring array with serial samplers (Honjo et al., 1999) at depths of ~0.8, 2.2, and 3.5 km was located approximately 50 km north of the central mooring site. The combined MVMS time series of derived primary productivity (Marra et al., 1998) and sediment trap time series (Honjo et al., 1999) indicate that the timing and amplitudes of the short-lived but intense phytoplankton blooms associated with seasonal stratification and eddies were well correlated (Figure 11). This suggests relatively rapid (days to weeks) export of organic carbon to the deep ocean. This is qualitatively consistent with the recent results of Conte et al. (2002) for the Bermuda area.

3.3 EQUATORIAL PACIFIC

The primary JGOFS equatorial Pacific process study (see Deep-Sea Research II, vols. 42 (2-3), 1995; 43(4-6), 1996, 44(9-10), 1997) took place in 1992 with some related studies preceding and extending after the main experiment. Field experiments were conducted in the central and eastern Pacific from 95^0W to 170^0W and between 10^0N and 10^0S. The equatorial Pacific is interesting because of its great variety of physical processes (including equatorial waves) and the El Niño-Southern Oscillation (ENSO) with its related global effects. The region is thought to play an important role in the global carbon cycle as it is estimated to supply roughly 1 gigaton of carbon (as CO_2) per year to the atmosphere by upwelling of carbon-rich deep water. It is also a high nutrient-low chlorophyll (HNLC) region, so that its productivity and carbon fluxes are relatively small considering the availability of nutrients. The physical dynamics of the equatorial Pacific have become increasingly well understood, largely because of the large number of measurements made from the Tropical Atmosphere Ocean (TAO) mooring array (McPhaden et al., 2001). However, understanding of biological and optical variability has been limited since few dedicated ship-based experiments could be performed in this remote oceanic region in the past. Only a few ship-based biological observations of chlorophyll and

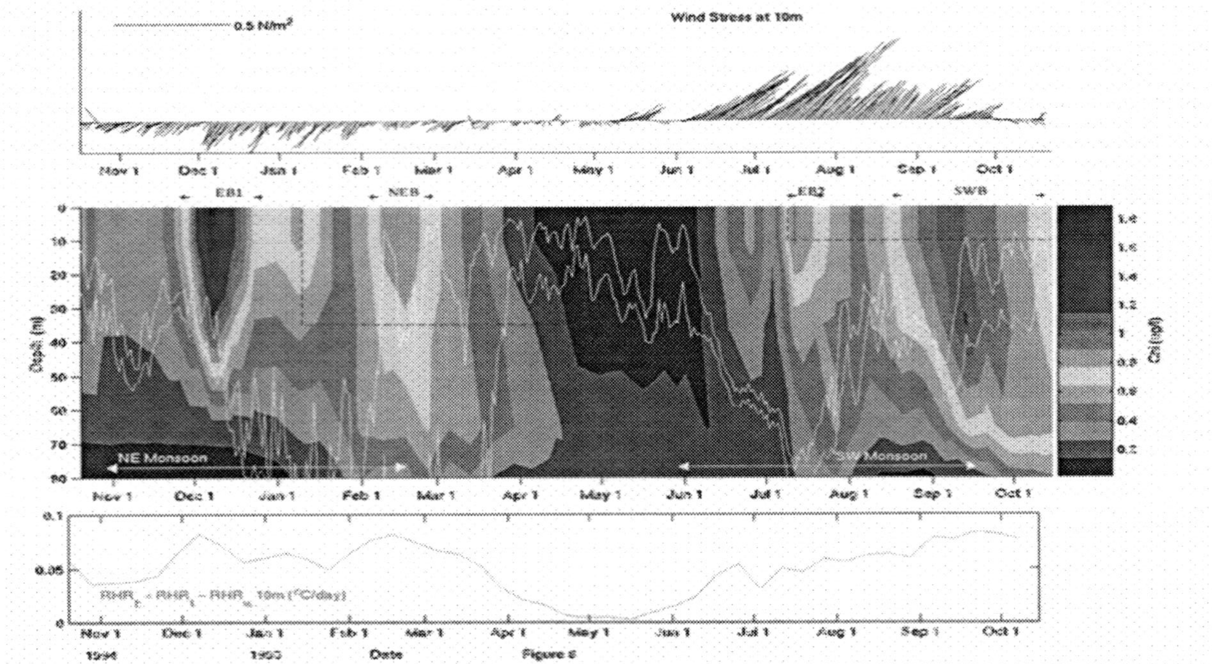

Figure 10. (Top) Time series of wind stress data based on Arabian Sea central mooring meteorological data indicating northeast and southwest monsoonal condition. (Bottom) Time-depth contour data showing chlorophyll data collected from MVMSs. Also overlain are mixed layer depths based on 0.1 and 1.0 °C temperature criteria. Designations EB1 indicate eddy bloom 1 and 2, respectively, and NEB and SWB indicate northeast monsoon and southwest monsoon blooms, respectively.

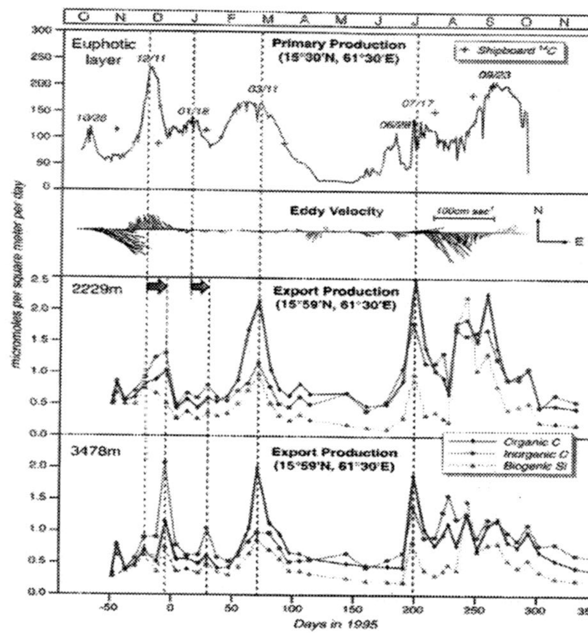

Figure 11. (Left) Picture of a mooring-based sediment trap used for measurements in the Arabian Sea near the central mooring indicated in Figure 9. (Right) Top to bottom: Time series of primary production based on MVMS and model formulations; time series of current vectors; export production (organic and inorganic carbon and biogenic silicate) measured in sediment traps at 2229 and 3478 m. MVMS data from Dickey et al. (1998b) and Marra et al. (1998). Sediment trap data provided by Sus Honjo (see Honjo et al., 1999). Vertical dashed lines indicate periods of high productivity in the upper layer as determined with moored bio-optical sensors and large export production fluxes of carbon and silicate at depths of 2229 and 3478 m depth sediment traps.

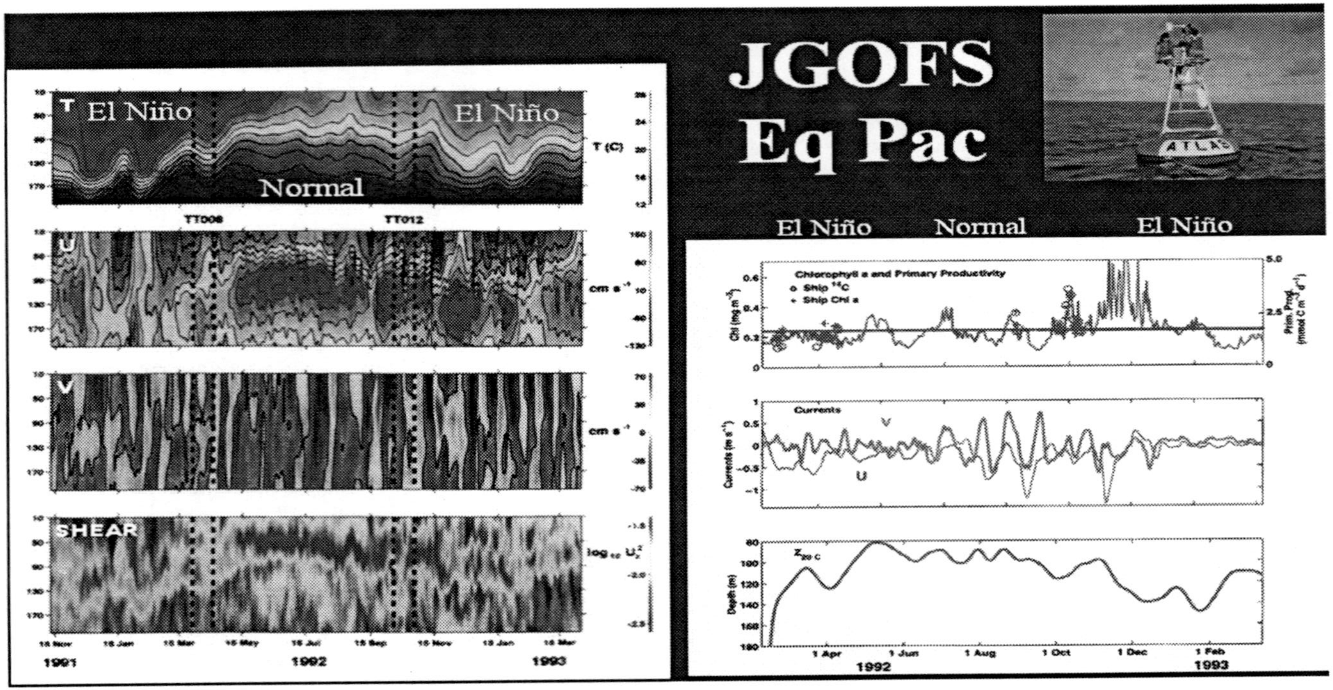

Figure 12. . Data collected during the JGOFS EqPac study using a mooring at 0°, 140°W (see Foley et al., 1998). (Left) Time versus depth contour plots of 1) temperature, 2) zonal currents, 3) meridional currents, and 4) current shear. Black vertical dashed lines indicate periods of concurrent shipboard sampling (TT008 and TT012) at the site by JGOFS investigators. (Right) Time series of 1) chlorophyll a and derived primary production at 10 m. (The same curve depicts chlorophyll a values using the left scale and primary productivity rates using the right vertical scale.), 2) zonal (red) and meridional (blue) currents at 10 m, and 3) the depth of the 20°C isotherm (note that it is deeper during the El Niño phase).

primary productivity were made each year prior to JGOFS; hence these were previously our only basis for annual estimates of chlorophyll and primary production for the expansive equatorial Pacific waveguide.

In order to study the temporal variability of the biogeochemistry of the central equatorial Pacific, MVMSs (described above) were added during the JGOFS intensive study to the TAO physical mooring at 0°, 140°W for an eighteen-month period in 1992 and 1993 (Foley et al., 1998). This sampling period was most opportune in that our observations spanned both El Niño and "normal" phases (Figure 12). During the El Niño phase, the mixed layer, the thermocline, and a very weak equatorial undercurrent (EUC) were very deep (at times greater than 150 m) and Kelvin waves (~60 day period) propagated eastward past the site (with depressions of the thermocline). Time series of key physical and bio-optical variables obtained from instruments deployed on the 0°, 140°W mooring are shown in Figure 12. Light levels were high. However, relatively high concentrations of nutrients including iron were deep while surface concentrations were low. As a result, measured chlorophyll *a* concentrations in the upper layer were low. However, as "normal conditions" returned, Kelvin waves ceased and the thermocline and a strong EUC shoaled and thus allowed the transport of nutrients into the euphotic layer. Importantly, westward propagating tropical instability waves (TIWs with periods of ~20 days) also contributed to large vertical upwelling cycles. TIWs are easily seen in the meridional current records and appear to be manifest in the chlorophyll *a* time series with values doubling and at times tripling those observed during the El Niño period (Figure 12). Importantly, strong, though highly complex, coupling is evident between the physical processes (El Niño, Kelvin waves, and TIWs) and the phytoplankton biomass and primary productivity of the equatorial Pacific. It is worth noting that the TAO array of physical instruments along with specially focused JGOFS ship-based sampling (ship–based mapping of biogeochemistry in space, ships on station for time series, sediment traps, etc.) provided a relatively comprehensive, though coarse 3-dimensional biogeochemical time series for the region. Subsequent time series studies in this region (Chavez et al., 1998, 1999) support the conclusions described here.

4. EMERGING RESEARCH APPROACHES: DATA SYNTHESIS AND MERGING OF OBSERVATIONS AND MODELS

4.1 DATA SYNTHESIS

A recurring theme of this review is that the oceans remain disturbingly undersampled in all variables, but especially biogeochemical and biological variables. New sensors, systems, and platforms are improving our ability to study and monitor the oceans. Yet, still too limited resources will be available, requiring judicious deployment of ocean assets and optimal utilization of the precious data sets. Application of the concept of nesting of platforms and model computational grids will be critical. Clearly, end-to-end data management systems will be required for multi-use data sets, quality control of data, data synthesis, and modeling. Ultimately, effective close-knit collaborations among observationalists and modelers will be needed to select key variables and sampling locations and to design sampling arrays. One of the challenging issues will revolve around the idea of more intensive sampling of features or locations, which are perceived (based on data and model simulations) to have extraordinarily great influence globally. Since much of the world ocean is still very poorly sampled (e.g., South Pacific and South Atlantic, Indian Ocean, Southern Ocean), mobile and stationary (mooring) platforms will be necessary even in areas that may presently be thought to be of lesser importance.

The oceanographic community is already beginning intensive efforts to design a global ocean observing system, which will need to satisfy a broad set of interdisciplinary needs and applications in coastal and open ocean environments (e.g., Dickey, 2001b, 2002b; Dickey et al, 2002). Future comprehensive networks must be carefully designed to take fullest advantage of the various platforms. This will likely require formation of cooperative partnerships, which involve governmental agencies at various levels (e.g., local, regional, national, and international), private industry, private foundations, and academic and research institutions.

One of the longstanding problems is sustainability of ocean sampling. An important factor in sustaining ocean observing networks is public interest and demonstration of utility. Again, a good example is the use of the TAO array and complementary satellite observations along with model predictions, which are contributing to the public's knowledge of ENSO and its societal and economic impacts (McPhaden et al., 2001). Clearly, the synthesis and visualization of these data sets and model simulations are critical elements.

4.2 MERGING OF OBSERVATIONS AND MODELS

Ocean models have become increasingly useful as new processes have been incorporated and parameterized in formulations. Also, numerical techniques improved, and more powerful computing capabilities have allowed increased spatial and temporal resolution and range as well as greater numbers of variables and balance equations. As more data have been collected, analyzed, and interpreted, and as computer model simulations have demonstrated more realistic results, observationalists and modelers have become more cognizant of and dependent upon each other for making scientific advances. The culmination of this cultural change is epitomized by two methodologies: inverse methods and data assimilation. Here we briefly outline some of the salient features of these, especially in regard to time series observations of biogeochemical studies.

The "ocean circulation inverse problem" has been defined by Wunsch (1996) as a problem of inferring the ocean circulation, understanding it dynamically, and forecasting it by quantitatively combining theory and observations. He defines "inverse methods" as techniques used for solving problems or systems of equations that are mathematically underdetermined. Most physical, and certainly all interdisciplinary, oceanographic problems lie within this "underdetermined" category. Again, we are almost always seriously limited in our studies by undersampling and aliasing. Hence, there is growing interest in application of inverse methods to many oceanographic problems. Inverse methods are often used in the context of data assimilative models as well.

The term data assimilation is defined by and Hofmann and Friedrichs (2001) as "the systematic use of data to constrain a mathematical model." In reference to data assimilation, they comment that: 1) it is assumed that dynamics responsible for processes and distributions of properties are inherent in the data, 2) by inputting data into a specific model, the model will produce a more accurate model simulation, and 3) hindcasts, nowcasts, and forecasts are improved through data assimilation. We can consider data assimilation to broadly comprise parameter estimation and setting/resetting of state variables. A thorough mathematical formulation of the biological (and chemical) data assimilation problem is presented in Robinson and Lermusiaux (2002) and Matear (this volume) discusses data assimilation methods and biogeochemical signatures of climate change.

The application of data assimilation methods to interdisciplinary oceanographic problems is non-trivial (e.g., Robinson et al., 1998, and Friedrichs, 2001, 2002; Robinson and Lermusiaux, 2001, 2002; Dickey, 2002; Matear, this volume). In particular, broad ranges of time and space scales need resolution to depict relevant processes for interdisciplinary problems. Additional considerations relevant for the interdisciplinary case include: 1) the number of important interdisciplinary processes and state variables is several times, if not orders of magnitudes, greater than those for physics alone, 2) biology involves complex physiological and behavioral effects that are either unknown or difficult to parameterize or model, 3) biological, chemical, and optical processes are often highly related: biological and chemical processes involve complex transformations (e.g., organic to inorganic forms of matter and vice versa), and require coupled sets of nonlinear equations, and 4) several biological, optical, and chemical processes have influential short time scales (minutes to days) and their rates are especially difficult to determine. Because of these many issues, interdisciplinary data assimilation is especially important for advancing interdisciplinary oceanography.

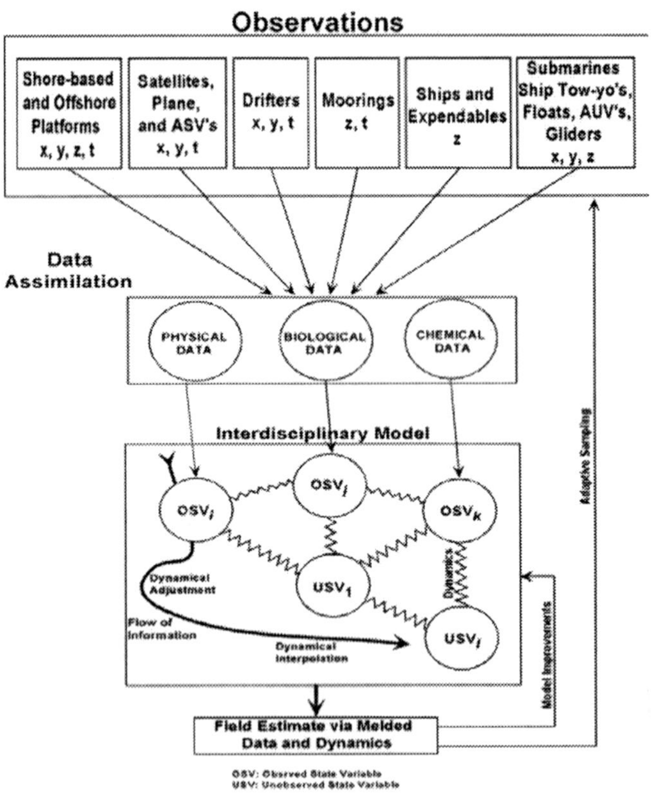

Figure 13. Schematic diagram showing data assimilation system components and their inter-relationships and feedbacks (based on Robinson et al., 1998). Network observational platforms and their primary spatial and temporal capabilities are shown at the top.

One example of a "data assimilative system" is depicted in Figure 13 (the lower portion follows Robinson and Lermusiaux, 2001). The system is composed of three fundamental elements: 1) an observational network with data telemetry, 2) an interdisciplinary model, and 3) a data assimilation scheme. Interdisciplinary data are collected and transmitted in real-time or near real-time from various platforms of the sampling network or array. Each platform has its own intrinsic capabilities and limitations in terms of types of sensors and systems that can be utilized and a defined spatial and temporal range of measurements. Data and error estimates are transferred for incorporation into the interdisciplinary model. The interdisciplinary dynamical model is composed of multiple modules (e.g., for physics, biology, and chemistry) and each module is represented with a set of equations with variables appropriate for describing the relevant processes. Importantly, these modules are coupled. The data-melding step produces state variable and parameter estimates of observed physical variables (OSVi), observed biological variables (OSVj), observed chemical variables (OSV_k), and unobserved variables (USVs) along with associated errors. This information feeds back to the data assimilative model for inclusion in the next model iteration and for model improvement and back to the sampling network enabling adaptive sampling (e.g., based on error simulation as discussed earlier). Adaptive sampling is accomplished using several possible modes: changing sensor sampling rates, changing gains of sensors and systems, and redirection of mobile sampling assets to locations of special informational value (e.g., fronts, upwelling sites, eddies, blooms, and regions where processes are leading to large model error estimates). Important goals of the data assimilation procedure are: 1) to develop and improve model formulations and functional dependencies, 2) to estimate parameters and variables as well as rates that cannot be measured, 3) to estimate or predict state variables and parameters on time and space scales inaccessible to some or all of the observational sampling assets (e.g., a smart interpolating function), 4) to provide model initializations, 5) to identify and elucidate ocean processes, 6) to design experimental or operational sampling networks using Observational System Simulation Experiments (OSSEs), 7) to optimize use of sampling assets through adaptive sampling, and 8) to make predictions for research and operational environmental management including uncertainty and error estimates. A general goal is to accurately predict the distributions of selected state variables given their values at some initial time.

It is worth noting the challenges of data assimilation and specifically inherent limitations in predictability in the form of nonlinear error propagation and error-scale transfers (Robinson and Lermusiaux, 2001). Other, observational and modeling problems remain for future work. For example, remote sensing of many physical and biogeochemical variables is most attractive for reasons described earlier. Unfortunately, many of the more important biological and chemical variables cannot be measured from space, even at the surface at present, and cloudy conditions negate ocean color (and temperature) data retrievals sporadically and regionally. In addition, measurements and models of the ocean at depth are vital because of the considerable vertical structure and variability, which cannot be inferred directly from remote sensing. This point was illustrated with the eddy studies described earlier. Another methodological problem concerns sampling of episodic events, some of which may contribute substantially if not dominantly to variance in physical and biogeochemical properties. Fortunately, more platforms and sensors will likely be available in the future to provide for high temporal and spatial resolution data at representative ocean sites; data obtained from these locations will be valuable for data assimilation systems. Data assimilative modeling using data collected from all of the available platforms (Figures 4a and 13) will be needed to characterize, quantify, and model the processes depicted in Figures 1, 2, and 3

5. FINAL COMMENTS

This chapter has concentrated largely on recent progress in quantifying and understanding global biogeochemical variability on a broad range of time scales. However, it is useful to take a look back and to consider some of the outstanding issues for future research. Interestingly, prior to the JGOFS-era, several elements and processes were yet to be identified or, at minimum, demonstrated to be important for biogeochemical cycling. These include the microbial loop/food web (and many heretofore unknown or undescribed forms of plankton), plankton community size and trophic structure, predators such as zooplankton (some with important diel migration behaviors), dissolved organic matter and dissolved organic carbon cycling, N_2 fixation, trace element (e.g., iron) and multi-nutrient limitation, episodic events and mesoscale eddies as they may affect primary production and carbon flux to the deep sea, particle transport and remineralization in the "twilight zone" (100-1500m depths), and the non-Redfield ratio nature of the sea's elemental stoichiometry (these are all discussed in papers in this volume and in papers Oceanography, 14(4), and several JGOFS Deep-Sea Research II volumes). Many of these were discovered through the intensive regional process and time series studies described above, yet it must be admitted that much remains to be learned and quantification is essential. Besides the obvious lack of data, one of the problems has been that different methodologies, which are suited for specific time windows and spatial domains, lead to divergent inferences and budget estimates (e.g., Berelson, 2001; Ducklow et al., 2001). Experiments that simultaneously utilize these various methodologies are needed for closure.

Observations have driven most of the new discoveries and paradigm shifts. However, biogeochemical models have rapidly evolved, generally becoming increasingly complex, to incorporate the plethora of new processes described in the previous paragraph. It is easier in many cases to add new model formulations than to develop and deploy new sensors to measure requisite variables and rates. Yet, models cannot advance without vital data sets for developing proper parameterizations that have universal applicability and for testing and evaluating models so that they be reformulated and improved (e.g., Doney, et al., 2001; Dickey, 2002). Nonetheless, it must be recognized that models are absolutely essential for synethsizing data sets and predicting biogeochemical variability and trends.

Despite the large number of challenging issues, there is cause for considerable optimism. For example, large numbers of sensors are being developed for many purposes outside of oceanography; likely some will be directly, or after modification, useful for ocean measurements. One possible benefit of advanced sensor technologies may well be more capable, smaller, simpler, less power-hungry, and less expensive sampling systems (e.g., "chip-based" chemical and biological sensors with automated data processing and transmission; Kaku, 1997), which will benefit oceanographic programs regardless of present technical capabilities and skill levels. There is increasing interest in making platforms available (e.g., platforms-of-opportunity) for a broader suite of interdisciplinary measurements. Also, there is growing support for systematic approaches that incorporate data acquisition, data telemetry, data assimilative modeling, prediction, and capability for adaptive sampling through redirection of assets.

There is now a strong international focus on global ocean observing systems for both the coastal and open oceans. These developments will likely reflect in improved sampling systems that will provide more variables and better spatial and temporal coverage. Importantly, data assimilation modeling is considered a key element of these systems (Le Traon et al., 2001; Matear and other chapters in this volume), and thus observational and modeling components are being developed synergistically. Joint instrumentation/data assimilation system testbeds need to be initiated as soon as possible to accelerate progress in this direction (e.g., Robinson and Dickey, 1997). There remain a myriad of challenges to be met in order for us to advance our knowledge of and predictive capabilities for biogeochemical variability on scales from minutes to the interannual. Nonetheless, we have great cause for optimism as the oceanographic community is in a very strong position to make major advances thanks to burgeoning technologies and increasing public and governmental concern for the global ocean environment and climate.

Suggested Further Reading

Readers interested in further related information at an introductory level are directed to other chapters within this volume and recent reviews by Dickey (2001a, 2002a), Dickey and Chang (2001), Hofmann and Friedrichs (2001), and Robinson and Lermusiaux (2001) and the JGOFS volume of Oceanography (vol. 14, no. 4, 2001). Those desiring more in-depth and detailed information are directed to reviews by Robinson et al. (1998, 1999), Dickey et al. (2000, 2001, 2002), Tokar and Dickey (2000), Chang and Dickey (2001), Dickey (2001b, 2002b), Griffiths et al. (2001), Koblinsky and Smith (2001), Send et al. (2001), Dickey and Falkowski (2002), Hofmann and Friedrichs (2002), Lewis (2002), Robinson and Lermusiaux (2002) and an extensive series of Deep-Sea Research II volumes dedicated to JGOFS research.

Griffiths et al. (2001), Send et al. (2001), and Dickey (2002a) provide comprehensive reviews on observing platforms. Detailed reviews concerning interdisciplinary sensors and systems may be found in Tokar and Dickey (2000), Dickey et al. (2000), Dickey (2001a,b; 2002a), and Dickey and Chang (2001). Reviews on merging of data sets and models are given elsewhere in this volume and by Robinson et al. (1998, 2001), Robinson and Lermusiaux (2002), Hofmann and Friedrichs (2001, 2002), Doney et al. (2001), and Dickey (2002).

ACKNOWLEDGEMENTS

I would like to thank Temel Oguz and Mick Follows for inviting me to lecture in the NATO ASI. NATO is acknowledged for their support of my travel to Ankara, Turkey for the ASI. Funding for research enabling the preparation of this chapter has been provided by several agencies, programs, and institutions including the National Science Foundation Ocean Technology and Interdisciplinary Coordination Program, Chemical Oceanography Program, and Oceanographic Instrumentation and Technical Service Program, the Office of Naval Research, the National Aeronautics and Space Administration, the National Oceanic and Atmospheric Administration, the National Ocean Partnership Program, and the University of California, Santa Barbara. Finally, I would like to thank my research team and the numerous collaborators and colleagues, who have generously shared ideas that have helped to shape my thinking about interdisciplinary oceanography for over two decades. This is JGOFS contribution Number 1025.

REFERENCES

Argo Science Team, 2001. Argo: the global array of profiling floats, in Observing the Oceans in the 21st Century, eds. C.J. Koblinsky and N.R. Smith, GODAE, Bureau of Meteorology, Melbourne, Australia, 213-247.

Behrenfeld and Falkowski, 1997a, Photosynthetic rates derived from satellite-based chlorophyll concentration, Limnol. Oceanogr., 42, 1479-1491.

Behrenfeld and Falkowski, 1997b, A consumer's guide to phytoplankton primary productivity models, Limnol. Oceanogr., 42, 1-20.

Berelson, W.M., 2001, The flux of particulate organic carbon into the ocean interior: A comparison of four U.S. JGOFS regional studies, Oceanography (14(4), 59-67.

Bidigare, R.R., J. Marra, T.D. Dickey, R. Iturriaga, K.S. Baker, R.C. Smith, and H. Pak, 1990. Evidence for phytoplankton succession and chromatic adaptation in the Sargasso Sea during springtime 1985. Mar. Ecol. Progr. Ser., 60, 113-122.

Bishop, J.K.B., 1989. Regional extremes in particulate matter composition and flux: effects on the chemistry of the ocean interior, in Productivity of the Ocean: Present and Past, eds. W.H. Berger, V.S. Smetacek, and G. Wefer, John Wiley and Sons, 117-137.

Bishop, J.K.B., 1999, Transmissometer measurement of POC, Deep-Sea Res. I, 46(2), 353-369.

Brewer, P.G., K.W. Bruland, R.W. Eppley, and J.J. McCarthy, 1986, The Global Ocean Flux Study (GOFS): Status of the U.S. GOFS Program, EOS, 67, 44.

Buesseler, K.O., 2001, Ocean biogeochemistry and the carbon cycle, Oceanography, 14(4), 5.

Chang, G.C., and T.D. Dickey, 2001, Optical and physical variability on time-scales from minutes to the seasonal cycle on the New England shelf: July 1996-June 1997, J. Geophys. Res., 106, 9435-9453.

Chavez, F., P.G. Strutton, and M.J. McPhaden, 1998. Biological-physical coupling in the central Pacific during the onset of the 1997-98 El Niño. Geophys. Res. Lett., 25, 3543-3546.

Chavez, F.P., P.G. Strutton, G.E. Friederich, R.A. Feely, G.C. Feldman, D.G. Foley, and M.J. McPhaden, 1999. Biological and chemical response of the equatorial Pacific Ocean to the 1997-98 El Niño. Science, 286, 2126-2131.

Colling, A. and Open University Course Team, 2001. Ocean Circulation, 2nd edition, Butterworth and Heinemann Publishing, Milton Keyes, UK, 286pp.,

Conte, M.H., T.D. Dickey, J.C. Weber, R. Johnson, and A. Knap, 2003. Transient physical forcing of pulsed export of bioreactive material to the deep Sargasso Sea, Deep-Sea Res. I, in press.

Dickey, T.D., 2001. New technologies and their roles in advancing biogeochemical research, Oceanography, 14(4), 108-120.

Dickey, T., 2001, Sensors: inherent and apparent optical properties, Encyclopedia of Ocean Science, eds. J.H. Steele, S.A. Thorpe, and K.K. Turekian, Academic Press, London, 1313-1323.

Dickey, T., 2002, Emerging ocean observations for interdisciplinary data assimilation systems, J. Mar. Sys., 40-41, 5-48.

Dickey, T., 2002, Instrumentation and new technologies, Oceans 2020: Science for Future Needs, eds. J.G. Field, G. Hempl, and C.P. Summerhayes, Island Press, Washington, DC, 213-256.

Dickey, T.D. and G.C. Chang, 2001, Recent advances and future visions: temporal variability of optical and bio-optical properties of the ocean, Oceanography, 14(3), 15-29.

Dickey, T. and P. Falkowski, 2002, Solar energy and its biological-physical interaction in the sea, eds. A.R. Robinson, J.J. McCarthy, and B.J. Rothschild, Chapter 10, The Sea, Vol. 12, John Wiley, New York, 401-440.

Dickey, T., D. Frye, H. Jannasch, E. Boyle, D. Manov, D. Sigurdson, J. McNeil, M. Stramska, A. Michaels, N. Nelson, D. Siegel, G. Chang, J. Wu, and A. Knap, 1998, Initial results from the Bermuda Testbed Mooring Program, Deep-Sea Res. I, 45, 771-794.

Dickey, T., D. Frye, J. McNeil, D. Manov, N. Nelson, D. Sigurdson, H. Jannasch, D. Siegel, T. Michaels, and R. Johnson, 1998, Upper-ocean temperature response to Hurricane Felix as measured by the Bermuda Testbed Mooring, Mon. Weather Rev., 126, 1195-1201.

Dickey, T., J. Marra, D.E. Sigurdson, R.A. Weller, C.S. Kinkade, .E. Zedler, J.D. Wiggert, and C. Langdon, 1998, Seasonal variability of bio-optical and physical properties in the Arabian Sea: October 1994 – October 1995, Deep Sea Res. II, 45, 2001-2025.

Dickey, T., M. Lewis, and G. Chang, 2003, Toward a global ocean system for measurements of optical properties using remote sensing and in situ Observations Manual of Remote Sensing: Marine Environment, ed. J. Gower, in review.

Dickey, T., N. Bates, R. Byrne, F. Chavez, R. Feely, C. Moore, and R. Wanninkhof, 2000, A Review of the NOPP Ocean-Systems for Chemical, Optical, and Physical Experiments (O-SCOPE) Project, Fifth Symposium on Integrated Observing Systems, Proc. Amer. Met. Soc.

Dickey, T., S. Zedler, X. Yu, S.C. Doney, D. Frye, H. Jannasch, D. Manov, D. Sigurdson, J.D. McNeil, L. Dobeck, T. Gilboy, C. Bravo, D.A. Siegel, and N. Nelson, 2001, Physical and biogeochemical variability from hours to years at the Bermuda Testbed Mooring site: June 1994 – March 1998, Deep-Sea Res. II, 48 (8-9), 2105-2140.

Doney, S.C., K. Lindsay, J.K. Moore, S. Dutkiewicz, M.A.M. Friedrichs, and R.J. Matear, 2001, Marine biogeochemical modeling: recent advances and future challenges, Oceanography, 14(4), 93-107.

Ducklow, H.W., D.K. Steinberg, and K.O. Buesseler, 2001, Upper ocean carbon export and the biological pump, Oceanography, 14(4), 50-58.

Flierl, G. and D. McGillicuddy, 2002, Mesoscale and submesoscale physical-biological interactions, in Biological-Physical Interactions in the Sea, eds. A.R. Robinson, J.J. McCarthy, and B.J. Rothschild, The Sea, Vol. 12, John Wiley and Sons, New York, 113-185.

Foley, D., T. Dickey, M. McPhaden, R. Bidigare, M. Lewis, R. Barber, C. Garside, and D. Manov, 1998: Longwaves and primary productivity variations in the equatorial Pacific at 0°, 140°W, Deep-Sea Res. II, 44, 1801-1826.

Griffiths, G., A. Knap, and T. Dickey, 2000, Autosub experiment near Bermuda, Sea Tech., 41(2), 35-45.

Griffiths, G., R. Davis, C. Eriksen, D. Frye, P. Marchand, T. Dickey, and R. Weller, 2001, Towards new platform technology for sustained observations, in Observing the Oceans in the 21st Century, eds. C.J. Koblinsky and N.R. Smith, Australian Bureau of Meteorology, 324-338.

Hansell, D.A. and C.A. Carlson, 2001, Marine dissolved organic matter and the carbon cycle, Oceanography, 14(4), 41-49.

Hofmann, E.E. and M.A.M. Friedrichs, 2001. Biogeochemical data assimilation, in Encyclopedia of Ocean Sciences, eds. J.H. Steele, S. Thorpe, and K. Turekian, Academic Press, London, 302-308.

Hoffman, E.E. and M.A.M. Friedrichs, 2002. Predictive modeling for marine ecosystems, The Sea, Volume 12, eds. A.R. Robinson, J.J. McCarthy, and B.J. Rothschild, Ch 13, John Wiley and Sons, New York, 537-566.

Honjo, S., J. Dymond, W. Prell, and V. Ittekot, 1999, Monsoon-controlled export fluxes to the interior of the Arabian Sea, Deep-Sea Res. II, 46(8-9), 1859-1902.

Kaku, M., 1997. Visions: How Science Will Revolutionize the 21st Century, Anchor Books Doubleday, New York, 403pp.

Karl, D.M., J.R. Dore, R. Lukas, A.F. Michaels, N.R. Bates, and A. Knap, 2001, Building the long-term picture: The U.S. JGOFS Time-series Programs, Oceanography, 14(4), 6-17.

Koblinsky, C.J. and N.R. Smith (eds.), 2001. Observing the Oceans in the 21st Century, GODAE, Bureau of Meteorology, Melbourne, Australia, 580pp.

Kolber, Z.S., O. Prasil, and P. Falkowski, 1998. Measurements of variable chlorophyll fluorescence using fast repetition rate techniques: defining methodology and experimental protocols, Biochim. Biophys. Acta, 1376, 88-106.

Lalli, C.M. and T.R. Parsons, 1993, Biological Oceanography: An Introduction, Pergamon Press, Oxford, 301pp.

Le Traon, P.Y., M. Reinecker, N.R. Smith, P. Bahurel, M. Bell, H. Hurlburt, and P. Dandin, 2001. Operational oceanography and prediction: a GODAE perspective, in Observing the Oceans in the 21st Century, eds. C.J. Koblinsky and N.R. Smith, GODAE, Bureau of Meteorology, Melbourne, Australia, 529-545.

Letelier, R.M., D.M. Karl, M.R. Abbott, P. Flament, M. Freilich, R. Lukas and T. Strub, 2000. Role of late winter mesoscale events in the biogeochemical variability of the upper water column of the North Pacific Subtropical Gyre, J. Geophys. Res., 105, 28,723-28,739.

Lewis, M.R., 2002, Variability of plankton and plankton processes on the mesoscale, in Phytoplankton Productivity: Carbon Assimilation in Marine and Freshwater Ecosystems, Ch. 6, eds. P.J. le B. Williams, D.N. Thomas, and C.S. Reynolds, Blackwell Science, London, 141-155.

Marra, J., T.D. Dickey, C. Ho, C.S. Kinkade, D.E. Sigurdson, R.A. Weller, and R.T. Barber, 1998, Variability in primary production as observed from moored sensors in the central Arabian Sea in 1995, Deep-Sea Res. II, 45, 2253-2267.

McGillicuddy, D.J., 2001. The internal weather of the sea and its influences on ocean biogeochemistry, Oceanography, 14(4), 78-92.

McGillicuddy, D.J., A.R. Robinson, D.A. Siegel, H.W. Jannasch, R. Johnson, T.D. Dickey, J.D. McNeil, A.F. Michaels, and A.H. Knap, 1998, Influence of mesoscale eddies on new production in the Sargasso Sea, Nature, 394, 263-266.

McNeil, J.D., H.W. Jannasch, T. Dickey, D. McGillicuddy, Jr., M. Brzezinski, and C.M. Sakamoto, 1999. New chemical, bio-optical, and physical observations of upper ocean response to the passage of a mesoscale eddy off Bermuda, J. Geophys. Res., 104, 15,537-15,548.

McPhaden, M.J., T. Delcroix, K. Hanawa, Y. Kuroda, G. Meyers, J. Picaut, and M. Swenson, 2001. The El Niño-Southern Oscillation (ENSO) Observing System, in Observing the Oceans in the 21st Century, eds. C.J. Koblinsky and N.R. Smith, GODAE, Bureau of Meteorology, Melbourne, Australia, 213-247.

Oschlies, A. and V. Garcon, 1998. Eddy-induced enhancement of primary production in a model of the North Atlantic Ocean, Nature, 266-269.

Robinson, A.R. and P.F.J. Lermusiaux, 2001. Data assimilation in models, in Encyclopedia of Ocean Sciences, eds. J.H. Steele, S. Thorpe, and K. Turekian, Academic Press, London, 623-634.

Robinson, A.R. and P.F.J. Lermusiaux, 2002. Data assimilation for modeling and predicting coupled physical-biological interactions in the sea, The Sea, Volume 12, eds. A.R. Robinson, J.J. McCarthy, and B.J. Rothschild, Ch 12, John Wiley and Sons, New York, 475-536.

Robinson, A.R. and T. Dickey (eds.), 1997. An Advanced Modeling/Observation System (AMOS) for physical-biological-chemical ecosystem research and monitoring (concepts and methodology), The GLOBEC International Working Groups on Numerical Modeling and Sampling Observational Systems, GLOBEC Special Contribution No. 2, GLOBEC International, Plymouth Marine Laboratory, Plymouth UK, 155pp.

Robinson, A.R., J.G. Bellingham, C. Chryssostomidis, T.D. Dickey, E. Levine, N. Petrikalakis, D.L. Porter, B.J. Rothschild, H. Schmidt, K. Sherman, D.V. Holliday, and D.K. Atwood, 1999, Real-time forecasting of the multidisciplinary coastal ocean with the Littoral Ocean Observing and Predicting System (LOOPS), Proceedings of the Third Conference on Coastal Atmospheric and Oceanic Prediction Processes, New Orleans, LA, American Meteorological Society.

Robinson, A.R., P.F.J. Lermusiaux, and N.Q. Sloan III, 1998. Data assimilation, The Sea, Volume 10, eds. K.H. Brink and A.R. Robinson, Ch 20, John Wiley and Sons, New York, 541-594.

Send, U., R. Weller, S. Cunningham, C. Eriksen, T. Dickey, M. Kawabe, R. Lukas, M. McCartney, and S. Osterhuis, 2001, Oceanographic time series observatories, Observing the Ocean for Climate in the 21st Century, eds. C.J. Koblinsky and N.R. Smith, Australian Bureau of Meteorology, 376-390.

Siegel, D.A., D.J. McGillicuddy, and E.A. Fields, 1999, Mesoscale eddies, satellite altimetry, and new production in the Sargasso Sea, J. Geophys. Res., 104, 13,359-13,379.

Tokar, J.M. and T.D. Dickey, 2000, Chemical sensor technology – Current and future applications, Chemical Sensors in Oceanography, Gordon & Breach Scientific Publishers, Amsterdam, 303-329, 2000.

Wunsch, C., 1996. The Ocean Circulation Inverse Problem, Cambridge University Press, Cambridge, UK, 442pp.

Yoder, J.A., J.K. Moore, and R.N. Swift, 2001, Putting together the big picture: remote sensing observations of ocean color, Oceanography, 14(4), 33-40.

Yu, X., T. Dickey, J. Bellingham, D. Manov, and K. Streilein, 2002, The application of autonomous underwater vehicles for interdisciplinary measurements in Massachusetts and Cape Cod Bays, Cont. Shelf Res, 22, 2225-2245.

REMOTE SENSING OF THE OCEAN FROM SPACE

GENNADY K. KOROTAEV
Marine Hydrophysical Institute, Sevastopol, Crimea, Ukraine,
ipdop@sebastopol.ua

1. PRINCIPLES AND METHODS OF SATELLITE OCEANOGRAPHY

1.1 SIGNIFICANCE OF REMOTE SENSING DATA FOR THE OCEAN OBSERVATIONS

The initial manifestations of climate change induced by the flow of anthropogenic carbon to the atmosphere are small. Extensive, regular and accurate observations are required to describe all of the important processes and to identify the long-term evolution of the climate system against a background of variability on a broad range of frequency. Due to its large inertia, the ocean plays a significant role in potential mechanisms of climate change and the ocean observing system should be expanded to provide continuous, long-term and basin-wide data sets.

The frequency and spatial resolution of observations should be appropriate for the spatial and temporal scales of oceanic processes. Consequently observations need to encompass typical scales of phenomena and temporal variations should be resolved in the development of the most energetic processes. Traditional oceanographic methods do not satisfy these requirements. Thus new technologies are being applied to perform basin-wide observations of the World Ocean.

M. Follows and T. Oguz (eds.), The Ocean Carbon Cycle and Climate, 189–216.

Remote sensing of the ocean from space now provides a major source of continuous data flow characterizing the state and variability of the World Ocean. Instruments placed on the spacecraft usually have a broad viewing angle and enable the observation of large areas of the sea surface simultaneously. The rotation of the Earth produces a successive shift of the spacecraft path projected on the Earth's surface so there can be whole ocean data coverage on timescales of a day or so. Long-term satellite datasets covering last 20 years are now available for studies of climate trends and climatic variability.

Ocean remote sensing is based on measurements of natural or artificial radiation propagating from the ocean surface. However, the electromagnetic radiation arriving at the spacecraft is transformed as it propagates through the overlying atmosphere. Therefore, remote sensing of the ocean relies upon the selection of appropriate electromagnetic wave bands. Millimetric, centimetric and decimetric radio waves with the wavelength from one millimeter to 100cm, natural light in the visible band (0.4 – 0.7μm), near infrared (0.7-3μm) and thermal infrared (3 – 14μm) ones are the most important for the remote sensing of the ocean from space.

1.2 FORMATION AND TRANSPORT OF THE RADIATION

Emission, absorption and scattering determine the formation and transport of radiation in all bands. Let us consider briefly all this processes (Stewart, 1985).

Absorption of radiation

Absorption by gases is induced by the energy exchange between a matter and radiation. The radiation can raise the energy level of orbital electrons or can cause vibration or rotational motion of the molecules. The energy exchange between radiation and matter is quantized: radiation can be absorbed if its energy is exactly equal to the difference of energy between two states of the matter. Since the energy quantum is proportional to the frequency of radiation, absorption as a function of frequency (an absorption spectrum) presents a set of narrow peaks or lines.

Emission of the radiation

The emission of the radiation by an idealised black body is defined by Planck's equation

$$B(\lambda) = \frac{2 \cdot h \cdot c^2}{\lambda^5} \cdot \frac{1}{\exp\left[h \cdot c / (k \cdot T \cdot \lambda)\right] - 1},$$

where λ is the wavelength of the emitted radiation, T is the temperature of the black body, c is the speed of light, and h and k are the Planck and Boltzmann constants respectively. Natural bodies, however, emit radiation at a somewhat reduced rate compared to a black body:

$$B^*(T) = \varepsilon \cdot B(T)$$

The ratio of the spectral radiance emitted by a real body relative to that emitted by a black body ε is named the spectral emissivity. An example of a natural source of radiation is the sun. The spectral radiation from the sun can be approximated as the radiation from a black body at 5900 K (Figure 1).

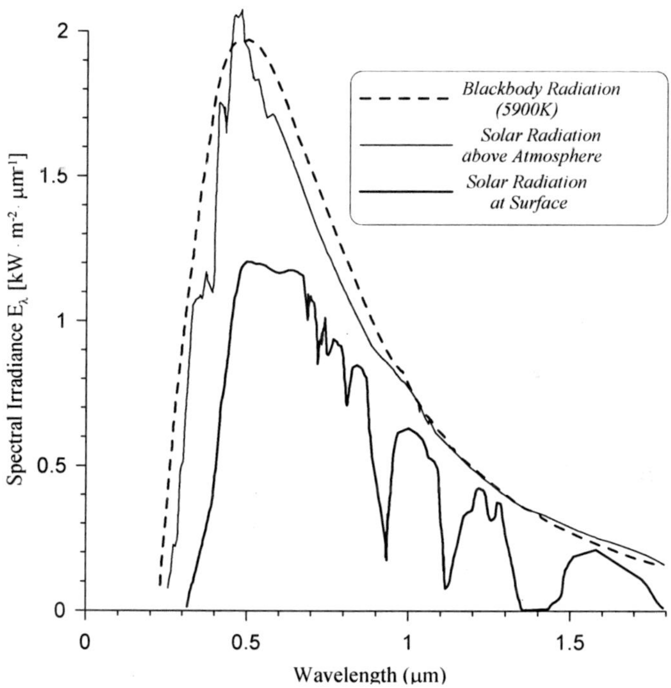

Figure 1.. The spectrum of solar radiation at the top of the atmosphere, its approximation by the Plank equation, and the spectrum of solar radiation at the bottom of the atmosphere.

Absorption and emission of radiation by a matter are not independent. Kirchoff's law states that the emission and absorption rates are the same for any material that is in the local thermodynamic equilibrium. This means that the spectral emissivity is equal to the spectral apsorptance.

Scattering of the radiation

Radiation propagating through the atmosphere is scattered by different molecules and particles including the water droplets in fog, clouds, rain and the solid particles of smoke and haze. The scattering is significant if the radiation has wavelength up to the same order of magnitude the particle size.

Scattering primarily changes the angle of propagation of radiation if particles are sparse. However the radiation is scattered again and again if particles are sufficiently dense. The angular distribution of the scattered radiance by a single particle, or ensemble of particles, is described by the phase function or the volume scattering function. The phase function is the volume scattering function normalized to the total volume scattering.

Radiative transfer; optical thickness

The propagation of radiation through the atmosphere with given properties of scattering and absorption can be simulated using the radiation transfer equation. In practice, such simulations are exceedingly difficult, and simplified approaches provide approximate descriptions of radiative transfer (Ishumaru, 1978).

The radiation passing through the atmosphere, along a selected ray, attenuates within an infinitesimally small volume due to absorption and scattering to different directions. It can also increase due to the scattering of other rays and local thermal emission. In general, coefficients of absorption and scattering depend on z , the distance along the ray. Therefore it is convenient to introduce a new variable, optical thickness, for the layer between z_1 and z_2 in the following way

$$\tau(z_1, z_2) = \int_{z_1}^{z_2} (k+s) \cdot dz$$

where k and s are absorption and scattering coefficients.

The radiation arriving at the top of the layer from above undergoes scattering and absorbtion so the downwelling radiation partially passes through the layer. Assume that at the bottom of the layer there is no upwelling radiation. The ratio of radiation, passing through the layer to that falling on its surface is the ``transmission coefficient''. The upwelling radiation on the top of the layer is non-zero due to the backscattering. The fraction of the upwelling radiation to the downwelling at the upper surface of the layer is the ``reflection coefficient''. Both coefficients are functions of wavelength.

The solar radiation spectrum at the Earth's surface; transparency windows in the atmosphere.

The atmosphere is the example of a layer that attenuates and reflects the radiation. The sunlight spectrum in particular, is significantly transformed as it passes through the atmosphere (Figure 1).

For some wavelengths, however, the attenuation of sunlight is not so strong (Figure 2). Hence we have some transparency windows in the atmosphere where the attenuation of the radiation is moderate. The transparency windows enable us to apply remote sensing methods to oceanographic investigations.

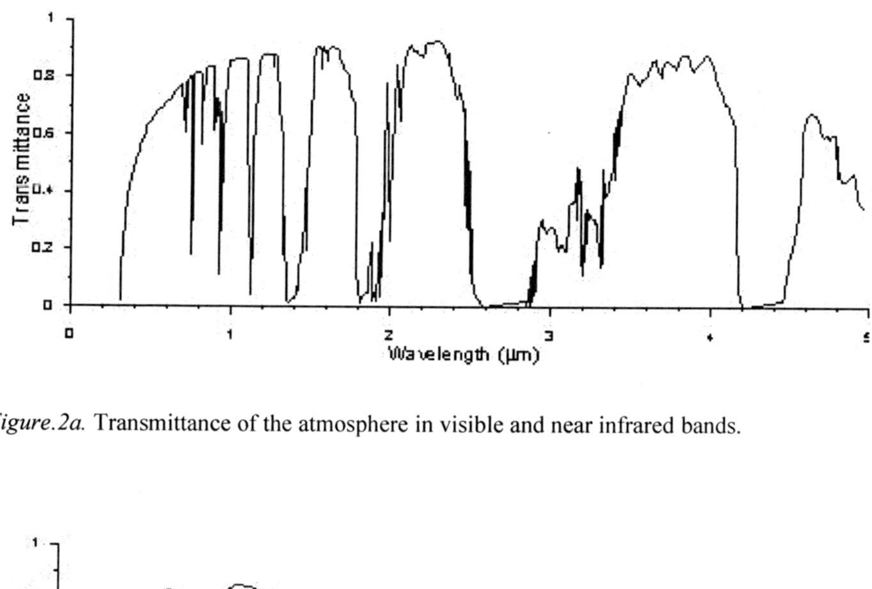

Figure.2a. Transmittance of the atmosphere in visible and near infrared bands.

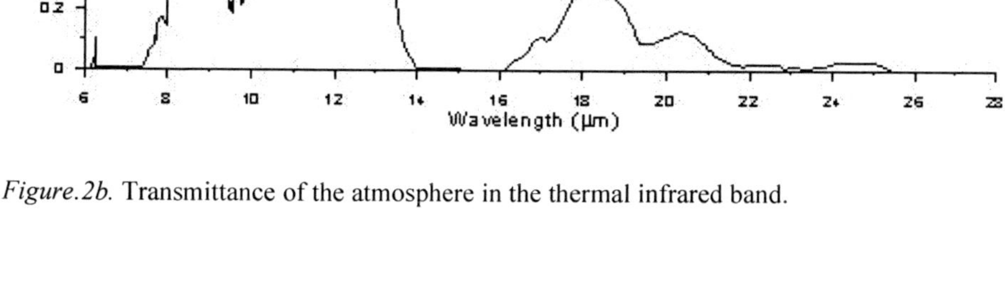

Figure.2b. Transmittance of the atmosphere in the thermal infrared band.

1.3 INTERACTION OF THE RADIATION WITH THE SEA SURFACE

Emission of the radiation by the sea surface

The radiation emitted by the sea surface depends on the sea surface temperature (SST) according to the Planck formula. However, the emission coefficient of the sea surface depends on temperature and salinity, sea surface roughness and the angle of observations. The dependence is weak in visible and IR bands where the emissivity coefficient is close to unity. However the emissivity coefficient is rather sensitive to variations of the sea surface state for radio waves. This dependence is also applied in the ocean remote sensing from space.

Scattering of light in the water.

A major part of the sunlight is reflected from the sea surface. The reflection occurs according to the rule of the mirror reflection stating that the incident angle is equal to the reflection angle. The waving surface also follows this rule but the reflection angle is modulated according to the slope of the sea surface. Observations of sun glint are used to determine the surface wave spectrum using aircraft observations. The sun glint area is simply excluded when observations are carried out from space.

The remaining sunlight is transmitted through the ocean interface and propagates to the depth. The propagation below the sea surface is described by the same radiation transfer equation that was considered earlier. Chlorophyll cells and dissolved matter absorb some fraction of the light while another part is scattered on molecules and suspended particles leading to upward propagating radiation. Ascending radiation can also be due to reflection from the bottom in shallow water. Backscattered light is partially transmitted through the sea surface and its spectrum contains information about the dissolved and suspended matter. This scattered light constitutes a few percent of the incident radiation.

Scattering of radio waves by the rough sea surface

The sea surface scatters radio waves in many directions. Mirror-like facets reflecting the energy approximately as a plane surface perform the scattering by the rough sea-surface for the incident angles that are close to vertical. The number of facets rapidly decreases with increasing incident angle but almost all waves have a slope of less than 25°. Thus a small portion of scattered energy, carrying the information on the phenomena at the surface, is eventually scattered backward. However surfaces with a regular lattice of roughness act like a diffraction grating at greater incident angles and the scatter is resonant. The resonant scatter from a single sinusoidal component of the spectrum of oceanic waves that matches the wavelength of the radio signal projected onto the sea surface reflects energy toward the receiver. The resonant scatter is often named ``Bragg scatter''. The Bragg equation

$$\vec{k} = 2 \cdot \vec{k}_r \cdot \sin \theta$$

relates the wave vector of the surface wave, $\vec{k}$, with the electromagnetic wave vector and the incident angle. The intensity of the backscattering is proportional to the energy of the wave. Thus processes at the sea surface which modulate the amplitude of surface gravity waves satisfying the Bragg equation can be seen in the backscattered signal.

1.4 MAJOR METHODS OF OCEAN REMOTE SENSING

Observations of the ocean from space are carried out using the wavelengths where the radiation is dependent on characteristics of the sea surface and the atmosphere is transparent or the atmospheric influence can be accurately accounted for. There are two broad classes of ocean remote sensing: The first is based on observations of natural radiation. The most popular is the use of the visible sunlight. Visible band observations may have very high spatial resolution (up to tens of centimeters) but, of course, can only be performed during daytime hours and in cloud-free conditions. Since the scattering of visible light in the atmosphere is significant, haze, fog and atmospheric aerosols contaminate visible band observations. Therefore the most efficient observations are made in clear atmosphere conditions and the major problem of quantitative description of the sea surface state is the correction of atmospheric scattering. Multi-channel measurements are used to reduce the atmospheric influence and to quantitatively characterize the ocean surface.

Infra-red (IR) radiation is used to infer thermal contrasts on the ocean surface. Space IR radiometers usually have resolution from a few hundred meters to one kilometer. Clouds also restrict IR observations, and absorption and emission of radiation in the atmosphere in general significantly contaminates the signal coming from the ocean. However using wavelengths in transparency windows keeps the atmospheric effect to a minimum and IR remote sensing is a very efficient method of observing the ocean. Multi-channel IR remote sensing is used to correct atmospheric contributions and to measure the sea surface temperature.

Radio frequencies are free from the major drawbacks of the visible band and IR remote sensing, as clouds are practically transparent for radio waves. However in order to achieve high spatial resolution the receiving antenna must be very large and its size grows with the wavelength. Therefore remote sensing of the ocean is usually restricted to millimetric to decimetric wavelengths. Typical spatial resolution of radio frequency radiometers is from a few to hundred kilometers. The received radiation depends from the state of atmosphere as clouds with high water content and rain distort the signal propagating from the oceanic surface. Therefore accurate interpretation of microwave imagery requires concomitant information, for example data from another channel. Measurements of temperature, salinity and wind at the sea surface are based on use of multi-channel data.

The second group of methods is based on the active radiolocation of the sea surface. Clouds are transparent for radiolocation and the ocean can be observed in practically all weather conditions.

The satellite-borne altimeter radiates radar pulses to the Earth's surface exactly downward and receives the reflected signal. The time required for the pulse transmission from the altimeter antenna to the Earth surface and back indicates the height of the satellite above the surface. Since the velocity of the light depends on the atmospheric state, the altimeter complex includes a microwave radiometer to estimate the water vapor content. Transmission of pulses at nadir means that altimetric observations occur only along the satellite track.

Real aperture and synthetic aperture radars depend upon pulse transmission at larger incident angles, so called sliding angles and reception of the back scattering signal from the ocean surface. Both instruments produce a scanning image of the sea surface with spatial resolution up to one km for real aperture radar. Synthetic aperture radar uses the coherent signal and measures the phase of the scattered signal. With synthetic aperture radar, motion of the satellite and processing of sequential observations of one area can increase the spatial resolution to ten meters.

The scatterometer is a radar designed to measure the surface wind. It measures the back scattering signal from two beams allowing determination of both the wind speed and the wind direction.

Three special types of orbit are used for the ocean remote sensing. Sun-synchronic orbits pass the Earth at a regular local time and are convenient for IR or visible band observations. A satellite with exact repeating orbit returns along each track during a certain time period. The repeat time is determined by a compromise between spatial and temporal resolution. Application of geostationary orbits, where the satellite hangs above the same point on the Earth's surface, is rather limited for ocean remote sensing as geostationary satellites observe the Earth from the height of about 36000 km. Their advantage is the opportunity to observe a large area of the ocean with high temporal resolution but it is difficult to achieve a useful spatial resolution at such great distance.

Remote sensing of the ocean has been intensively developed during the last two decades. Some particularly significant missions are presented in Table 1. The last decade is characterized by transition from experimental remote sensing of the ocean to operational observations.

2. OBSERVATION OF OCEANIC PROCESSES BY SCANNING INSTRUMENTS

Scanning imagery from space provides a unique opportunity to observe large areas of sea surface and to track their temporal evolution. Even without complicated preprocessing of the raw data the observations can be used to reveal specific ocean processes and aid physical understanding of the underlying mechanisms.

Table 1. Some important remote sensing missions.

NOAA meteorological satellites: TIROS 11, NOAA-6,7,8,... Since 1978 Major device – AVHRR (Advanced Very High Resolution Radiometer)	*Seasat oceanographic satellite* Launch in1978 Major devices SMMR (Scanning Multifrequency Microwave Radiometer) SAR (synthetic aperture radar) SCAT (scatterometer) ALT/2 (altimeter)	*Nimbus-7* Launch in1978 Major devices CZCS (Coastal Zone Color Scanner) SMMR (Scanning Multifrequency Microwave Radiometer)	*Kosmos 1151 oceanographic satellite* Launch in 1980 Major devices Along track multichannel IR radiometer Along-track multichanel microwave radiometer Along-track visible band spectrometer
Interkosmos-21 Launch in 1981 Major device – MCS (Multi Channel Spectrometer)	*Kosmos 1500, Okean 2, 4, 6, 7, Sych-1* Beginning from 1985 SLR (Side Looking, real aperture radar)	*ERS-1,2* Launch in 1991 ATSR (Along Track Scanning Radiometer) SAR (Synthetic Aperture Radar) SCAT (Scatterometer) ALT (Altimeter)	Geosat Launch in 1986 Altimeter RADARSAR SAR (Synthetic Aperture Radar)
Topex/Poseidon Launch in 1992 Topex and Poseidon (altimeters)	*IRS* Launch in 1996 MOS (Multichannel Optical Scanner)	*ADEOS-1* Launch in 1996 OCTS (Sea color scanning spectrometer) NSCAT (Scatterometer)	*OrbView-2* Launch in 1998 SeaWiFS (Sea color scanning spectrometer)

2.1 NATURAL EMISSION OF RADIATION BY THE OCEAN

Seawater emits radiation almost like a black body in the visible and infrared bands (emissivity is about 0.98 – 0.99). Therefore IR observations in the atmosphere transparency windows are used for observing thermal contrasts on the sea surface. For example, sea surface temperature (SST) images allow us to identify jets and eddies. Figure 3 presents the image of the Black Sea surface obtained by AVHRR scanner in the thermal IR band. The cyclonic Rim Current encircling the sea along the coast is visible on Figure 3 as the boundary between the light and dark tones. The light tone corresponds to relatively warm water and the dark indicates cold water near the coast. Large mesoscale eddies are also evident on this figure to the east and to the west of the Crimea peninsula. Intense meandering of the Rim Current jet and some smaller scale eddies are seen to the north of the Anatolian coast.

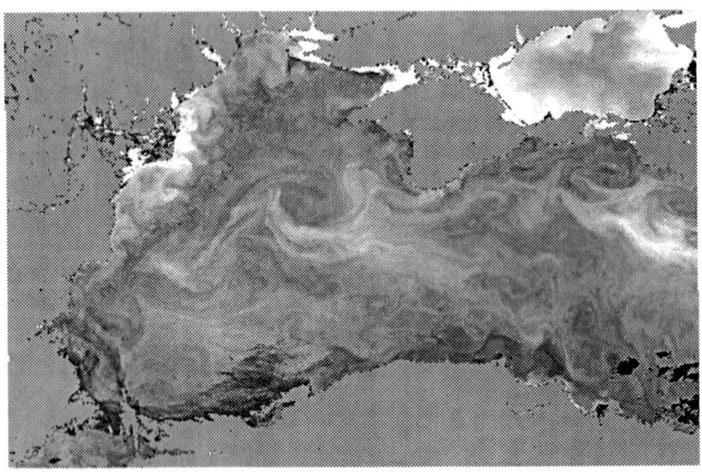

Figure3. Infrared image of the Black Sea. Light tone corresponds to warm water.

Another example is the infrared image of the Gulf Stream (Figure 4). The Gulf Stream jet is seen on the image as a warm river, transporting warm water polewards from the Gulf of Mexico, and restricted to the north by a sharp thermal front. The thermal front is a basic feature of any strong current subjected to the geostrophic balance (i.e. the balance between pressure gradient and the Coriolis force) driving the flow along the front.

The sharp thermal front is unstable, inducing meanders of the jet. Strong meanders of the Gulf Stream sometimes detach from the flow and form intense mesoscale eddies, so called cold and warm Gulf Stream rings. They are seen on infrared images as a circle of warmer or cooler water since the eddy traps the water from other side of the thermal front. Detachment of a Gulf Stream meander and formation of a cold ring is evident in the left corner of Figure 4. Well-developed cold and warm rings are seen in the center of Figure 4. Eddies similar to the Gulf Stream rings are often observed near other strong jets.

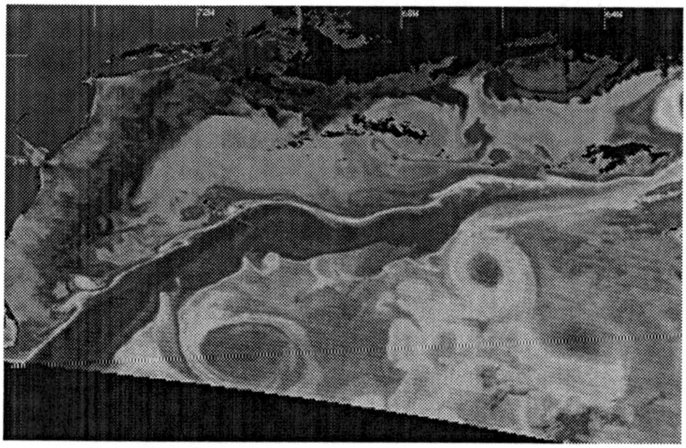

Figure 4. Thermal infrared image of the Gulf Stream by MODIS. Cold and warm Gulf Stream rings are seen to the right and to the left of the warm current jet.

Thermal contrasts in the IR imagery can be observed also in the open ocean, far from intense currents. Such contrasts arise because of the fine features of the upper ocean thermal structure and reflect certain peculiarities of the air-sea interaction. Usually the instability of drift currents, break down of surface and internal gravity waves or convection induces turbulent mixing of the quasi-homogeneous, surface ocean layer that is known as the mixed-layer. The temperature within the layer is vertically almost uniform. The seasonal thermocline, with a low level of turbulent mixing, is situated just below the quasi-homogeneous layer which is typically on the order of 100 meters deep. IR radiation, however, is emitted by a very thin layer situated just below the surface. This layer has thickness on the same order of magnitude as the wavelength of thermal IR radiation. Therefore IR imagery reflects the non-uniformity of small-scale processes developing near the ocean surface.

One such phenomena is the skin layer which is located in the top few millimeters below the sea surface. The temperature within the skin layer decreases toward the surface, as the evaporation and the emission of long-wave radiation are concentrated in a few upper millimeters. There are three regimes of the ocean skin layer. The first is observed when the surface wind is very weak and the surface waves are stable. The skin layer in this case is well pronounced and the temperature jump across the layer can achieve 1K (Emery, Castro, Wick, Schluessel, Donlon, 2001). The second regime is developed under moderate winds when the turbulence reduces both the temperature gradient and the thickness of the layer. The third case corresponds to strong winds when the breaking down of short gravity waves destroys the skin layer and the surface temperature becomes the same as in the upper few meters. Thus the wind modulates the temperature of the skin layer. Since IR radiation is formed within the skin layer, non-uniform wind can lead to spatial contrasts in IR images.

Another effect is connected with the diurnal cycle of temperature in the upper layer of the ocean. A diurnal thermocline, which forms within the first meter below the sea surface at the daytime under the calm weather conditions, can cause a difference between SST and the mixed layer temperature in a few degrees. The diurnal cycle of SST is clearly seen in continuous observations from the IR radiometer on a geostationary satellite (Legeckis, Tong Zhu, 1997). This example also shows one advantage of geostationary satellites for ocean remote sensing as polar orbit satellites are unable to accurately observe the temporal variability of SST over a full day.

However the diurnal thermocline is generated only under low wind conditions when the wind-induced turbulence cannot mix the whole quasi-homogeneous layer. Stronger wind destroys the diurnal thermocline and distributes heat through the whole depth of the mixed-layer. Thus a spatially non-homogeneous surface wind can again produce spatial contrasts on the IR images.

Strong hurricanes excite very intense surface winds also giving rise to SST contrasts. A hurricane wind increases the level of turbulent energy in the mixed layer and the cold water from below of the seasonal thermocline is entrained to the mixed layer, reducing SST (Figure 5). Such a pool of cold water is observed by IR radiometer along the trajectory of the hurricane (Monaldo et al., 1997).

A wind blowing along the coast can lead to a signature in the SST distribution which is captured on IR images. An example of the synoptic upwelling near the Crimean coast, induced by the north-westerly winds, is presented on Figure 6. A patch of cold water near the coast is well seen as a dark spot on the image.

The wind blowing along the coast induces an offshore Ekman drift and warm surface water is replaced by cold water coming from below the mixed layer due to the intense vertical motion (Figure 7).

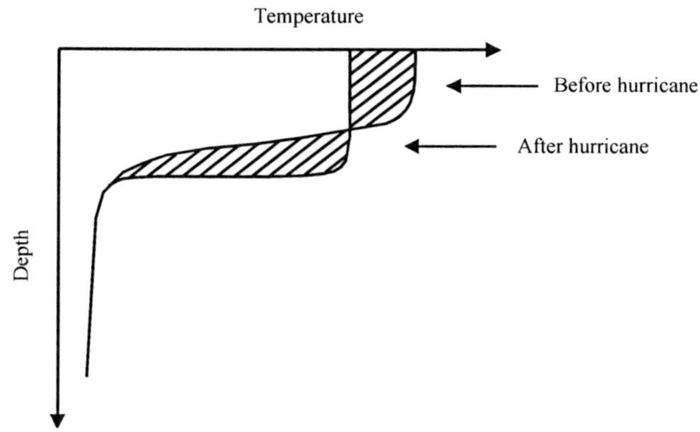

Figure 5. Sea surface temperature decrease along a hurricane path.

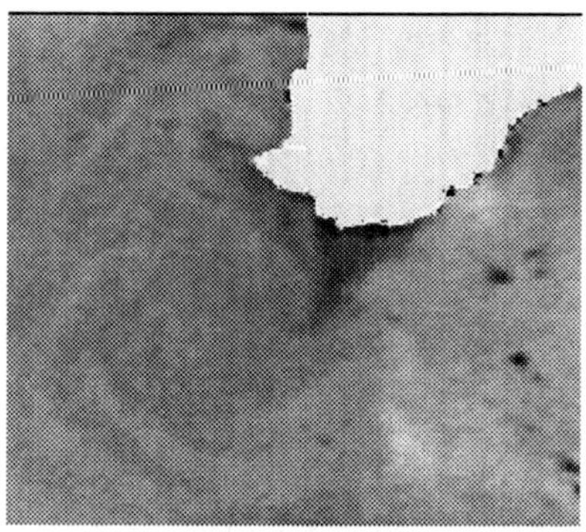

Figure 6. Synoptic coastal upwelling near the Crimea coast.

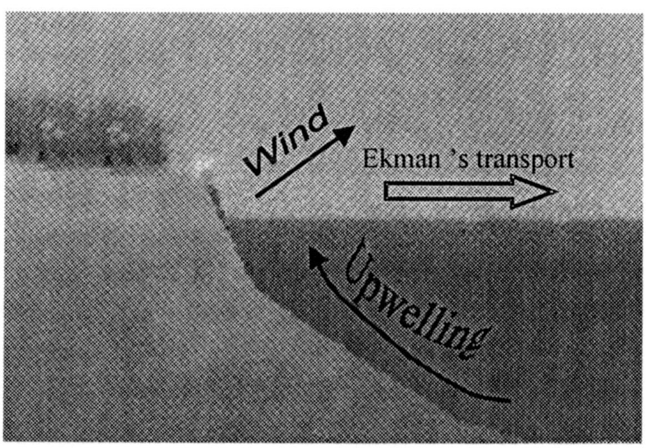

Figure 7. Schematic of upwelling induced by the wind blowing along the coast.

2.2 OBSERVATIONS USING VISIBLE LIGHT.

Visible band scanning radiometers use the sun as the source of radiation and observations are available only at the daytime. Spatial variations in the absorption and scattering of sunlight forms the image of the sea surface. For example, snow and ice weakly absorb visible light. Therefore a considerable contrast is observed on the satellite image if a part of the sea surface is covered by ice.

A sequence of visible band images obtained by the scanner MOS in April 1998 shows the dynamics of the ice field in the area of Riga Bay (Figure 8). Ice sheets appear as white patches surrounded by water, which is black on the visible band image.

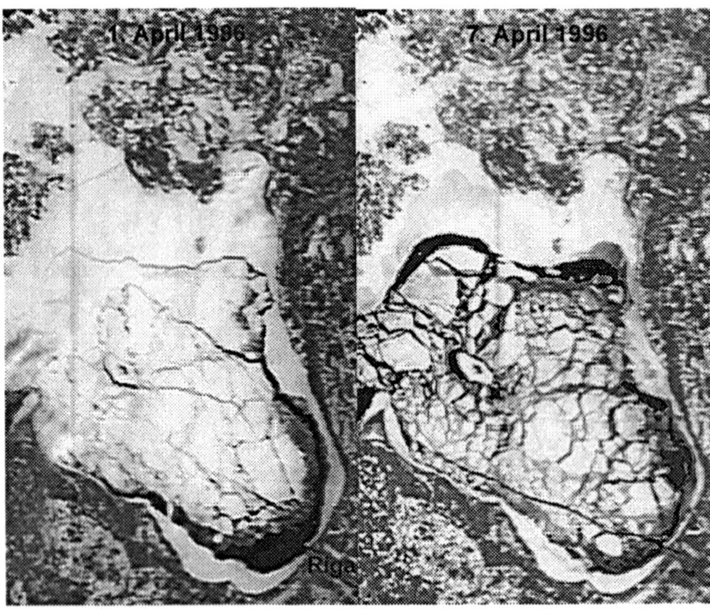

Figure 8. Visible band observation of sea ice field dynamics in Riga Bay.

The visible band signal in the open sea depends on the concentration of absorbing and scattering substances in the seawater. The example in Figure 9 shows a set of structures observed on the AVHRR image of the Black Sea obtained at the wavelength 0.55 μm. Backscattering of light prevails in formation of a signal on this wavelength. Sunlight is back-scattered from suspended particles of different origin, though in May and June the open Black Sea is characterized by coccolith blooms. The patterns are determined by dynamical structures which modulate the growth and transport of the particles.

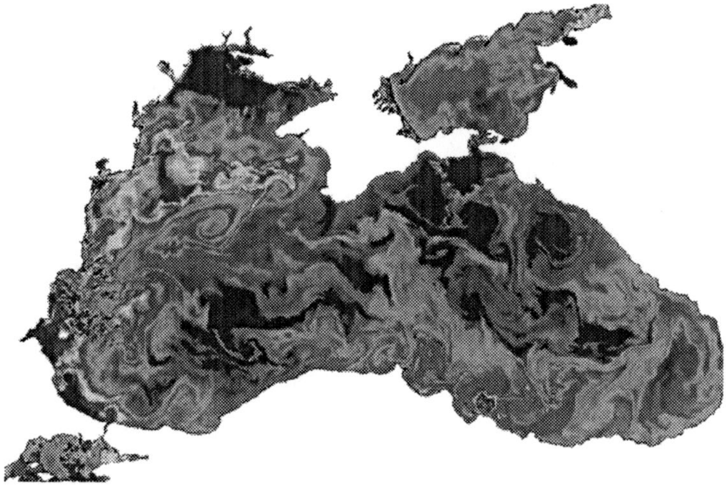

Figure 9. Coccolithophores bloom in the open Black Sea, observed in AVHRR visible band image in June 1993.

In another example (Figure 10) a jet structure is observed in the contrast of visible light absorption due to a phytoplankton bloom which reflects the dynamical structure.

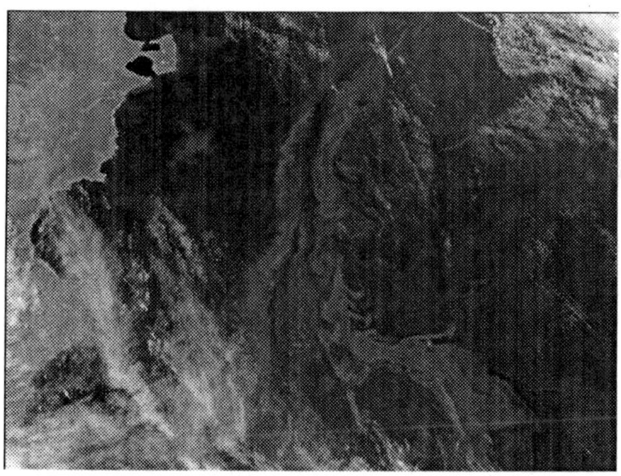

Figure10 . Jet near Patagonia in the MODIS visible band image

Sometimes intense synoptic coastal upwelling gives rise to a local plankton bloom. The upwelling of deep water to the surface supplies nutrients to the euphotic zone stimulating the growth of phytoplankton. Such an upwelling event has been observed in the southern part of the Black Sea near the Turkish coast. The upwelling is seen as a black strip of cold water on the thermal AVHRR image (Figure 11). The simultaneous SeaWiFS image shows high chlorophyll concentration (light colour) in the same strip along the coast.

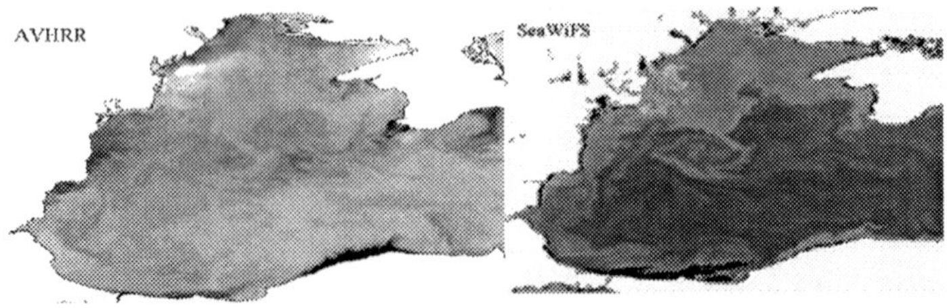

Figure 11. Phytoplankton bloom induced by coastal upwelling.

2.3 RADIOLOCATION OF THE OCEAN SURFACE UNDER SLIDING ANGLES

Surface waves and wind: Cyclones in radar imagery

The spectrum of short surface waves is sensible to wind intensity and surface currents. Therefore intense atmospheric structures, particularly, intense cyclones, and strong currents should be well pronounced on images of synthetic aperture radar (SAR) or side looking real aperture radar (SLR) working with radio waves of a few centimeter wavelengths.

The typical structure of an atmospheric cyclone is shown on the visible image of clouds (Figure 12). Strong winds occupy the periphery of the cyclone. The non-uniform structure of clouds along the radius of the cyclone indicates the complex structure of the wind field. The cloud-free zone near the center of the cyclone ("eye of the cyclone") that is well seen on the satellite image is an area of low winds. Thus we can anticipate low backscatter near the center of cyclone in the radar image of the sea-surface, and non-uniform increase of the scattering along the radius.

Figure 13 shows the image of the tropical cyclone obtained by the SLR of the satellite "Kosmos-1500". The eye of the cyclone is well presented on the image as a dark region. The increased roughness of the sea surface, manifested in the radar signal, is observed further from the center.

Sikora et al. (2000) present an interesting analysis of RADARSAT SAR image of the polar mesoscale cyclone (PMC). As in Figure 13, the area of low backscatter or low winds is observed in the center of PMC.

It also has a sharp frontal zone which appears on the radar image as a boundary between high and low backscatter due to the strong wind speed shear. Wave-like features with the wavelength of about 20km are probably the expression of wind shear instability and can be seen along the frontal zone on the radar image. The authors also found a trace of atmospheric gravity waves and cellular convection in the SAR image.

Figure 12. Visible band image of a tropical cyclone.

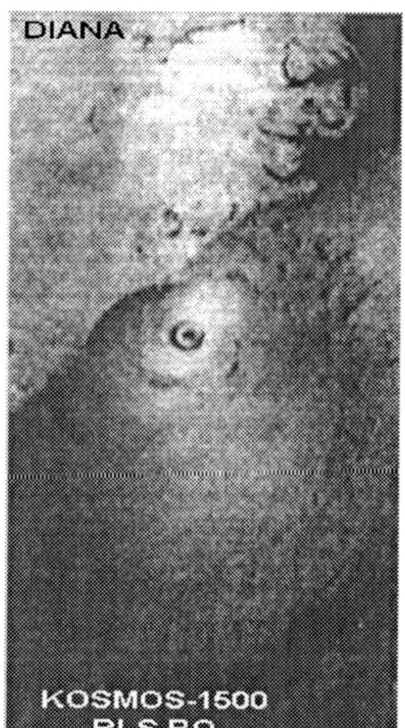

Figure 13. Image of the tropical cyclone "Diana" obtained by SLR

Internal waves in SAR imagery

SAR imagery also reveals manifestations of deep-sea processes such as internal waves. These waves displace the permanent or seasonal pycnocline and stimulate convergence or divergence of horizontal currents at the sea surface. Variable surface currents modulate the steepness of surface waves and the intensity of SAR signal. Examples of the observation of internal waves with SAR are to be found in Stewart (1985) and Porter and Thompson (1999).

Observation of sea ice in SLR images.

The difference in scattering properties of the open water and ice, or ice covered by snow, enables observation of the evolution of ice sheets in radar imagery. The example of the ice sheets evolution is shown on the SLR images of the Sea of Azov obtained from the «Sych-1» satellite in February – April 1996 (Figure 14).

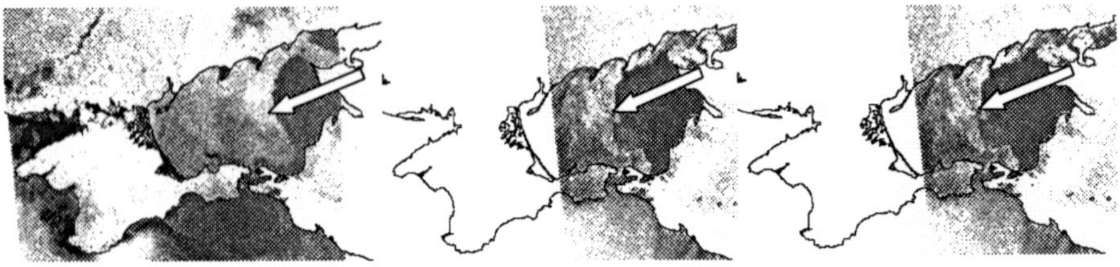

Figure 14. Evolution of the ice field in the Sea of Azov based on SLR observations.

These examples of the analysis of the ocean imagery from space show that the repeated survey of large oceanic areas provides an efficient tool for observing many oceanic phenomena and studying their dynamics. Joint application of different instruments enables the characterization of some complex processes and establishes interdisciplinary links.

3. QUANTITATIVE AND MODELLING APPLICATIONS OF REMOTE SENSING

3.1 MULTI-CHANNEL SYSTEMS AS A MEANS OF MEASUREMENTS FROM SPACE

Observations of the ocean from space provide qualitative description of many oceanic phenomena and processes. However practical problems require quantitative characterization of oceanic fields for now casting and forecasting of its state. Therefore the development of instruments, methods and algorithms capable to provide quantitative measurements is one of the major tasks of the ocean remote sensing.

We have mentioned that the scattering and absorption of radiation at each wavelength depends on various atmospheric and oceanic factors. A thorough investigation of the formation and transport of radiation makes it possible to select wavelengths where the contribution of a desirable parameter dominates. Then we can use data of other wavelengths to correct the measurement of the basic channel and to reach the required accuracy. Thus, difficulties of measurement from space can be overcome by the use of multi-channel systems. All current measurements from space are based on this principle.

Phytoplankton pigment determination in visible band spectrometry

The major final product of multi-channel visible band measurements is the chlorophyll concentration, which is a direct characteristic of the phytoplankton, or low level trophic level, of the marine ecosystem. The absorption of light by the phytoplankton pigments determines the color of the open ocean. Therefore the spectral reflectance of the ocean depends on the concentration of phytoplankton pigments. This dependence is used to estimate the chlorophyll concentration from the remote measurements.

However direct measurements of the spectral reflectance from space are impossible due to the scattering of the light by molecules and aerosols in the atmosphere which obscures the true color of the water body. Typically, the atmospheric contribution to the radiance measured from space constitutes more than 90% of the signal. The contribution to scattering by gas molecules can estimated theoretically, whereas aerosols effects must be estimated remote measurements on pixel-by-pixel basis.

There are a currently several algorithms for atmospheric correction based on multi-channel measurements. The most widely used are algorithms based on the evaluation of the aerosol optical model from the data of two near-infrared channels and then its application to the visible bands (Gordon and Wang, 1994). Such an approach relies on the fact that the sea surface reflectance is negligible in the near-infrared. The knowledge from the aerosol optical model makes possible to evaluate the atmospheric transparency in the visible band and to simulate the water-leaving radiance on the sea surface.

The next step is the estimation of phytoplankton pigment concentration using the known values of the water-leaving radiance. The concentration of phytoplankton pigments is determined using empirical regressions of the ratio of the water-leaving radiance at two wavelengths in the visible range. Usually regression has the form (O'Reilly et al., 1998):

$$C = 10^{f(R)} + a$$

where C is the concentration of phytoplankton pigments, $R=\log(L_1/L_2)$, L_1 and L_2 are the water-leaving radiance measured on two wavelengths, $f(R)$ is a linear, quadratic or cubic polynomial. Coefficients of the polynomial are found statistically using the huge sets of *in situ* data collected in different regions of the ocean (O'Reilly et al., 1998). The constant a is either zero or is also found statistically.

The algorithm described above is applied to regular mapping of phytoplankton pigments field in the World Ocean. Composite maps are used to fill gaps caused by clouds on each image. An example of a monthly composite map is presented on Figure 15 where the major features of the global phytoplankton pigments distribution are clearly seen. The red color corresponds to the waters richest in chlorophyll and the dark blue to waters with very little phytoplankton content.

Figure 15. Monthly global phytoplankton pigment map reconstructed using SeaWiFS data.

Phytoplankton pigments determine the spectral reflectance of the open-sea waters (so called Case 1 water). Water near the coasts (Case 2 water) has more complicated optical properties since light is also scattered add by terrigenic suspended particles and absorbed by CDOM (coloured, dissolved organic matter). Therefore the correspondence between phytoplankton pigments and the spectral reflectance of the sea is distorted. Accurate separation of terrigenic particles and biogenic materials in Case 2 water is a complicated problem discussed, for example, by Carder et al. (1991).

Retrieving of the sea surface temperature using IR observations

The emissivity coefficient of water is almost a constant in the thermal IR band. Therefore SST determines the radiance emitted by the sea surface.

SST measurements from space are carried out in the transparency windows where the atmospheric influence is rather low. However the absorption in the atmosphere (mainly by the water vapor) and the radiance emitted by the atmosphere directly upward, or reflected from the sea surface, still distort the signal received on the satellite. Fortunately a simple linear algorithm that uses two or three measuring channels excludes distortions (Stewart, 1985). The accuracy of the SST determination using linear algorithms is about 0.5 degree C. Clouds restrict the mapping of SST using IR radiometry so a snapshot of the ocean surface contains many gaps and the SST map is constructed by means of spatial averaging or data collection over a few days. An example of global SST map resulting from a one-day survey is presented in Figure 16.

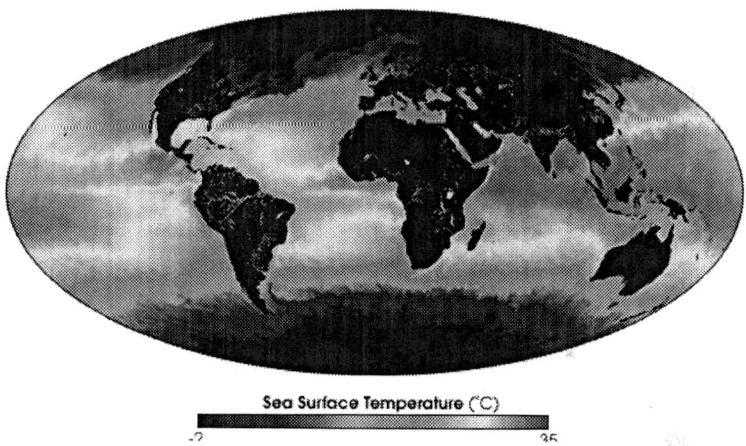

Figure 16. Global SST map obtained from MODIS data

Absorption of solar short-wave radiation, which peaks at about 0.5 μm, heats the surface layer of the ocean. The visible and IR band radiometers installed on geostationary satellites are used to estimate the short-wave radiation flux to the ocean as the difference between the incoming solar radiation and that reflected back by clouds and atmosphere (Stewart, 1985).

Multi-parametric microwave radiometry

Radio frequencies have an advantage over the visible and IR bands since even a cloudy atmosphere is almost transparent in this range. Nevertheless the total water vapor content and water content in clouds should be taken into consideration for accurate estimation of the sea surface state. The emissivity of the sea surface depends on the surface temperature, salinity (for decimetric waves) and the foam coverage, which is a function of the surface wind. Therefore the sea surface wind, temperature and salinity can be retrieved from the multi-channel microwave remote sensing.

Bernstein (1982) has demonstrated the determination of SST using multi-channel microwave measurements. The accuracy of microwave SST is about 1K, i.e. it is lower than that obtained from IR data. The spatial resolution of microwave radiometery is also rather coarse compared to infrared. However microwave observations are available in the areas covered by clouds.

Examples of sea surface winds determined using passive microwave measurements are presented by Atlas et al. (1996). The surface wind derived from the microwave data, together with climatological fields, are presented in Figure 3 and Figure 5 of their paper. The maps show good consistency testifying to the quality of the remote sensing data.

Spaced-based measurements of the sea surface salinity are not yet available. However the brightness temperature at wavelengths of about 70 – 80 cm is sensitive to the salinity variations. Therefore there is a real possibility to develop an appropriate instrument for measuring the sea surface salinity from space. Encouraging experiments are currently being conducted from aircraft (Le Vine, et al., 1998). Le Vine et al. (1998) show a good correspondence between aircraft and *in situ* observations of sea surface salinity. Therefore there is a hope that the salinity measurements from space soon will be available for the oceanographic community.

The major problem of the centimetric and decimetric passive radiometry is the spatial resolution on the sea surface. Now it is about ten km for centimetric waves, that is insufficient to resolve mesoscale features of SST field. Measurements of the surface salinity are based on use of decimetric waves and the spatial resolution is anticipated to be about 50 km.

Scatterometric three beam system

The amplitude of surface waves at the centimetric wavelength is sensitive to the variability of the surface wind. Therefore measurements of the backscatter by centimetric waves provides information about the intensity of the surface wind field. Centimetric waves propagate mainly along the wind and their two-dimensional spectrum is almost symmetric with respect to the wind direction. The perfect scattering diagram of radio waves is also symmetric relative to the wind direction, whereas the scattering intensity is a function of the wind speed. Thus measurements of backscatter at a few angles makes it possible to reconstruct the scattering diagram and to evaluate a wind velocity. This principle is used by scatterometers to map the surface wind field. Modern scatterometers are based on a three-beam system to reduce ambiguity in the reconstruction of the scattering diagram. Real scattering diagrams are contaminated by multiple systems of wind waves or by swell. Nevertheless, surface wind determination is rather accurate.

Accurate altimetric measurements

Satellite-borne altimeters measure the distance from a satellite to the ocean surface from a height of 1200km. A set of corrections must be applied (Table 2) to evaluate the distance from the satellite to the sea surface and to simulate the sea surface topography.

Table 2. List of corrections applied for altimeter measurements.

Source of Error	Uncorrected (Seasat) (cm)	Corrected (Seasat) (cm)	Corrected (Topex) (cm)	Wavelength (km)
Geoid	100(m)	100-200	10-50	200-40,000
Orbits	5 (km)	100-200	5-10	10,000
Coordinate system	100-200	100-200	10	10,000
Ionosphere	0.2-20	0.2-5	1.3	20-10,000
Mass of air	230	0.7	0.7	1000
Water vapor	6-30	2.0	1.2	50-1000
Atmospheric pressure	4	2	2.0	100-1000
Surface waves	5	5	2.0	0.005-0.1
Tides	10	4	10	100-1000
Steric oscillations	50	5-10	2.0	1000
Altimeter(noise)	5	5	2.0	6-20

Unfortunately, absolute determination of the sea surface elevation is currently impossible due to the low accuracy of knowledge of the geoid. However the anomaly of the sea surface elevation, relative to the time mean, is estimated from altimeter measurements with rms accuracy of about 4cm, permitting measurement of tides and the sea level oscillations in open ocean (Koblinsky et al. 1999, Chernyavsky, Foreman, Crawford and Henry, 2001) and in semi-enclosed basins (Ducet et al. 1999).

However, the major advantage of the altimeter measurements is the ability to map sea surface elevation from which surface currents can be inferred. The sea level gradient at low frequencies characterizes so-called gradient currents. The geostrophic relations

$$-f \cdot v = g \cdot \frac{\partial \varsigma}{\partial x} \qquad f \cdot u = g \cdot \frac{\partial \varsigma}{\partial y},$$

where u, v are components of the current velocity, f is the parameter of Coriolis, g is the gravitational constant, and ς is the sea level, connects the slope of the sea level and the surface gradient currents everywhere excluding the narrow band near the equator. Thus sea level is related to the large-scale dynamic topography of the sea surface. The altimeter measurements can be used to infer the surface flow. On the large scale, they reproduce major oceanic gyres (Fu and Smith 1996).

Application of the dynamic method shows that

$$\zeta \approx \frac{1}{\rho_0} \int_{-D}^{0} \rho \cdot dz ,$$

where under the integral is the anomaly of the water density. Therefore the sea level reflects the deep-ocean processes, such as mesoscale eddies, meanders of currents and planetary waves. The root-mean-square elevation of the sea surface shows the spatial distribution of energy on the mesoscale.

3.2 DATA TRANSMISSION THROUGH SATELLITES

Application of satellites as data transmitters constitutes a significant contribution to the development of oceanographic observing system. Data is received from the anchored buoys, ecological observing stations, automatic meteorological stations and other autonomous devices, and relayed to ground stations.

Surface drifting buoys

One application involves drifters, i.e. buoys free-drifting on the sea surface. The essence of the method consists in via-satellite transmission of the buoy coordinates, barometric pressure, temperature of air and sea, optical properties and biological characteristics measured by various sensors. Meteorological drifters are equipped with a tall mast with meteorological gauges that distort the Lagrangian transport of the buoy. In contrast oceanographic drifters have minimal above-water parts. Some of them are equipped with an underwater sail and are transported by surface currents.

The positioning of the buoy at two consecutive moments of time makes it possible to evaluate the average velocity of its motion. More than thousand of surface drifting buoys are now launched in the oceans beginning with the TOGA and WOCE programs. Observations by drifting buoys are carried out also in the marginal seas. Figure 17 shows trajectories of surface drifting buoys in the Black Sea collected during international experiments in 1999-2001. Data from drifters is used to study currents and heat transport in the surface layer of the sea.

Profiling floats

A new possibility for the observation of deep-sea processes is provided by profiling floats which drift at a prescribed depth or density and rise to the surface at prescribed time intervals. As they ascend they measure profiles of temperature and salinity. This data, and the co-ordinates of the float, are transmitted via satellite. The knowledge of coordinates at two consecutive moments allows estimation of the average velocity on the float parking depth. A few hundred of such profiling floats are currently launched in the World Ocean in framework of the ARGO project.

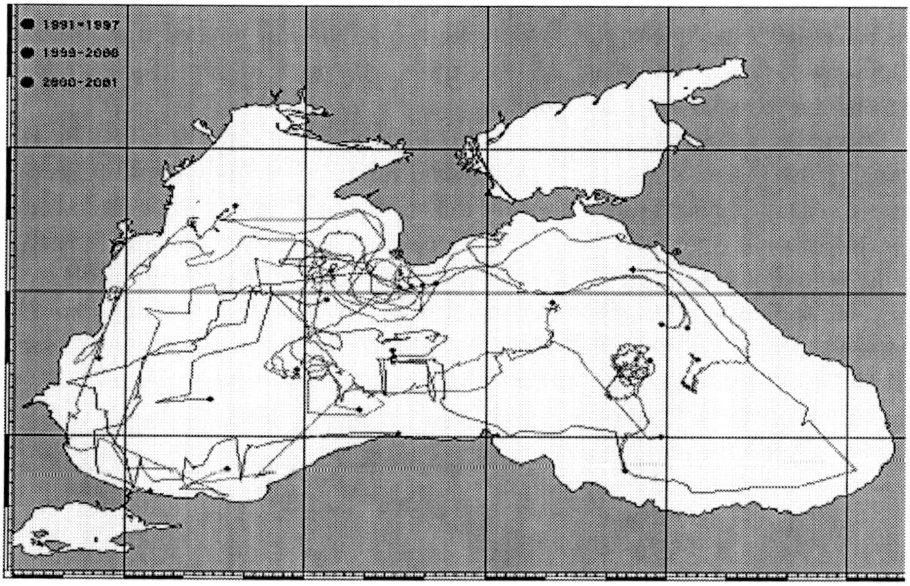

Figure 17. Surface drifting buoy trajectories in the Black Sea.

3.3 APPLICATION OF REMOTE SENSING TO THE MODELING OF OCEANIC DYNAMICS

Remote measurements from space provide information on practically all boundary conditions on the ocean surface that are necessary for running of global ocean circulation models. Scatterometers or passive microwave radiometers measure the momentum fluxes at the sea surface that are major driving force of the oceanic circulation. Measurements of surface temperature and salinity from space provide thermodynamic boundary conditions for the global circulation models. Altimeter data characterize the large-scale and mesoscale processes occurring deeper in the ocean. The assimilation of altimeter measurements in global circulation models increases accuracy of simulations and reproduces a more realistic circulation. Additional information is transmitted via satellite from different platforms. Among these, profiling floats are the most important supplementing space-based altimetry since they also register the variability of temperature and salinity that is not reflected in the topography of the sea surface.

Models can reproduce continuous evolution of currents and deep-sea hydrography using realistic boundary conditions measured from spacecrafts and assimilating deep-sea data of profiling buoys. The Global Ocean Data Assimilation Experiment (GODAE) has a goal to run general circulation models with assimilation of all available data and to reproduce three-dimensional circulation in real-time.

Due to the simplicity of their lateral boundary conditions, the land-locked seas like the Black Sea are good candidate regions for models assimilating remote sensing data. Figure 18 shows the results of altimetry assimilation in a basin-scale model of the Black Sea circulation. The cyclonic Rim Current, being the major feature of the Black Sea circulation, is clearly seen in Figure 18. The application of realistic boundary conditions and assimilation of deep-sea data in the model make it possible to continuously monitor mesoscale eddies and waves (Figure 18).

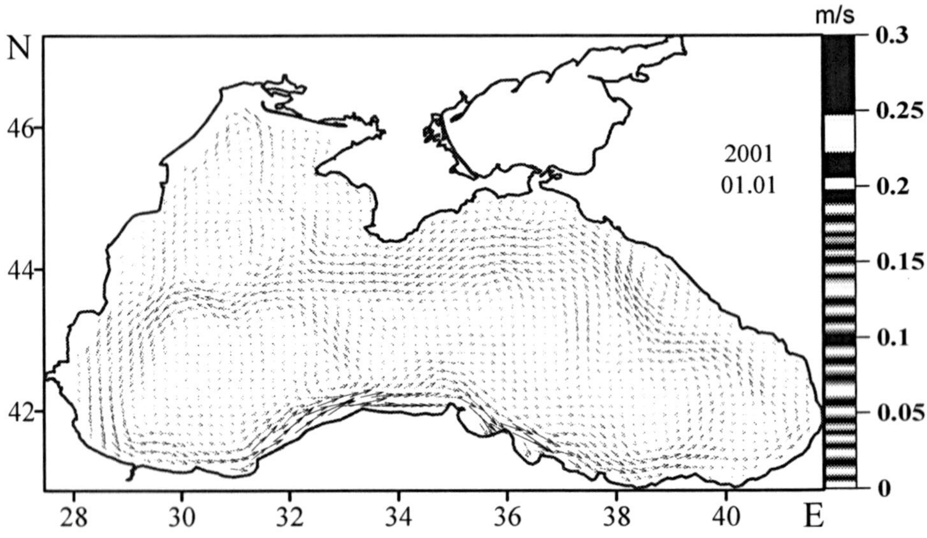

Figure 18. Surface circulation in the Black Sea estimated by assimilation of Topex/Poseidon and ERS altimeter data in a numerical ocean model.

The fields obtained after assimilation ensure a good model initialization for the following forecast of the basin dynamics.

Thus remote sensing data, together with data transmitted via satellites, provide the required boundary conditions and certain constraints on the deep-sea circulation that are sufficient for realistic simulation of the three-dimensional sea state. Realistic circulation and three-dimensional fields of temperature and salinity also constitute the base for the development of the pollution transport models, ecosystem models and coupled models of air-sea interaction.

4. SUMMARY

This brief overview shows that remote sensing methods are now a powerful tool for oceanographic research. The launch of near operational oceanographic spacecraft makes it possible to develop a new ocean observing system, which is similar to, and may later even exceed in quality, the standard meteorological network. The development of the new observing system is coordinated by the international program GOOS (Global Ocean Observing System). Improvement of the circulation models and assimilation technique is being carried out by the GODAE project.

The new observing system allows continuous monitoring of the three-dimensional ocean dynamics. Archiving of the system output makes it possible to trace climatic changes of the World Ocean. Accurate knowledge of the global-scale circulation is the base of the development of ocean ecosystem and biogeochemistry models; a key issue in the CO_2 problem.

However the quantitative measurement of oceanic fields is not the only application of remote sensing data which contains very rich qualitative information about the World Ocean, and is the basis for new discoveries.

REFERENCES

Alves, J. O. S., K. Haines, D. L. T. Anderson, Sea Level Assimilation Experiments in the Tropical Pacific. J. Phys. Oceanogr., 31, 2001.

Atlas, R., R. N. Hoffman, S. C. Bloom, J. C. Jusem, and J. Ardizzone, A Multiyear Global Surface Wind Velocity Dataset Using SSM/I Wind Observations. Bull. Amer. Meteor. Soc., 77, 1996.

Bernstein R. L., Sea Surface Temperature Mapping with the SEASAT Microwave Radiometer. J. Geophys. Res., 87, 7865 – 7872, 1982.

Qiu, B. Seasonal Eddy Field Modulation of the North Pacific Subtropical Countercurrent:TOPEX/Poseidon Observations and Theory. J. Phys. Oceanogr., 29, 1999.

Carder,K.L., S.K.Hawes, K.A.Baker, R.C.Smith, R.G. Steward, and B.G. Mitchell, Reflectance model for quantifying chlorophyll a in the presence of productivity degradation products. J. Geophys. Res., 96, 20599-20611, 1991.

Chernyavsky, J. Y., M. G. G. Foreman, W. R. Crawford, R. F. Henry, Ocean Tides from TOPEX/Poseidon Sea Level Data. J. Atm. Ocean. Tech., 18, 2001.

Dorandeu, J., P. Y. Le Traon, Effects of Global Mean Atmospheric Pressure Variations on Mean Sea Level Changes from TOPEX/Poseidon. J. Atm. Ocean. Tech., 16, 1999.

Ducet, N., Le Traon, P.Y., Gauzelin, P., Response of the Black Sea level to atmospheric pressure and wind forcing. J. Mar. Syst., 22, 311-327, 1999.

Emery, W. J., S. Castro, G. A. Wick, P. Schluessel, and C. Donlon, Estimating Sea Surface Temperature from Infrared Satellite and In Situ Temperature Data. Bull. Amer. Meteor. Soc., 82, 2001.

Fu, L.-L. and R. D. Smith, Global ocean circulation from satellite altimetry and high-resolution computer simulation. Bull. Amer. Meteor. Soc., 77, 2625–2636, 1996.

Gordon H.R and M. Wang, Retrieval of water-leaving radiance and aerosol optical thickness over the oceans with SeaWiFS: a preliminary algorithm. Appl. Optics. 33, 443 – 452, 1994.

Ishumaru A., Wave Propagation and Scattering in Random Media Vol. 1 Single Scattering and Transport Theory. Acad. Press, 1978.

Legeckis R. and T. Zhu, Sea Surface Temperatures from the GOES-8 Geostationary Satellite. Bull. Amer. Meteor. Soc., 79, 1997.

Le, D. M. Vine, M. Kao, R. W. Garvine,T. Sanders, Remote Sensing of Ocean Salinity: Results from the Delaware Coastal Current Experiment. J. Atm. Ocean. Tech., 15, 1998.

Monaldo F. M., T. D. Sikora S. M. Babin and R. E. Sterner, Satellite Imagery of Sea Surface Temperature Cooling in the Wake of Hurricane Edouard, 1996. Monthly Weather Review, 125, 1997.

O'Reilly, J.E., Maritorena S., Mitchell B.G., Siegel D.A., Carder K.L., Garver S.A., Kahru M., and C. McClain, Ocean color chlorophyll algorithms for SeaWiFS. J. Geophys. Res., 103, 24937 – 24953, 1998.

Porter D. L. and D. R. Thompson, Continental Shelf Parameters Inferred from SAR Internal Wave Observations. . J. Atm. Ocean. Tech., 16, 1999.

Sikora T. D., K. S. Friedman W. G. Pichel and P. Clement-Colon, Synthetic Aperture Radar as a Tool for Investigating Polar Mesoscale Cyclones. Weather and Forecasting, 15, 2000.

Stammer D., Global Characteristics of Ocean Variability Estimated from Regional TOPEX/POSEIDON Altimeter Measurements. J. Phys. Oceanogr., 27, 1997.

Stewart R., Methods of Satellite Oceanography. Univ. of California Press, 1985

White W. B., Evidence for Coupled Rossby Waves in the Annual Cycle of the Indo–Pacific Ocean. J. Phys. Oceanogr., 31, 2001.

Wu, X., W. P. Menzel, and G. S. Wade, Estimation of Sea SurfaceTemperatures Using GOES-8/9 Radiance Measurements. Bull. Amer. Meteor. Soc.,10, 1999.

MECHANISMS CONTROLLING THE AIR-SEA FLUX OF CO_2 IN THE NORTH ATLANTIC

MICK FOLLOWS
Program in Atmospheres, Oceans and Climate
Department of Earth Atmosphere and Planetary Sciences
Massachusetts Institute of Technology, Cambridge, MA, USA
mick@ocean.mit.edu

RICHARD G. WILLIAMS
Oceanography Laboratories
Department of Earth and Ocean Sciences
University of Liverpool, Liverpool, England
ric@liv.ac.uk

INTRODUCTION

The air-sea flux of carbon is controlled by the disequilibrium in partial pressure of carbon dioxide between the atmosphere and surface ocean. This disequilibrium is a consequence of the interactions of physical, chemical and biological processes in the ocean and, today, includes a response to the anthropogenic increase of atmospheric pCO_2. Fig. 1 illustrates the annual mean air-sea flux of carbon, $\mathcal{F}$, estimated from a knowledge of the atmospheric partial pressure, pCO_2^{at}, and compilation of surface pCO_2 observations by Takahashi *et al.* (1999). The air-sea flux of carbon is determined by

$$\mathcal{F} = -K_g K_0 (pCO_2 - pCO_2^{at}), \tag{1}$$

where K_0 is the solubility of CO_2 at local temperature and salinity. K_g is the air-sea gas transfer coefficient, which is dependent on local environmental conditions and is usually parameterized as a function of wind speed, sea-surface temperature and sea-surface salinity (Wanninkhof, 1992). The major global scale features in Fig. 1 are the outgassing of CO_2 from the tropical oceans, and the influx at mid and high latitudes. In this chapter we focus on understanding what sets the basin wide, and regional patterns of air-sea carbon flux in the North Atlantic basin. While we focus on the North Atlantic, some of the concepts and discussions are also relevent to other regions of the ocean.

M. Follows and T. Oguz (eds.), The Ocean Carbon Cycle and Climate, 217–249.

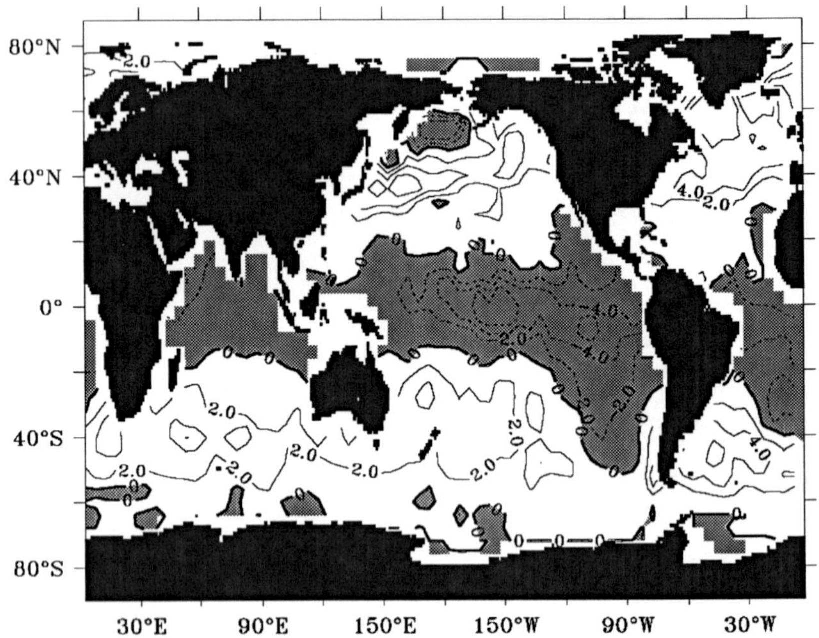

Figure 1. Air-sea flux of carbon, $mol\ m^{-2}\ yr^{-1}$ estimated from a global compilation of surface pCO_2 observations by Takahashi *et al.* (1999). Shaded areas indicate outgassing. The map is derived from a compilation of pCO_2 observations made over a number of years and the pattern has been normalized to the situation in 1990. However, the map still reflects a transient situation due to fossil fuel burning and the rapid increase in pCO_2^{at} in the past century (Takahashi *et al.*, 1999).

In Section 1, we briefly describe the distribution of carbon in the Atlantic basin and its relation to other water-mass properties. Then, in Section 2, we discuss how the large scale overturning circulation helps to drive a net basin flux of carbon into the ocean over the basin. We discuss the connections between the sub basin-scale patterns of the air-sea carbon flux, heat flux, and biological production in Section 3 and formalize these relationships in Section 4, where we describe the surface ocean carbon system following a water parcel moving through the surface Atlantic Ocean. In Section 5, we illustrate how the long timescale for equilibration of mixed-layer carbon with the atmosphere impacts the air-sea flux patterns using a numerical model. We discuss the role of the subduction process and the importance of the reservoir of carbon within the ventilated thermocline in Section 6. In Section 7, we speculate on the importance of time-varying, eddy circulations for the air-sea flux of carbon and, in Section 8, we briefly discuss the role of interannual variations.

1. THE DISTRIBUTION OF DISSOLVED INORGANIC CARBON IN THE ATLANTIC

There is a general increase in the concentration of dissolved inorganic carbon (DIC) with depth in the oceans (Fig. 2a). This storage in the thermocline and deep waters, away from the surface ocean and atmosphere, is achieved through a combination of physical and biological processes, referred to as 'pumps'. The total amount of dissolved carbon in a water parcel is enhanced due to the chemistry of the carbonate system in seawater (Murray, this volume). At equilibrium with a fixed atmospheric partial pressure, cooler waters can hold more dissolved inorganic carbon, DIC, than warm waters. Hence there is some increase in DIC with depth because cool, dense high latitude surface waters form the deep waters of the ocean (Fig. 2b). This effect is referred to as the *solubility pump* (Volk and Hoffert, 1985).

While the DIC distribution does reflect the undulations of the thermocline, marked by the enhanced vertical gradient in T over the upper 1 km, there are very large-scale interleaving patterns of DIC in the deep waters which are not prominent in T. However, similar patterns are revealed in salinity, S (Fig. 2c); a conservative tracer away from the surface of the ocean. For example, salty North Atlantic deep water spreads southwards from the deep water formation regions in the Labrador and Greenland seas at depths between 1000 and 3000 m. Fresher water spreads northwards at intermediate depths, as Antarctic Intermediate Water (around 1000 m), and as Antarctic Bottom Water along the seafloor. The thicker subtropical thermoclines are also highlighted by their saline character, due to strong net evaporation from the warm surface waters. These features are echoed by the DIC distribution.

This general correspondence between the distributions of DIC and S in the deep waters reflects transport of water masses from their point of subduction (see Section 6) and ventilation through the global oceans. In contrast to salinity, however, there is an interior source of DIC leading to enhanced vertical, and basin to basin, gradients. This interior source is due to biological processes and is termed the *biological pump*. In the process of photosynthesis, phytoplankton convert dissolved inorganic carbon and nutrients (such as nitrate, phosphate and iron) into organic matter (Anderson and Totterdell, this volume). This conversion occurs in the euphotic zone, typically the upper 100 m or so of the ocean where there is sufficient solar radiation. While the organic matter is cycled and transformed within the pelagic ecosystem, ultimately a fraction either sinks to depth as detritus or fecal matter, or is transported downwards in dissolved organic form (Hansell and Carlson, 1999). The exported organic matter is remineralized to inorganic form at depth, thus increasing the vertical contrast in DIC. This transfer is the *soft tissue pump*. There is an additional contribution from the calcium carbonate structural material formed by certain planktonic species termed the *carbonate pump* (Volk and Hoffert, 1985).

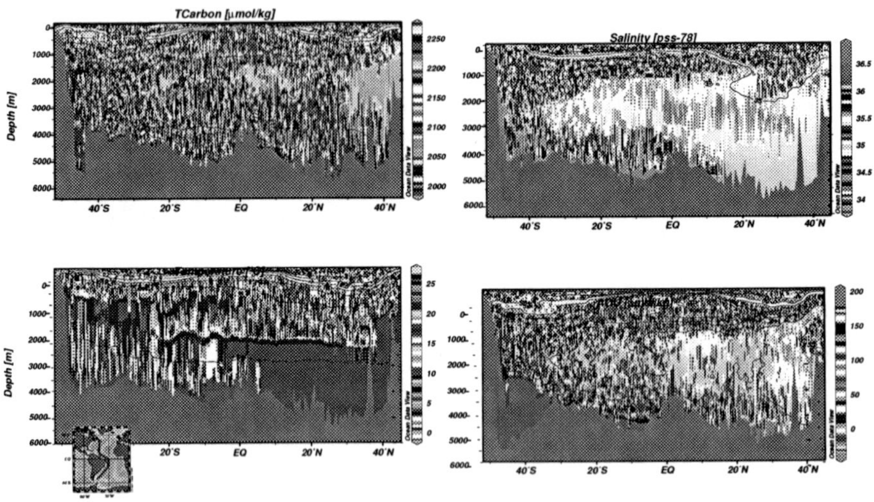

Figure 2. Atlantic meridional distributions of **(a)** dissolved inorganic carbon ($\mu mol\ kg^{-1}$), **(b)** temperature (^oC), **(c)** salinity (psu), and **(d)** apparent oxygen utilization ($\mu mol\ kg^{-1}$). Data are from the WOCE-JGOFS (World Ocean Circulation Experiment-Joint Global Ocean Flux Study) global survey. (Plotted using Ocean Data View.)

The remineralization of organic matter is accompanied by a consumption of oxygen. The stoichiometry of this conversion is reasonably well defined and dissolved oxygen is generally close to saturation in the surface waters. Hence the amount of organic matter which has been remineralized in a water parcel can be inferred from the "apparent oxygen utilization", defined $AOU = [O_2]_{sat}(T,S) - [O_2]$. Here $[O_2]$ is the *in situ* concentration of dissolved oxygen and $[O_2]_{sat}(T,S)$ is the saturated concentration, the concentration the water parcel is assumed to have had at the time it left the surface waters and inferred from its T and S. AOU has a strong maximum between 200 and 2000 m (Fig. 2d), the layer where most of the sinking organic matter is remineralized. AOU is also high in the older, deeper waters of the ocean due to the long term accumulation of remineralized carbon. The DIC and AOU distributions in Figs. 2a and d show a strong similarity due to the significant influence of the soft tissue pump on the DIC distribution.

In brief, the ocean DIC distribution has strong relationships with temperature, conservative water mass tracers like salinity, and indicators of biological processes such as AOU. It is clear that the solubility and biological pumps both are significant in setting the carbon distribution in the ocean. Both pumps are intimately linked with water masses and transport by the large-scale ocean circulation.

2. THE NET AIR-SEA FLUX OF CARBON OVER THE NORTH ATLANTIC

In the annual mean there is a net air-to-sea flux of carbon over most of the North Atlantic, with a small region of net outflux in the tropics, as depicted in (Fig. 1). Takahashi *et al.* (1999) estimated a net, basin wide air-sea flux of CO_2 (normalized to 1990) of about 0.7 $GtCy^{-1}$ into the Atlantic Ocean north of the Equator. The surface waters do not generally reach equilibrium with the overlying atmosphere, $pCO_2 \neq pCO_2^{at}$, due to the oceanic transport of DIC and the continual experience of thermal and biological forcing, including the seasonal cycle (e.g. Brostrom, 2000).

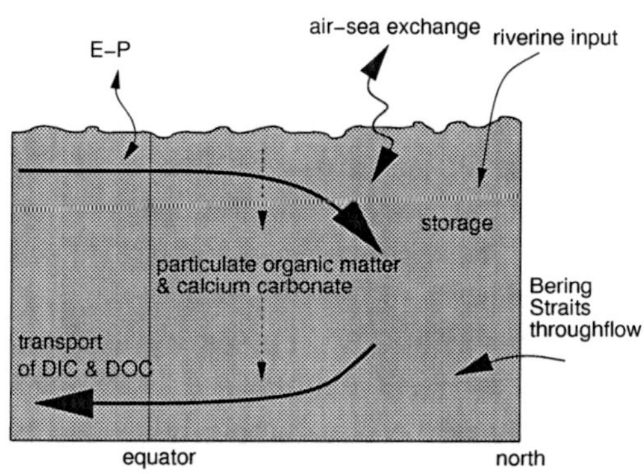

Figure 3. Schematic figure indicating the processes and fluxes which control the carbon distribution and net air-sea flux in the North Atlantic basin. Consider the box in the schematic to represent the North Atlantic basin from equator to its Northern extent. There is a net flux of carbon across the sea surface which is balanced by a combination of local storage, large scale transport of DIC and other processes, including riverine inputs, flow through the Bering Straits and transport of dissolved organic carbon, DOC.

The basin scale uptake of carbon is tied to the thermal and freshwater forcing of the region. Warm, low solubility surface waters are transported Northwards in the Atlantic and cooled, inducing the oceanic uptake of carbon, reinforced by biological transfer carbon and nutrients to the deep ocean. The contrast in thermal and freshwater forcing over the basin also supports the overturning circulation so the cold, carbon-rich deep water is transported southwards out of the North Atlantic (Fig. 3). Hence there is a net transport of carbon to the south at low latitudes. Holfort *et al.* (1998) estimated the net southwards transport of carbon at 20°S to be 2150 $\pm$ 200$kmol\ s^{-1}$ (or about 0.8 $Gt\ C\ yr^{-1}$) from several observed transects of hydrography and DIC between 11°S and 30°S. Fig. 4 illustrates the depth dependence of the carbon transport (at 30°S) which, since the vertical change in DIC is small relative to the ocean mean value, closely follows the structure of the mass flux associated with the overturning circulation. The estimate of net southward transport is of comparable magnitude to the basin wide air-sea flux estimated by Takahashi *et al.* (1999). However Lundberg and Haugan (1996) estimate a flux of carbon into the Atlantic from the Bering Straits and through the Arctic ocean also of similar magnitude, about 0.8 $Gt\ C\ yr^{-1}$, associated with a net volume flux of about $1Sv$. Hence the flow out of the North Atlantic to the south seems to be roughly balanced by flow in from the north and other processes must be invoked to balance the air-sea exchange in the budget including the accumulation of carbon within the basin, the effect of net surface freshwater fluxes, and riverine inputs of carbon and freshwater.

In addition, the transport and transformations of particulate and dissolved organic carbon (POC and DOC respectively) may be significant but are not yet quantified. Surface DOC concentration varies seasonally, between 40 and 60 $\mu mol\ kg^{-1}$ near Bermuda (Hansell and Carlson, 1999). At depth, it is relatively uniform at about 40 $\mu mol\ kg^{-1}$. The opposing vertical gradient suggests a DOC transport associated with the overturning circulation which may oppose that of DIC. The regional impact of DOC transport is, however, yet to be quantified.

The observed air-sea flux (Fig. 1) includes a contribution due to the air-sea disequilibrium caused by fossil fuel burning, on top of the pre-industrial pattern. The signature of the anthropogenic perturbation on oceanic DIC has been estimated (Gruber, 1998; Sabine et al. 1999). The role of the North Atlantic in responding to the atmospheric increase is currently unclear. As the North Atlantic is generally a sink of carbon from the atmosphere we might expect that is absorbs anthropogenic carbon accordingly. However, inversions of transport estimates (Wallace, 2001) and direct flux determinations (Lefevre et al., 2003) suggest that there might be a local loss of anthropogenic carbon from ocean to atmosphere. This unexpected inference could be due to a horizontal supply of anthropogenic carbon, absorbed elsewhere in the global ocean, and transported into the North Atlantic. Sarmiento *et al.* (2003) illustrate the importance of Southern Ocean mode waters as a source of nutrients to the North Atlantic basin: transport in this water mass might also be significant for the anthropogenic carbon budget.

In summary, while there is uncertainty in terms of the anthropogenic carbon budget and the cycling of DOC, the climatological picture of the basin-scale uptake of CO_2 from the atmosphere is broadly in accord with what is expected from the surface thermal and biological forcing and the DIC transport by the overturning circulation. In the following sections we discuss what sets the regional, sub-basin scale patterns of the air-sea flux (as shown in Fig. 5a) which reflect the impact of the gyre-scale circulation.

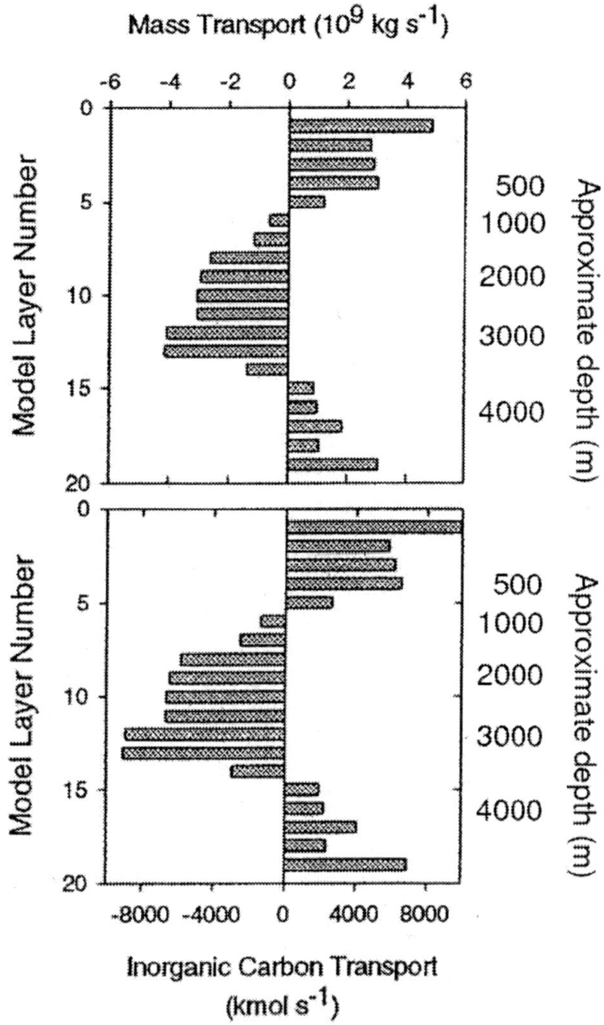

Figure 4. Zonally-integrated fluxes of mass and dissolved inorganic carbon (upper and lower panels) as a function of depth at 30°S in the Atlantic basin from Holfort *et al.* (1998). Note how the mass and carbon transports have a similar vertical structure with a northwards flux of surface and bottom waters, and a southwards flux of North Atlantic Deep Water. Figure adapted from Holfort *et al.* (1998).

3. REGIONAL PATTERN OF THE AIR-SEA CARBON FLUX

The regional variation of the air-sea flux of CO_2 in the North Atlantic basin is shown in Fig. 5a from the estimate of Takahashi *et al.* (1999) - a detail of Fig. 1. As noted earlier, there is an uptake of CO_2 by the oceans over almost all of the basin northwards of 10°N with outgassing in the tropics. This pattern reflects the interaction of solubility effects, biological processes and the gyre scale circulation of the North Atlantic. Here we discuss these sometimes competing influences.

3.1 AIR-SEA HEAT FLUX AND SOLUBILITY FORCING

The regional pattern of the air-sea flux of CO_2 partly reflects that of the air-sea heat flux in which an uptake of CO_2 corresponds with a surface cooling, compare Fig. 5a and b (the latter as compiled by Josey *et al.*, 1998). The equilibrium surface ocean concentration of DIC, for a given atmospheric pCO_2 varies linearly with sea surface temperature. Hence the air-sea flux of heat drives changes in the surface equilibrium DIC concentration. The two maps reveal strong ocean uptake of CO_2 (unshaded) in the regions of strong oceanic heat loss (unshaded) over the Gulf Stream system and the subpolar gyre. On the integrated basin scale, this connection has been exploited by Watson *et al.* (1995) to estimate the air-sea CO_2 flux from the observed basin scale heat transport (together with estimates of nutrient and alkalinity changes accounting for the biological contributions).

However, the North Atlantic air-sea flux of CO_2 differs in detail from the pattern of the surface heat flux. For example, over the subtropical ocean, there is an oceanic uptake of CO_2, even in regions of a weak surface heat uptake (which is acting to decrease the solubility).

So why does the subtropical gyre of the North Atlantic absorb CO_2 since it is not a direct response to local heat forcing? We will discuss several possible explanations: Biological drawdown, the increase of atmospheric CO_2, and the role of physical transport of carbon in the ocean.

3.2 BIOLOGICAL DRAWDOWN OF CO_2

The distribution of chlorophyll-a in the surface North Atlantic is shown in Fig. 5c, revealing vigorous biological production at high latitudes and close to the eastern boundary. Much of the associated primary production (a measure of the rate of formation of organic matter by photosynthesis) is associated with locally recycled carbon and nutrients. A small fraction however, termed export production, is transferred into the deep ocean through a combination of sinking of organic particles and advection of dissolved organic matter. Export production reduces the surface DIC concentration and is expected to lead to an uptake of CO_2 over the whole basin. This contribution reinforces the local cooling and increase in solubility at higher latitudes and may be significant in the tropics and coastal waters, where it opposes the local tendency in solubility due to local warming. Due to variations in the rate of local recycling, export production is probably less variable across the basin relative to primary pro-

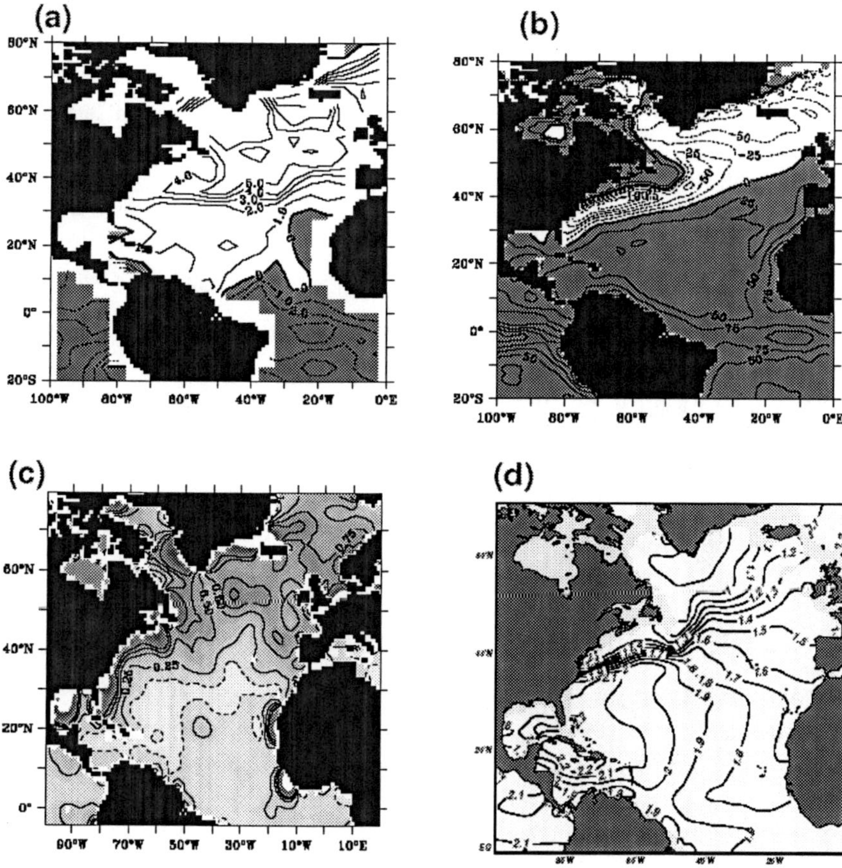

Figure 5. (a) Air-sea flux of CO_2, $mol\ m^{-1}\ yr^{-1}$ (Data from Takahashi *et al., 1999*). Shaded regions correspond to a loss of CO_2 to the atmosphere. (b) Annual mean air-sea flux of heat, Wm^{-2}, in the North Atlantic (Data from Josey *et al.*, 1998). Negative values (unshaded) denote a loss of heat to the atmosphere and a corresponding increase in solubility of CO_2. The difference in the extent of the shaded regions between (a) and (b) indicates the regions where carbon and heat flux are decoupled. (c) Climatological surface ocean chlorophyll-a concentration ($\mu g\ l^{-1}$) from the World Ocean Atlas (Conkright *et al.*, 2001). The chlorophyll distribution suggests vigorous biological productivity and the strong spring bloom in the subpolar regions, and the nutrient limited subtropical gyre, broadly corresponding to the patterns of gyre-scale upwelling and downwelling. (d) Dynamic height relative to 2000m (dyn m) in the North Atlantic, derived from the climatological hydrography of the WOA01 atlas (Conkright et al., 2002). Upper ocean geostrophic circulation approximately follows the contours and is inversely proportional to the distance between them. The figure reveals the anti-cyclonic flow of the subtropical gyre, the swift flow close to the western boundary (tight contours) and the cyclonic subpolar circulation.

duction (Emerson et al., 1997). In the subtropics the rate of export production is ultimately limited by the slow rate of nutrient supply to the surface. In addition, nutrients transported from the deep ocean to the surface may bring with them excess carbon of biogenic origin, sufficient to provide the necessary carbon for organic matter without requiring a flux across the sea surface (Lewis 1992). This transfer of biogenic carbon is tied up with the concept of fixed stoichiometry for biological transformations and the Redfield ratio (see Anderson and Totterdell, 2003, this volume) and we discuss this further in the next section.

3.3 ATMOSPHERIC TRANSIENT OF CO_2

Over the past century atmospheric pCO_2 has been increasing due to fossil fuel emissions at a (variable) rate of about 1 $ppmv\ yr^{-1}$. To what extent is the uptake of CO_2 by the subtropical gyres due to the rapid increase in atmospheric pCO_2^{at}? While this factor might increase the air-sea flux in those regions, numerical models consistently suggest there would have been a flux of CO_2 into the pre-industrial subtropical gyres. For example, the preindustrial simulations in all of the OCMIP (Ocean Carbon-cycle Model Intercomparison Project) models summarized by Orr (2002).

3.4 DELAYED RESPONSE OF CO_2 FLUX

In our view there is another important factor driving the uptake of CO_2 over the subtropical gyres; the long equilibration timescale for mixed layer carbon which causes the air-sea flux response to take place downstream of the regions of strong solubility and biological forcing.

The upper ocean circulation of the North Atlantic can be inferred (Fig.5d) from the dynamic height of the water column relative to 2000m. The figure reveals the anti-cyclonic, subtropical gyre and the intense western boundary current including tight, inertial recirculations. Part of the fluid in the western boundary recirculates around the subtropical gyre, while the rest is swept further north into the cyclonic, subpolar gyre, where it is cooled and converted to a denser water mass. With respect to the subtropical gyre, the regions of most intense solublility and biological forcing are associated with the region of the Gulf Stream, with intense cooling, and the gyre margins, where there is significant biological drawdown. The anomalies generated by these processes in the margins are transported into the interior of the gyre in the anti-cyclonic gyre flow where they are gradually eroded by the air-sea flux.

The significance of this delayed response is supported by numerical models which have a lagged response to cooling over the Gulf Stream (Sarmiento *et al.*, 1995) which extends downstream into the interior of the subtropical gyre (Follows *et al.*, 1996; 2002). In the next Section we will explore in detail the implications of this delayed response of the air-sea carbon flux to surface forcing. We illustrate the concepts using a Lagrangian frame of reference, where one follows a hypothetical water column as it moves around the North Atlantic.

4. A LAGRANGIAN VIEW OF AIR-SEA CARBON DIOXIDE FLUXES

We now consider the carbon cycle in the surface mixed-layer, and its response to surface solubility and biological forcing in a Lagrangian frame, following a moving water column. From this point of view, the development of properties of the water parcel can be described without explicit account for transport. (In contrast, in the more usual Eulerian framework, temporal changes are evaluated at a fixed point in space and the advective transfer of properties must be explicitly accounted for). This Lagrangian view is somewhat idealized; we ignore the effects of a vertical shear in the horizontal velocity. Here we will consider idealized trajectories of water columns starting in the tropics, following the Gulf Stream system and then circuiting around the subtropical gyre or passing into the subpolar gyre.

We schematically depict a water column moving within an ocean basin in Fig. 6, illustrating how its change in temperature, T, and nutrient, N, properties affect the air-sea flux of carbon dioxide (see caption). How does the air-sea flux of CO_2 relate to Lagrangian changes in temperature, mixed-layer thickness and nutrient concentration? Significant surface temperature changes can be induced by air-sea heat exchange on timescales as short as a few weeks. Biological drawdown may occur on even shorter timescales. In contrast, the subsequent equilibration of mixed-layer carbon with the overlying atmosphere is relatively slow, occurring on a timescale on the order of a year (Broecker and Peng, 1974). This mismatch in forcing and response timescales means that changes in surface carbon concentration occur downstream of surface temperature and nutrient changes.

4.1 A LAGRANGIAN DESCRIPTION OF THE EVOLUTION OF SURFACE DISSOLVED INORGANIC CARBON

Here we outline an idealized description of the upper ocean carbon cycle which clearly reflects the temporal and spatial separation of forcing and response. We then go on to illustrate this separation in examples using a numerical model. Related Lagrangian approaches to the carbon system have previously been applied by Xu (1990), Follows *et al.* (1996), Brostrom (2000) and Andersson and Olsen (2002).

The Lagrangian change in the surface dissolved inorganic carbon concentration, C, is controlled by the air-sea CO_2 flux, $\mathcal{F}$, the entrainment of carbon,

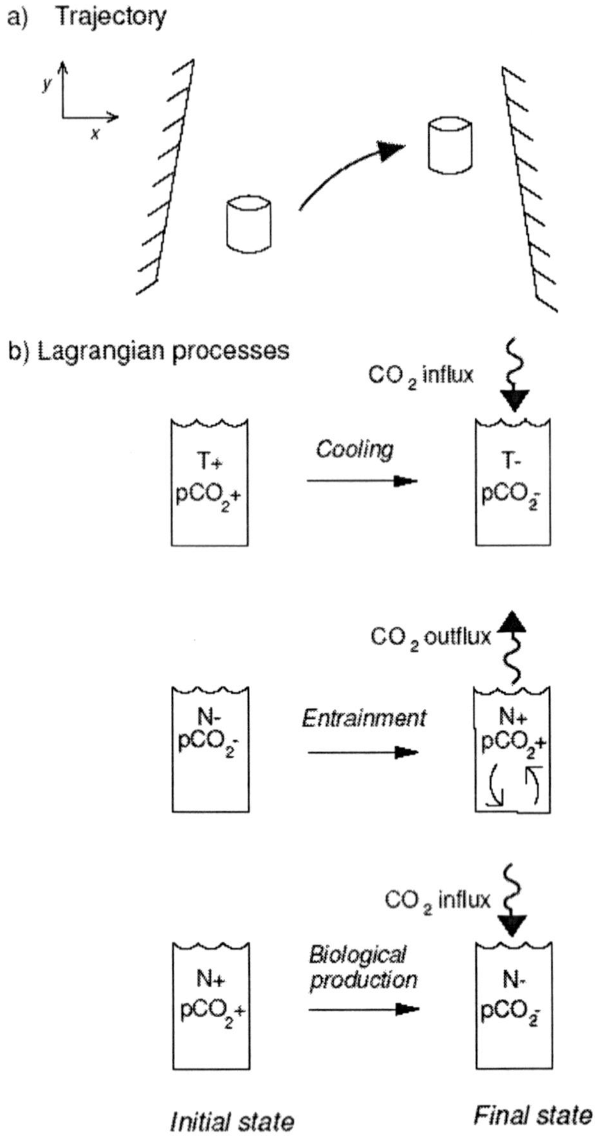

Figure 6. Schematic depiction of a water column moving within an ocean basin and how changes in its properties influence the air-sea flux of carbon dioxide. Cooling and a reduction of surface T increases the solubility and drives an influx of CO_2. However, cooling the surface may also lead to a thickening of the mixed layer and a vertical supply of thermocline waters, nutrient and carbon rich, due to entrainment. Entrained DIC will drive outgassing (or reduce uptake). The presence of nutrients may lead to biological production and reduction of surface DIC, driving an influx of CO_2.

E_C, and the net biological source-sink term, S_C:

$$\underset{storage}{\frac{DC}{Dt}} = \underset{air-sea}{\frac{\mathcal{F}}{h}} + \underset{entrain}{E_C} + \underset{biology}{S_C}. \tag{2}$$

Here the Lagrangian derivative, $D/Dt = \partial/\partial t + \mathbf{u}.\nabla$, denotes the rate of change following the movement of a water column, where $\mathbf{u}$ represents the horizontal velocity vector. $\mathcal{F}$ is the air-sea flux of CO_2 given by (1) and is defined to be positive when the flux is directed into the ocean. h is the thickness of the surface ocean mixed layer. The supply of carbon into the mixed layer through entrainment and upwelling from the thermocline is represented by E_C where $E_C > 0$ for an input of carbon-rich, thermocline waters. The term S_C represents the net supply of carbon due to biological processes. $S_C < 0$ indicates the dominance of consumption of inorganic carbon, and export of the organic matter to the deep ocean, over local remineralization.

4.2 THE EVOLUTION OF SURFACE NUTRIENTS

The governing equation for a nutrient, here phosphorus, in the mixed layer is analogous to that of carbon (2), but the air-sea exchange of nutrients is generally a negligible term (for phosphorus; not necessarily so for nitrogen or iron).

$$\underset{storage}{\frac{DN}{Dt}} = \underset{entrain}{E_N} + \underset{biology}{S_N}. \tag{3}$$

Here, E_N represents the supply of inorganic nutrients through entrainment and upwelling, and S_N the influence of biological processes. The carbon and nutrient relations, (2) and (3), can be combined after the following assumptions: Biological transformations of carbon and nutrients are assumed to occur with fixed stoichiometry; the Redfield ratio, $\mathcal{R}$ (see, for example, Anderson and Totterdell, 2003, this volume). Hence

$$S_C = \mathcal{R} S_N. \tag{4}$$

We relate the entrainment of C and N by

$$E_C = \mathcal{R}_{th} E_N, \tag{5}$$

where $\mathcal{R}_{th}$ is the ratio of the vertical contrast in C and N between the mixed layer and thermocline:

$$\mathcal{R}_{th} = \frac{C_m - C_{th}}{N_m - N_{th}},$$

where subscripts m and th refer to mixed layer and thermocline values.

Thus, the evolution of surface carbon (2) can be re-expressed in terms of nutrient changes by combining it with (3), (4) and (5), and rearranging to give

$$\underset{storage}{\frac{DC}{Dt}} = \underset{air-sea}{\frac{\mathcal{F}}{h}} + \underset{biology}{\mathcal{R}\left(\frac{DN}{Dt} - E_N\right)} + \underset{entrain}{\mathcal{R}_{th}E_N}. \tag{6}$$

Hence, the carbon evolution is now defined in terms of the air-sea flux, the input of carbon from entrainment, and the biological cycling defined in terms of Lagrangian changes in nutrient concentration and entrainment.

Consider an artificial limit where the vertical gradients of carbon and nutrients in the thermocline are set entirely by Redfieldian biological processes, $\mathcal{R}_{th} = \mathcal{R}$. In this case, entrainment has no net effect, since upwelled "biogenic carbon" is simply utilized again as nutrients are consumed (Lewis, 1992). Biological processes can then be described solely in terms of DN/Dt and an increase in surface N implies an accompanying increase in surface C driving an outgassing of CO_2. However, in practice, the vertical gradients of carbon and nutrients are not always in a Redfield ratio due to the air-sea exchange of CO_2 leading to differences in the ratio of the surface C and N, which are then subducted into the interior.

4.3 A LAGRANGIAN DESCRIPTION FOR THE EVOLUTION OF A CARBON ANOMALY

It is convenient to express these relationships in terms of C', the disequilibrium of surface ocean and atmospheric carbon

$$C' = C - C^{eq}, \tag{7}$$

where C is the dissolved inorganic carbon concentration and C^{eq} the local equilibrium value; the concentration at which the water parcel would be in balance with the pCO_2 in the overlying atmosphere at the current temperature, T, salinity, S, and alkalinity. It can be shown that the air-sea flux of carbon may be approximated as a function of C' (see Appendix A):

$$\frac{\mathcal{F}}{h} = -\frac{C'}{\tau}, \tag{8}$$

where τ represents an equilibration timescale for carbon in the mixed layer which is of the order of 1 year; $\tau^{-1} = K_g\alpha_o\beta/h$; here, β is the buffer factor and $\alpha_o = [CO_2]/C$ at local equilibrium (Appendix A).

Substituting for C and $\mathcal{F}/h$from (7) and (8) in (6) we rewrite the governing equation for surface carbon in terms of the disequilibrium:

$$\frac{DC'}{Dt} = -\frac{C'}{\tau} - \frac{DC^{eq}}{Dt} + \mathcal{R}\frac{DN}{Dt} + (\mathcal{R}_{th} - \mathcal{R})E_N. \tag{9}$$

Following a water column, (9) describes the development of the disequilibrium (i.e. difference from equilibrium with overlying atmosphere) of dissolved inorganic carbon in the surface waters. The first term on the right represents the decay of the disequilibrium due to air-sea gas transfer. The second term represents the forcing due to changes in the local equilibrium carbon concentration; a function of temperature, salinity and alkalinity which is largely determined by physical influences, though the production of calcium carbonate by organisms does affect alkalinity (for simplicity, we do not address this issue further here). The second term might, for example, reflect the increase in equilibrium carbon concentration as a water parcel cools during a poleward transit. The third and fourth terms together represent the combined effects of entrainment and biological cycling as in (6).

We can rearrange (9) in terms of the carbon anomaly to give

$$\frac{DC'}{Dt} + \frac{C'}{\tau} = \mathbf{F(t)}, \tag{10}$$

where $\mathbf{F(t)}$ represents the sum of the forcing terms following the parcel: $\mathbf{F(t)} \equiv -\frac{DC^{eq}}{Dt} + \mathcal{R}\frac{DN}{Dt} + (\mathcal{R}_{th} - \mathcal{R})E_N$.
Since the air-sea flux, $\mathcal{F} = -K_0K_gp' = -hC'/\tau$, the relationship in (9) can also be expressed in terms of the air-sea partial pressure difference, p';

$$\frac{D(K_gK_0p')}{Dt} + \frac{K_0K_gp'}{h} = \mathbf{F(t)}\,. \tag{11}$$

Though the air-sea partial pressure difference, p', is perhaps a more tangible quantity than the disequilibrium expressed as a DIC anomaly, C', the prognostic equation for p', (11), is not so clean since the solubility coefficient, also a function of temperature and salinity, K_0 now appears explicitly inside the derivative.

4.4 LAGGED RESPONSE OF THE CARBON SYSTEM TO EXTERNAL FORCING

In this framework, it is easy to illustrate the lagged response of the carbon system to external physical or biological forcing if we study the case where the forcing terms, $\mathbf{F}$, and the equilibration timescale, τ, are constant over a period of time, t. Assuming that the water parcel starts out at time $t = 0$ with initial carbon anomaly, $C'(0)$, and that the forcing, $\mathbf{F}$, is constant in time, we can solve (10) for the temporal development of the disequilibrium:

$$C'(t) = C'(0)e^{-t/\tau} + \tau \mathbf{F}\left(1 - e^{-t/\tau}\right). \quad (12)$$

The carbon anomaly, $C'(t)$, depends on (i) the initial disequilibrium with the atmosphere, $C'(0)$, which decreases with time as the air-sea flux erodes the anomaly, and (ii) the additional disequilibrium caused by the physical or biological forcing over time t, inducing a further air-sea flux. Notice that the timescale for the decay of the initial anomaly, and the subsequently induced anomalies, is τ which is typically on the order of a year or more and comparable with the residence time of a water parcel in the surface waters with respect to subduction into the thermocline.

Again, since $\mathcal{F} = -hC'/\tau$, the expression for the change in C' can be expressed in terms of the air-sea flux, $\mathcal{F}$,

$$\mathcal{F}(t) = \mathcal{F}(0)e^{-t/\tau} - h\mathbf{F}\left(1 - e^{-t/\tau}\right). \quad (13)$$

The air-sea flux at a time t is likewise controlled by (i) the initial air-sea flux at time $t = 0$, in response to the initial disequilibrium, which decays as the anomaly is eroded, and (ii) the additional flux created in response to a change in the equilibrium carbon concentration due to external forcing over the time t.

In summary, we have presented an idealized description of the surface ocean carbon system in the Lagrangian frame which brings out clearly the lagged response of the surface disequilibrium and air-sea flux to physical and biological forcing. In the following section we illustrate the consequences of this lagged response, in a qualitative sense, using a numerical model.

5. AN UPPER OCEAN WATER COLUMN MODEL.

We now illustrate this point, and the relationships between the air-sea flux and forcing using a more detailed one-dimensional model with which we can explore more complex situations; for example, with time varying forcing. The model resolves the vertical structure of a water column moving around the surface ocean. It includes explicit descriptions of the carbonate chemistry system and biological export of nutrients and carbon.

5.1 FORMULATION OF THE ONE-DIMENSIONAL MODEL

The model represents a one-dimensional (depth) column of water which we imagine to be swept around the surface of the North Atlantic. Though the model is more detailed than the simplified mathematical description above it is still highly idealized and assumes that the water column is isolated from the waters surrounding it and that there is no vertical shear in the flow; in other words the water column remains vertical during its passage around the gyres. We resolve the vertical structure of tracers, including temperature, salinity, dissolved inorganic carbon and nutrients. They are transported within the column by an imposed "diffusive" mixing or convective adjustment. If the water column becomes unstable due to surface cooling, i.e. if the surface density exceeds that at depth, then the water column and its properties are vigorously mixed until a homogeneous vertical density profile is achieved. Conversely, if we warm the surface ocean, the the mixed-layer depth shallows and water is subducted into the thermocline. We solve for the full carbonate chemistry system (see Murray, 2003, this volume) and specify the biological consumption and export of nutrients in the euphotic zone (upper 100m). CO_2 is exchanged with the atmosphere which has a fixed pCO_2 of $280ppmv$ according to the changes in the carbonate chemistry sytem and an imposed gas transfer coefficient, K_g.

Using this simple, but explicit model, we illustrate the respsonse of the surface carbon system and air-sea flux to imposed changes in physical conditions, consistent with idealized trajectories around the surface North Atlantic ocean. The key point illustrated is the lag in the response of the air-sea flux of carbon, and decay of induced dissolved inorganic carbon anomalies, induced by physical or biological forcing. We have applied a similar approach, in a abiotic framework, in Follows *et al.* (1996) and some relevant detail may be found in that article.

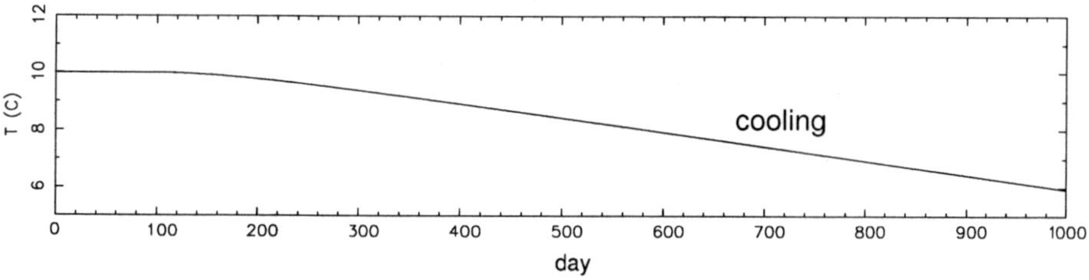

Figure 7. Change in surface properties and air-sea carbon flux following an idealized water column moving through the surface ocean. In this case the mixed layer depth is unchanging (120 m) and the water parcel is initially at equilibrium with the overlying atmosphere for the local temperature and salinity. The panels show: (a) C, solid line, and C^{eq}, dashed line ($mol\ m^{-3}$); (b) mixed-layer temperature (^oC) following the water parcel.

5.2 RESPONSE TO COOLING: THE ABIOTIC LIMIT WITH NO ENTRAINMENT OR UPWELLING

If the water column is cooled C^{eq} increases, creating a negative surface DIC anomaly, C', (and an air-sea difference in pCO_2) inducing a flux of carbon into the ocean. The example in Fig. 7 clearly illustrates the lagged response to a sustained cooling of the mixed layer: C^{eq} increases ahead of C (dashed and full lines respectively in Fig. 7a). The forcing leads the response by about a year in accord with the interpretation of the simplified theory in (12) and (13).

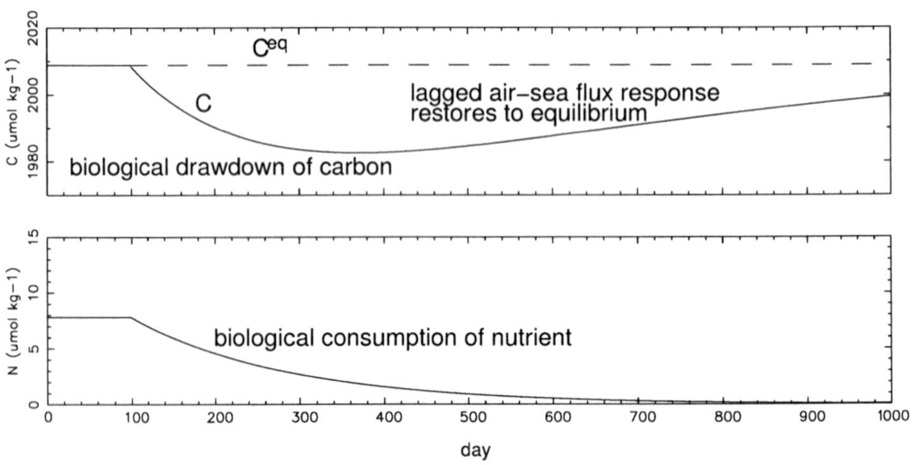

Figure 8. Change in surface properties and air-sea carbon flux following a water column moving through the surface ocean. In this case the mixed layer depth is unchanging (100 m) and the water parcel is initially at equilibrium with the overlying atmosphere for the local temperature and salinity. The panels show: (a) C, solid line, and C^{eq}, dashed line ($mol\ m^{-3}$); (b) nitrate concentration ($mol\ m^{-3}$) following the water parcel.

5.3 RESPONSE TO BIOLOGICAL DRAWDOWN

If surface nutrient, N, decreases through biological consumption and formation of organic matter, then a negative surface C' (i.e. undersaturation) is formed driving an air-sea flux of CO_2 into the ocean. In contrast with the cooling case above, the anomaly is driven by a loss of C from the water column as opposed to change in the equilibrium value, C_{EQ}. We illustrate a simple case from the simplified numerical model in which the mixed-layer temperature and thickness are held constant, but nutrients and carbon and biologically consumed. Since the equilibrium carbon concentration, C^{eq}, is a function only of temperature and salinity here (and alkalinity is fixed), it does not change in response to the nutrient drawdown and there is a subsequent air-sea flux of carbon compensating for the biological drawdown. However, the air-sea flux occurs with a time lag on the order of a year, consistent with the theory outlined in the previous section.

5.4 WITH THERMAL AND BIOLOGICAL FORCING AND ENTRAINMENT: SUBPOLAR AND SUBTROPICAL PATHWAYS

Now we consider two more realistic, though still idealized, trajectories; the first passing through the Gulf Stream and circuiting into the subpolar gyre, and the second passing through the Gulf Stream and circuiting into the subtropical gyre (Fig. 9). We will examine these pathways in the context of the theory and the numerical model. In order to clearly demonstrate the key gyre-scale interactions we do not resolve the seasonal cycle though this could modify details of the results (Brostrom, 1997). Figures 10 and 11 illustrate the surface ocean carbon cycle processes in these idealized pathways and again we emphasize the impact of thc lagged response.

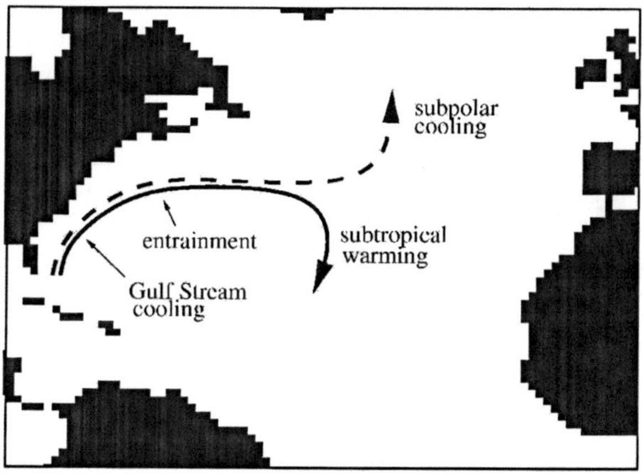

Figure 9. Idealized pathways for surface water parcels in the North Atlantic basin. Both parcels start out moving northwards through the Gulf Stream system where they cool strongly. At the inter-gyre boundary the water columns form deep winter mixed-layers, as a result of the cooling, and move into the interior of the basin. One water column is caught in the anti-cyclonic flow of the subtropical gyre, returning southwards with a warming and shoaling mixed layer. The other water column is caught in the cyclonic subpolar flow and continues to move polewards with increasing mixed layer thickness and surface cooling.

In the subpolar example, the water column starts its transit close to equilibrium with the overlying atmosphere and the surface depleted of nutrients (Fig. 10). It transits through the Gulf Stream system experiencing cooling, increased solubility, but also thickening of the mixed layer and entrainment of nutrients and carbon to the surface. There is a large air-sea flux of CO_2 directed into the ocean over the subpolar gyre (reflected in the increased disequilibrium), which is due to downstream cooling increasing C^{eq} and reinforced by the reduction in C due to biological consumption. In this ideal trajectory, the water column continues to cool as it circulates in the subpolar gyre, and so continues to absorb atmospheric CO_2 at a greater rate.

In this particular experiment, the downstream entrainment and subsequent biological export of N enhances drawdown of CO_2; C decreases even though C^{eq} is increasing because $(\mathcal{R}_{th} - \mathcal{R}) > 0$. If, in contrast to the example shown, $(\mathcal{R}_{th} - \mathcal{R}) < 0$, then entrainment of thermocline waters will bring excess carbon, over and above that required to support biological export of the entrained nutrients, and partially compensate for the increased solubility due to cooling, reducing the uptake of CO_2 from the atmosphere. Mahadevan *et al.* (2002) have mapped $\mathcal{R}/\mathcal{R}_{th}$ in the global oceans. Their analysis suggests that regions of both positive and negative $(\mathcal{R}_{th} - \mathcal{R})$ are present in the subpolar North Atlantic implying that entrainment can alter both the sign and magnitude of the air-sea flux of CO_2.

In the subtropical example, the water column follows a similar pathway through the Gulf Stream, but then recirculates and moves into the interior of the subtropical gyre where the surface water is warmed and the mixed layer shoals (Fig. 11). Despite the warming, the surface water remains under-saturated, C is always below C^{eq}. There is an uptake of CO_2 from the atmosphere in the interior of the subtropical gyre due to both the drawdown of nutrients and, significantly, the lagged response to the upstream cooling as illustrated in Fig. 7. The subtropical and subpolar responses differ in the subduction peak associated with the rapid mixed-layer shoaling in the subtropical case (Fig. 11). The shoaling leads to a decrease in the mixed-layer equilibration timescale (Appendix A) and thus downstream decrease in disequilibrium: C approaches C^{eq}.

In summary, as the idealized examples suggest, the surface carbon concentration rarely reaches a local equilibrium with the overlying atmosphere (witnessed by Fig. 1). This disequilibrium is due to the continual forcing induced by the physical and biological processes, changing the equilibrium carbon concentration (by changes in T, S and alkalinity) and changes in nutrient concentrations. Incorporating a seasonal cycle introduces more rapid changes in the equilibrium conditions. The exchange timescale, τ, seasonally varies from a few months over summer to several years in winter due to the thickening of the mixed layer which extends the delay in the response of the carbon flux to large scale changes in forcing. This mismatch in forcing and response timescales leads to changes in surface carbon concentration occurring downstream of surface temperature and nutrient changes.

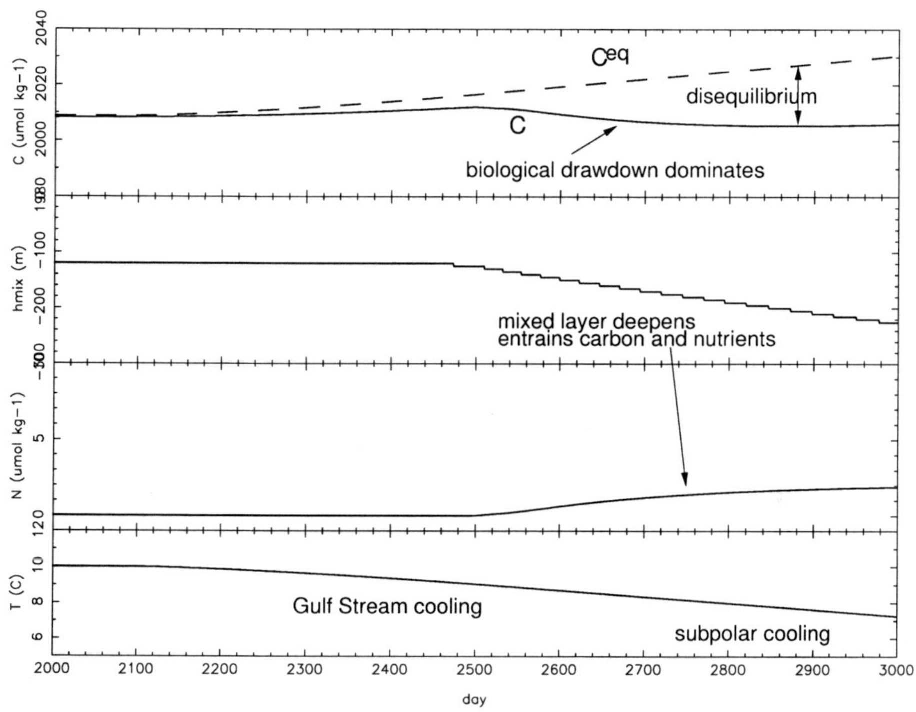

Figure 10. Subpolar example: evolution of surface carbon system following water column on an idealized trajectory through the Gulf Stream system and into the subpolar gyre (dashed line Fig. 9). While cooling increases solubility and promotes an uptake of CO_2 entrainment of nutrients promotes biological drawdown which dominates the trend of C in the subpolar gyre. **(a)** C, solid line, and C^{eq}, dashed line ($mol\ m^{-3}$); **(b)** mixed-layer thickness (m); **(c)** nitrate concentration ($mol\ m^{-3}$); **(d)** mixed-layer temperature (^oC), and **(e)** cumulative flux of carbon into the ocean ($mol\ m^{-2}$), following the water parcel.

6. SUBDUCTION OF CARBON INTO THE THERMOCLINE

The ocean interior acquires its water-mass structure, such as temperature, salinity, and stratification, principally through the ventilation process, where fluid in the surface mixed layer (in contact with the atmosphere) is transferred into the main thermocline and deep ocean. In addition the properties at the point of ventilation also affect the interior carbon, nutrient and dissolved gas distributions.

Ventilation over the open ocean can be separated into deep convection and the subduction processes. Both processes occur preferentially at the end of winter when the mixed layer is at its maximum thickness and density. Deep convection occurs where there is strong surface buoyancy loss and favourable pre-conditioning with a doming up of isopycnals. The surface buoyancy loss in

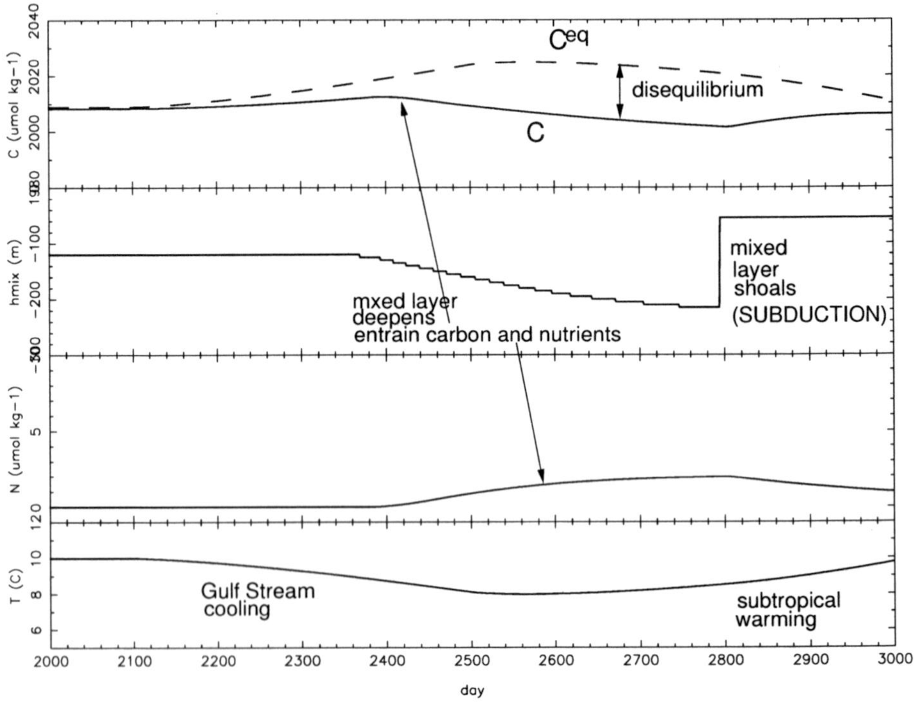

Figure 11. Subtropical example: evolution of surface carbon system following a water column on an idealized trajectory through the Gulf Stream system and into the subtropical gyre (solid line, Fig. 9). Note that the surface waters are undersaturated, and drawing CO_2 from the atmosphere even in the subtropical gyre where the water column is warming. This is, in part, due to the lagged response to cooling upstream in the Gulf Stream system. Panels are as in Fig. 10.

winter erodes the relatively thin, thermocline and lead to striking overturning (see review by Marshall and Schott, 1999). The overturned fluid then spreads from the convective sites through boundary currents if there are topographic barriers or through geostrophic eddies. In the North Atlantic deep convection occurs in very localized regions of the Labrador and Greenland Seas.

Subduction occurs principally over the interior of the subtropical gyres, where fluid is transferred into the main thermocline through a combination of downwelling and a lateral transfer through a shoaling mixed layer (Fig. 12a); see North Atlantic study by Marshall et al. (1993) and review by Williams (2001). Over the North Atlantic, the subduction rate estimated from climatology reveals a high band downstream of the Gulf Stream in the subtropical gyre and lower values elsewhere over the subtropical gyre (Fig. 12b). Conversely, fluid is transferred from the thermocline into the mixed layer over much of

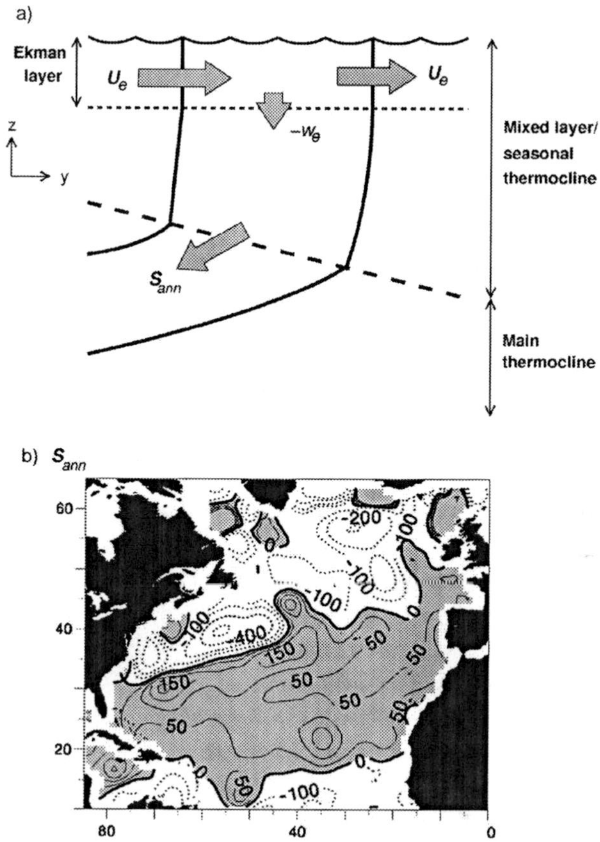

Figure 12. **(a)** Schematic representation of the subduction process where fluid is transferred from the mixed layer into the stratified thermocline. **(b)** subduction rate or volume flux per unit horizontal area into the main thermocline ($m\ y^{-1}$) over the North Atlantic; see Marshall *et al.* (1993) for details.

the subpolar gyre. This gyre-scale contribution to subduction can also be augmented by eddy or frontal-scale contributions (Marshall, 1997).

The ventilation process peaks over certain regions and leads to the formation of 'mode waters', characterized by their weak stratification and extensive volume. For example, the band of enhanced subduction downstream of the Gulf Stream is associated with the formation of 18°C Water, and the continual entrainment and thickening of the mixed layer as fluid moves anti-cyclonically around the subpolar gyre eventually leads to the formation of Labrador Sea Water. These 'mode waters' and their passage around the ocean represent the dominant physical pathway for carbon to be transferred into the ocean interior; see the review on mode waters by Hanawa and Talley (2001) and the nutrient-related study of Sarmiento *et al.* (2003).

6.1 CONSEQUENCES FOR ATMOSPHERIC CO_2

The ventilated thermocline is on the order of a few hundred metres thick throughout the large subtropical gyres of the world's oceans. Consequently, the thermocline represents significant reservoir of carbon which, if changed on a global scale, may have a significant impact on atmospheric pCO_2 (Follows *et al.*, 2002; Ito and Follows, 2003). The carbon properties of the thermocline, with respect to the solubility pump, are set at the point of subduction. The biological pump subsequently modifies these waters as they transit through the interior. The region of maximum subduction rate in the North Atlantic is also a place where the surface waters are strongly undersaturated and absorbing CO_2 following recent vigorous cooling in the Gulf Stream (Fig. 5a, Fig. 12). Also note the undersaturation at the point of subduction in the idealized water column model shown in Figure 11. Hence, subducted waters are undersaturated and the thermocline stores less carbon than if full saturation were achieved before subduction.

For a closed ocean-atmosphere system, thickening the thermocline can lead to an increase in the volume of warm waters, a reduction in the mean solubility of CO_2 in the ocean, and a net transfer of carbon from ocean to atmosphere. Reducing the saturation state of the subducted waters will have the same effect. The saturation state depends upon how long it takes water parcels to be advected through the Western Boundary Current system relative to the air-sea equilibration timescale. The wind-stress forcing on the ocean has a significant influence on both the thickness of the ventilated thermocline and the swiftness of the boundary currents. Theoretical arguments and numerical model experiments (Ito and Follows, 2003) suggest that a factor of two change in wind-stress forcing on the ocean could lead to changes in atmospheric pCO_2 of as much as $30ppmv$ by modulating the volume and saturation state of the thermocline waters (see Fig. 13).

7. EFFECT OF TIME-VARYING, EDDY CIRCULATIONS

We have discussed processes and interactions on the basin and gyre scales, but there is significant energy in motions on much smaller scales in the ocean; geostrophic eddies on scales of tens to one or two hundred kilometres, and even smaller frontal scales. In principle, such time-varying, finer-scale circulations alter the structure of the upper ocean and thus the exchange of heat and dissolved gases with the atmosphere. With respect to biological productivity, there has been much discussion of the importance of mesoscale eddies (McGillicuddy et al., 1998; Oschlies, 2002), propagation of planetary waves for biological production (Uz et al., 2001; Cipollini et al., 2001), and frontal scale circulations (Mahadevan and Archer, 2000; Levy et al., 2001). Across

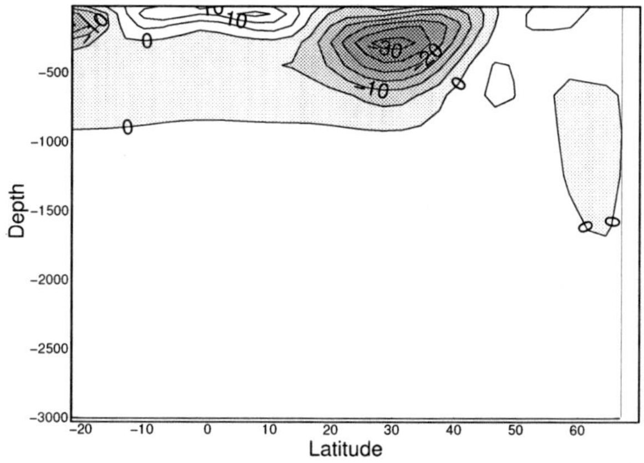

Figure 13. Difference in zonally averaged DIC of two numerical models of the solubility pump in a single ocean basin; one with and one without wind forcing. The basin is about the size of the North Atlantic ocean and is coupled to an atmospheric reservoir of CO_2 through air-sea exchange. Introducing wind-stress forcing has developed thick, warm thermocline in the upper few hundred metres. Due to the swiftness of the Western Boundary Current which feeds the region of subduction, the waters of the thermocline are significantly under-saturated with DIC concentrations as much as $30\mu M$ less than the case without wind forcing. The deep ocean DIC is largely unchanged. In this simplified model a doubling of wind-stress forcing can lead to a greater outgassing of carbon from the ocean and an increase in atmospheric pCO_2 of as much as $30ppmv$.

this range of scales there are deformations of the near-surface isopycnals which may be reversible or irreversible.

The extent to which these time-varying circulations are significant for the air-sea exchange of CO_2 is, as yet, unquantified. However it is clear that their importance for the air-sea exchange of CO_2 partly depends on how persistent these phenomena are relative to the air-sea equilibration timescale, whether there is associated biological activity, and whether there is any modification of the surface mixed-layer.

In the following thought experiments we consider how the air-sea flux might change in response to idealized, eddy-induced changes in surface ocean DIC and N concentrations. In these examples we consider the impact of a persistent eddy or "ring" as it moves through the ocean. (We note that there may be associated perturbations of temperature and alkalinity which are significant for the carbon system but, for simplicity, we will not discuss them further here).

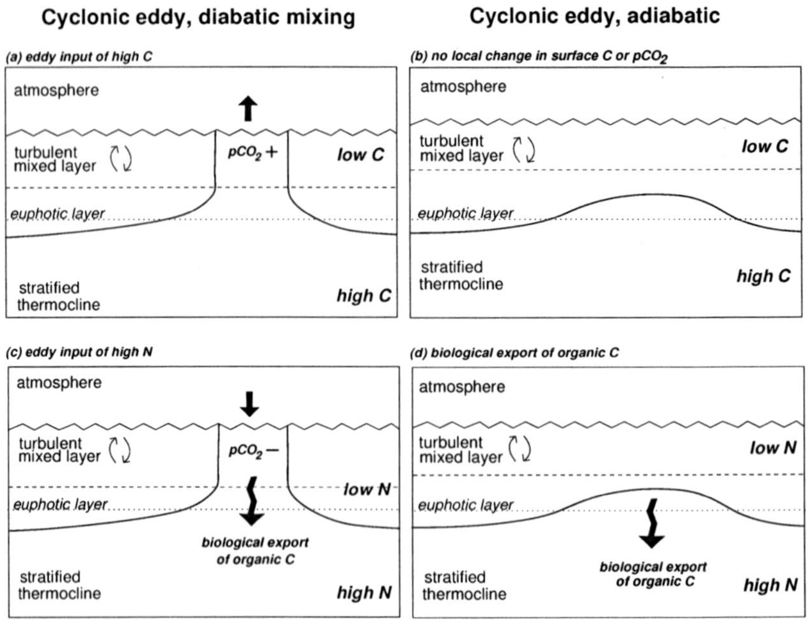

Figure 14. Schematic depiction of the biogeochemical influences of cyclonic eddies in circumstances where either isopycnal surfaces are lifted and mixed into the surface mixed-layer (left panels), or they are not mixed to the surface but are deformed into the sunlit euphotic layer (right panels). The turbulent mixed layer (with its base denoted by a dashed line) overlies a stratified thermocline. Here we examine cases where the euphotic layer (its base denoted by dotted line) is deeper than the mixed-layer. The full line denotes an isoline separating low C and N surface waters from the thermocline, richer in C and N.
In the left panels, a cyclonic eddy has a raised thermocline and nutricline in contact with the surface mixed-layer where convection bring anomalies of C and N to the surface. The enrichment of surface C promotes outgassing (top left) while the entrainment of nutrients leads to enhanced biological drawdown with the opposite effect (bottom left).
The right panels depict a case where the thermocline undulations do not reach the surface mixed-layer. There is no direct carbon anomaly at the surface (top right) but nutrients may be lifted into the euphotic layer (bottom right), promoting production and export of organic matter. In the latter case, even though the carbon profile is changed below the mixed-layer, there is not a local, contemporary expression in surface pCO_2 or air-sea exchange.
In contrast to these cases, anti-cyclonic eddies have a depressed thermocline and nutricline, thus having little effect on the air-sea flux of CO_2 or in biological production.

7.1 IRREVERSIBLY RAISED THERMOCLINE AND INJECTION OF DIC AND N INTO THE MIXED LAYER

Injection of DIC and N into the mixed layer from below might occur within a travelling cyclonic eddy (Fig. 14a). Thermocline waters are raised into, and irreversibly mixed with, the mixed layer, injecting DIC from the cooler thermocline. This promotes outgassing of CO_2 into the atmosphere *if* other mixed layer properties are constant. Equilibration with the atmosphere takes several seasons to a year (see Fig. 7).

A supply of nutrient, N, into the mixed layer by raising the nutricline (Fig. 14b), promotes phytoplankton growth within the euphotic zone and, ultimately, the export of nutrients and carbon to the deep ocean in organic form. The sign of the induced air-sea exchange of carbon depends on whether the $C : N$ ratio of the thermocline waters raised into the mixed layer is greater or lesser than that in the organic matter (Lewis, 1992; Mahadevan *et al.*, 2002). This is illustrated in the biology and entrainment terms in (6) where the net forcing due to entrainment, $E_N(R_{th} - R)$, changes sign depending on these relative $C : N$ ratios.

The counter-example of an anti-cyclonic eddy with a depressed thermocline is unlikely to lead to changes in air-sea flux or enhancement of biological production since there will be no surface DIC perturbation and the nutricline is pushed downwards, away from the sunlit euphotic zone and the surface mixed layer.

7.2 REVERSIBLE OSCILLATIONS OF THE THERMOCLINE

An undulating thermocline signal, possibly associated with the passage of eddies or planetary waves, need not penetrate into the surface mixed layer. In this case, we regard these thermocline undulations as reversible.

In this case there will be no concurrent, local surface expression of the eddy in DIC or N and no immediate response in surface pCO_2 and gas exchange (Fig. 14c,d). However, there may still be a rectified biological response to such perturbations. During the upwelling phase nutrient rich waters are uplifted into the illuminated, euphotic zone, assuming it extends below the surface mixed-layer. Here photosynthesis can consume N and DIC as long as there is sufficient time for phytoplankton to grow; typically on the order of days (McGillicuddy and Robinson *et al.*, 1997). Thus, the oscillating signal can lead to an enhancement of export of organic carbon to the deep ocean (Fig. 14d). This situation might occur in the tropical oceans, or summer subtropics, where the euphotic zone is deep and mixed-layers are shallow. While these anomalies in DIC created below the mixed layer and in the upper thermocline will not lead to an immediate response in the air-sea flux, there will be a delayed response if the DIC anomaly becomes entrained into the mixed layer in the following winter.

In addition to inducing direct changes in the air-sea flux of carbon there are more subtle and indirect effects of the smaller-scale, time-varying motions (see review by Williams and Follows, 2003). For example, eddies lead to slantwise exchange of fluid parcels on frontal and eddy scales, increasing the stratification and suppressing convection (Levy et al., 1998; Nurser and Zhang, 2000) and impacting upon surface DIC distribution and air-sea fluxes. Likewise, eddies provide an additional transport (in addition to the time-mean circulation), which can alter the subduction process (Marshall, 1997) and the transport of carbon and nutrients within the thermocline (Lee and Williams, 2000), which is particularly important in the Southern Ocean.

In contrast with the recent focus on the role of vertical meso-scale and submeso-scale motions, Mahadevan *et al.* (2002) suggest that lateral displacements of surface properties by eddies probably play a greater role in modulating surface pCO_2 than the associated vertical excursions.

In summary, it is presently unclear as to the systematic response of the carbon system to the presence of eddies and time-varying circulations. Their importance depends on how persistent the phenomena are relative to the air-sea equilibration timescale, whether there is associated biological activity, and whether there is any modification of the surface mixed-layer. The long term response for the air-sea exchange of CO_2 is also sensitive to the way in which eddies systematically alter the stratification, and vertical and horizontal transports of carbon and nutrients.

8. SEASONAL AND INTERANNUAL VARIATIONS IN THE NORTH ATLANTIC

So far, we have focussed on the large scale structures which control the regional pattern of the annual mean air-sea carbon flux, and the potential of smaller scale motions to influence those patterns. There is a vigorous seasonal cycle in the North Atlantic air-sea carbon flux driven by seasonal sea-surface temperature change and the annual cycle in mixed-layer depth. There is also interannual variability in the air-sea flux of CO_2 driven by changes in the physical environment and biological community.

The variability in the subtropical North Atlantic, has been characterized at the Bermuda Atlantic Time-Series Station (BATS), summarized and discussed by Bates (2000). Seasonal and interannual variability in the CO_2 flux at that site is strongly related to temperature changes and the variability in winter-time convective mixing. Such physical variations can be related to the North Atlantic Oscillation (NAO) index (Hurrell *et al.*, 2003); a measure of the sea level pressure difference between Iceland and Portugal. The NAO index captures a significant fraction of the regional variability in the atmosphere and surface ocean physical properties. A characteristic of the variability often associated with the NAO is the change in winter-time convective mixing. During the high index phase of the NAO, winter mixed layers in the subpolar North Atlantic are anomalously deep, and those in the subtropics anomalously shallow (Dickson *et al.*, 1996). The opposite is true during low NAO index winters. The variability in winter mixing can have a direct impact on regional productivity, nutrient and gas fluxes (Bates, 2000; Williams *et al., 2000*; McKinley *et al.*, 2003). Interannual variations in the meteorological forcing can also lead to changes in the rate of subduction and the ventilation of mode waters which can impact on the longer term role of the carbon reservoir in the North Atlantic subtropical thermocline (Bates *et al.*, 2003).

An outstanding question is the role of the North Atlantic as a modulator of the oceanic uptake of fossil fuel CO_2 on interannual timescales. The observations at Bermuda show local variations in the annual flux which, if extrapolated over the whole basin, or whole subtropical gyre, suggest a significant role for the Atlantic in the global variability (Gruber *et al.*, 2002). Atmospheric inverse models concur with this extrapolation. In contrast, experiments with global ocean circulation and biogeochemistry models, which do capture the amplitude of local variability at BATS, do not find significant basin integral variations in the annual air-sea flux (McKinley *et al.*, 2003). Hence, the variability at BATS might in practice not reflect changes over the whole basin, since there can be cancellation from opposing changes in different parts of the gyre.

9. SUMMARY

The air-sea flux of carbon is driven by the disequilibrium between the partial pressure of CO_2 in the surface ocean and atmosphere. This disequilibrium is maintained by a combination of the physical, chemical and biological processes within the ocean.

The extent to which physical or biological processes keep the surface ocean from equilibration, and drive air-sea fluxes of CO_2, depends upon the relative timescale of each process. The air-sea equilibration timescale for CO_2 is relatively long, on the order of one year, and this has significant consequences:

(i) The location of physical or biological forcing and the consequent air-sea flux of CO_2 may be significantly separated in space and time. There is only a simple one to one correspondence between the forcing patterns and the carbon response *if* the air-sea flux response timescale is relatively short compared with the forcing timescale.

(ii) The western boundary currents are swift, the physical forcing (cooling) is strong, and surface water parcels may be subducted shortly after passing through them. Consequently the waters of the main thermocline are probably undersaturated and the ocean's solubility pump is not working at full efficiency. The importance of the thermocline thickness and boundary current speed makes the carbon reservoir of the thermocline sensitive to the wind-stress forcing on the ocean.

(iii) On the basin scale, the disequilibrium between ocean and atmosphere is maintained by the net loss of heat from the North Atlantic ocean and the biological transfer of nutrients from upper waters to depth. The thermal forcing also drives the overturning circulation which facilitates a net oceanic transport of carbon out of the basin to the south.
(iv) The large scale impact of small scale, time-vary structures, such as eddies and fronts, in the ocean is, as yet, unclear. Their affect on the carbon system and air-sea exchange depends on the timescale of the small-scale structures compared with the air-sea equilibration timescale, whether biological production occurs and whether the mixed layer properties are irreversibly modified.
(v) There is significant interannual variability in the surface ocean physical environment which promotes variability of biological productivity and air-sea fluxes. The global significance of North Atlantic variability is yet to be fully quantified.

ACKNOWLEDGMENTS

MJF is grateful for support from NSF grant OCE-0136609. RGW is grateful for support from the UK NERC AMT Consortium Programme NER/O/S/2001. We thank Alison McLaren for discussions and assistance in developing the Lagrangian theory. We thank Amala Mahadevan for an insightful and helpful review

APPENDIX: DEFINING THE TIMESCALE FOR CARBON EQUILIBRATION BETWEEN SURFACE OCEAN AND ATMOSPHERE

In the limit of no entrainment, upwelling or biological processes, the carbon evolution following a water column is controlled by the air-sea exchange:

$$\frac{DC}{Dt} = \frac{\mathcal{F}}{h} = \frac{K_0 K_g}{h}(pCO_2^{at} - pCO_2), \tag{A1}$$

where pCO_2^{atmos} and pCO_2 are the partial pressures of CO_2 in the atmosphere and oceanic mixed layer, K_0 is the solubility of CO_2, K_g is the gas transfer coefficient. The air-sea flux, $\mathcal{F}$, in (A1) may be written in terms of $C' = C - C^{eq}$, the mismatch in total inorganic carbon concentration and its equilibrium with the partial pressure in the atmosphere, pCO_2^{at} at local T and S. Using the buffer factor, β, defined

$$\beta \equiv \frac{pCO_2'/pCO_2^{at}}{C'/C^{eq}}, \tag{A2}$$

and the flux is expressed

$$\mathcal{F} = K_0 K_g \beta \frac{pCO_2^{at}}{C^{eq}}(C^{eq} - C), \tag{A3}$$

The ionization fraction (Stumm and Morgan, 1996) at equilibrium for local conditions, α_0, may be defined

$$\alpha_0 = \frac{K_0\, pCO_2^{at}}{C^{eq}} = \frac{[CO_2]^{eq}}{C^{eq}}. \tag{A4}$$

Substituting from (A4) into (A1) and (A3) relates the carbon evolution to the air-sea forcing expressed in terms of a mismatch between the surface carbon and equilibrium carbon, C^{eq}:

$$\frac{DC}{Dt} = \frac{\mathcal{F}}{h} = \frac{1}{\tau}(C^{eq} - C), \qquad (A5)$$

where the exchange timescale for carbon is defined by $\tau^{-1} = K_g \alpha_0 \beta / h$.

Typically for seawater, $\alpha_0 \sim 0.005$ and $\beta \sim 10$ and, if $K_g \sim 10^-4\ m\ s^{-1}$ and $h \sim 100m$, then the timescale τ is estimated to be about 1 $year$.

REFERENCES

Anderson, T. and I. Totterdell (2003) Modelling the response of the oceans biological pump to climate change. This volume.

Andersson, L.G. and A. Olsen (2002) Air-sea flux of anthropogenic carbon dioxide in the North Atlantic *Geophys. Res. Lett.*, **29(19)**, 10.1029/2002GL014820.

Bates, N.R. (2000) Interannual Variability of Oceanic CO2 and Biogeochemical Properties in the Western North Atlantic Subtropical Gyre *Deep-Sea Research II*, **48**, 1507-1528.

Bates N.R., A. C. Pequignet, R. J. Johnson and N. Gruber (2002) A Variable Sink for Atmospheric CO_2 in Subtropical Mode Water of the North Atlantic Ocean, *Nature*, **420**, 489-493.

Broecker, W. S., Peng, T-H., (1974). Gas exchange rates between air and sea. *Tellus*, 26, 21-35.

Brostrom, G. (1997) Air-sea flux of CO_2 - can we shortcut the annual cycle? A Norwegian Sea case study. *Phys. Chem. Earth*, **21**, 517-522.

Brostrom, G. (2000) The role of annual cycles for the air-sea exchange of CO_2. *Marine Chem.*, **72**, 151-169.

Cipollini, P., D. Cromwell, P.G. Challenor and S. Raffaglio, (2001). Rossby waves detected in global ocean colour data. *Geophys. Res. Lett.*, **28**, 323-326.

Conkright, M.E., R.A. Locarnini, H.E. Garcia, T.D. O'Brien, T.P. Boyer, C. Stephens, J.I. Antonov (2002) *World Ocean Atlas 2001: Objective analyses, Data Statistics and Figures. CD-ROM Documentation.* National Oceanographic Data Center, Silver Spring, Maryland.

Dickson, R., J. Lazier, J. Meinke, P. Rhines and J. Swift (1996) Long term coordinated changes in the convective activity of the North Atlantic. *Prog. Oceanogr.*, **38**, 241-295.

Emerson, S., P. Quay, D. Karl, C. Winn, L. Tupas, and M. Landry (1997) The carbon pump in the Subtropical Pacific Ocean: Implications for the Global Carbon Cycle, *Nature* , **389**, 951-954.

Follows, M.J.,T. Ito and J. Marotzke (2002) The ocean's subtropical gyres and Atmospheric pCO2. *Global Biogeochem. Cycles*, **16**, 1113, doi:10.1029/2001GB001786.

Follows, M.J., Williams, R.G., Marshall, J.C., (1996). The solubility pump of carbon in the subtropical gyre of the North Atlantic. *J. Mar. Res.*, **54**, 605–630.

Gruber, N., (1998). Anthropogenic CO_2 in the Atlantic Ocean. *Glob. Biogeochem. Cycles*, **12**, 165-191.

Gruber, N., N.Bates, and C.D. Keeling (2002) Interannual variability in the North Atlantic Ocean carbon sink, *Science*, **298**, 2374-2378.

Hanawa, K. and L.D. Talley (2001) Mode Waters. In *Ocean Circulation and Climate: Observing and Modelling the Global Ocean*, Eds. G. Siedler, J. Church and J. Gould, 373-387.

Hansell, D.A. and C.A. Carlson (1998) Deep ocean gradients in dissolved organic carbon concentrations. *Nature*, **395**, 263-266.

Holfort, J., K.M. Johnson, B. Schneider, G. Siedler and D.W.R. Wallace, (1998) Meridional transport of dissolved inorganic carbon in the South Atlantic Ocean. *Glob. Biogeochem. Cycles*, **12**, 479-499.

Hurrell, J.W., Y. Kushnir, M. Visbeck, and G. Ottersen (2003) An Overview of the North Atlantic Oscillation. in *The North Atlantic Oscillation: Climate Significance and Environmental Impact* J.W. Hurrell, Y. Kushnir, G. Ottersen, and M. Visbeck, Eds. Geophysical Monograph Series, **134**, pp. 1-35.

Ito, T., and M.J. Follows (2003) Upper ocean control of the solubility pump of CO_2. *J. Marine Res.*, accepted for publication.

Josey, S.A., E.C. Kent and P.K. Taylor, (1998). The Southampton Oceanography Centre (SOC) Ocean-Atmosphere Heat, Momentum and Freshwater Flux Atlas. Southampton Oceanography Centre, Reports 6, 30pp.

Lee, M.-M. and R.G. Williams, (2000). The role of eddies in the isopycnic transfer of nutrients and their impact on biological production. *J. Mar. Res.*, **58**, 895-917.

Lefevre, N, A.J. Watson, A. Olsen, A. Rios, F. Perez, T. Johannessen, R. Bellerby, I. Skjelvan, (2003). A decrease in the sink for atmospheric CO_2 in the North Atlantic, To be submitted to *Geophs. Res. Lett.*

Levy, M., L. Memery and G. Madec, (1998). The onset of a bloom after deep winter convection in the northwest Mediterranean sea: mesoscale process study with a primitive equation model. *J. Mar. Sys.*, **16**, 7-21.

Levy, M., P. Klein and A-M. Treguier, (2001). Impact of sub-mesoscale physics on production and subduction of phytoplankton in an oligotrophic regime. *J. Mar. Res.*, **59**, 535-565.

Lewis, M.R. (1992) Satellite ocean color observations of global biogeochemical cycles. In *Primary Productivity and Biogeochemical Cycles in the Sea*. Ed. P.G. Falkowski and A.D. Woodhead, Plenum Press, New York.

Lundberg, L. and P.M. Haugan (1996) A Nordic Seas - Arctic Ocean Carbon Budget from Volume Flows and Inorganic Carbon Data. *Global Biogeochem. Cycles*, **10**, 493-510.

Mahadevan, A. and D. Archer, (2000). Modelling the impact of fronts and mesoscale circulation on the nutrient supply and biogeochemistry of the upper ocean. *J. Geophys. Res.*, **105**, 1209-1225.

Mahadevan, A., M. Levy, and L. Memery (2002) Mesoscale variability of sea surface pCO_2: What does it respond to? *Global Biogeochem. Cycles*, submitted.

Marshall, D., (1997). Subduction of water masses in an eddying ocean. *J. Mar. Res.*, **55**, 201-222.

Marshall, J.C., A.J.G. Nurser and R.G. Williams, (1993). Inferring the subduction rate and period over the North Atlantic. *J. Phys. Oceanogr.*, **23**, 1315-1329.

Marshall, J. and F. Schott, (1999). Open-ocean convection: observations, theory and models. *Rev. Geophysics*, **37**, 1, 1-64.

McGillicuddy, D.J. and A.R. Robinson, (1997). Eddy-induced nutrient supply and new production in the Sargasso Sea. *Deep-Sea Res. I*, **44**, 1427-1449.

McGillicuddy, D.J., A.R. Robinson, D.A. Siegel, H.W. Jannasch, R. Johnson, T. Dickeys, J. McNeil, A.F. Michaels and A.H. Knap, (1998), New evidence for the impact of mesoscale eddies on biogeochemical cycling in the Sargasso Sea. *Nature*, **394**, 263-266.

McKinley, G.A. (2002) Interannual variability of the air-sea fluxes of carbon and oxygen. Ph.D. thesis. Massachusetts Institute of Technology.

McKinley, G.A., M.J. Follows and J.C. Marshall (2003) Mechanisms of interannual variability of the air-sea flux of CO_2. Submitted to *Global Biogeochem. Cycles.*

Murray, J. (2003) Ocean carbonate chemistry. This volume.

Nurser, A.J.G. and J.W. Zhang, (2000), Eddy-induced mixed-layer shallowing and mixed-layer/thermocline exchange. *J. Geophys. Res.*, **105**, 21851-21868.

Orr, J. (2002) Global Ocean Storage of Anthropogenic CO_2: Final Report. EC Environment and Climate Program.

Oschlies, A., (2002) Can eddies make ocean deserts bloom? *Global Biogeochem. Cycles*, **16**, 1106, doi:10.1029/2001GB001830.

Sabine, C.L., R.M. Key, K. Johnson, F.J. Millero, J. Sarmiento, D. Wallace and C. Winn (1999) Anthropogenic CO2 inventory of the Indian Ocean, *Global Biogeochem. Cycles*, **13**, 179-198.

Sarmiento, J.L., Murnane, R., Le Quere, C., (1995). Air-sea CO_2 transfer and the carbon budget of the North Atlantic. *Phil. Trans. Roy. Soc. Lond.*, **B348**, 211–219.

Sarmiento, J.L., N. Gruber, M.A. Brzezinski, and J.P. Dunne (2003) High latitude controls of the global nutricline and low latitude biological productivity. Nature, in press.

Sathyendranath, S., R.S.A. Longhurst, C.M. Caverhill, and T. Platt, (1995). Regionally and seasonally differentiated primary production in the North Atlantic. *Deep-Sea Res. I.*, **42**, 1773-1802.

Schmitz, W.J. and M.S. McCartney, (1993). On the North Atlantic circulation. *Rev. Geophysics.*, **31**, 29-49.

Stumm, W. and J. J. Morgan (1996) *Aquatic Chemistry.* Third Edition. John Wiley and Sons., N.Y., 1022pp.

Takahashi, T., Olafsson, J., Goddard, J.G., Chipman, D.W., Sutherland, S.C., 1993. Seasonal variation of CO_2 and nutrients in the high-latitude surface oceans: a comparative study. *Global Biogeochemical Cycles*, **7**, 843-878.

Takahashi, T., Wanninkhof, R.H., Feely, R.A., Weiss, R.F., Chipman, D.W., Bates, N., Olafsson, J., Sabine, C., Sutherland, S.C., (1999). Net sea-air CO_2 flux over the global ocean: An improved estimate based on the sea-air pCO_2 difference. *Proceedings of the 2nd International Symposium CO_2 in the Oceans*, Ed. Y. Nojiri, Center for Global Environmental Research, CGER-IO37-'99, Tsukuba, Japan, 9-15.

Uz, B.M, J.A. Yoder and V. Osychny, (2001). Pumping of nutrients to ocean surface waters by the action of propagating planetary waves, *Nature*, **409**, 597-600.

Volk T., and M.I. Hoffert (1985) Ocean carbon pumps: Analysis of relative strengths and efficiencies in ocean-driven atmospheric CO2 changes. *Geophys. Monogr.*, **32**, 99-110.

Wallace, D.W.R., (2001). Storage and transport of excess CO_2 in the oceans: the JGOFS/WOCE Global CO_2 survey. In *Ocean Circulation and Climate: Observing and Modelling the Global Ocean*, Eds. G. Siedler, J. Church and J. Gould, 489-521.

Wanninkhof, R. (1992) Relationship between wind speed and gas exchange over the ocean. *Journal of Geophysical Research*, **97**, 7373-7382.

Watson, A.J., P.D. Nightingale and D.J. Cooper, (1995) Modelling atmosphere-ocean CO_2 transfer. *Phil. Trans. Roy. Soc. Lond.*, **B348**, 125-132.

Williams, R.G., (2001). Ocean Subduction. In *Encyclopedia of Ocean Sciences*, Eds. J.H. Steele, K.K. Turekian and S.A. Thorpe, Academic Press, 1982-1992.

Williams R.G. and M.J. Follows, (2003). Physical transport of nutrients and the maintenance of biological production. In *Ocean Biogeochemistry: a JGOFS synthesis*. Ed. M. Fasham, Springer, pp. 19-51.

Williams, R.G., McLaren, A.J., Follows, M.J., (2000). Estimating the convective supply of nitrate and implied variability in export production over the North Atlantic. *Global Biogeochem. Cycles*, **14**, 1299-1313.

Xu, Y.-F. (1990) *A Study of the Biogeochemical Cycle of CO_2 in the Ocean Using a Parcel Model.* Ph.D. thesis. 160 pp. University of East Anglia, Norwich, U.K.

THE ROLE OF THE OCEANS IN THE GLOBAL CARBON CYCLE: AN OVERVIEW

IRINA MARINOV & JORGE L. SARMIENTO
Program in Atmospheric and Oceanic Sciences
Princeton University, Sayre Hall, Forrestal Campus
Princeton, New Jersey, USA
imarinov@princeton.edu, jls@splash.princeton.edu

INTRODUCTION

Typical gases such as oxygen tend to be distributed mostly in the atmosphere (99.4%) and less so in the ocean (0.6%). Carbon dioxide is vastly different from such gases, with 98.5% of the pre-industrial CO_2 in the ocean and only 1.5% in the atmosphere. The reason that CO_2 is found preferentially in the ocean is because of its high solubility (thirty times that of oxygen) and because of the hydrolysis reaction it undergoes to form carbonate and bicarbonate ions. The oceanic carbon inventory of total dissolved inorganic carbon (DIC) is also influenced by a combination of biologically mediated processes (the biological pump) and physical and chemical processes (the solubility pump). Both the biological and solubility pumps contribute to a higher concentration of DIC in the deep ocean, which reduces atmospheric CO_2 relative to what it would be otherwise.

The fact that the largest part of the total carbon inventory is in the ocean suggests that the ocean exerts a dominant control on atmospheric CO_2. Section 1 introduces the carbon pumps and their role in modulating $p\text{CO}_{2\,atm}$ on the glacial-interglacial time scale.

Human activities such as fossil fuel burning, production of cement, and land use change (e.g., tropical deforestation, land conversion for crops) have led to an increase in emissions of carbon dioxide, three quarters of which is due to fossil fuel burning. As a consequence, the partial pressure of carbon dioxide in the atmosphere has increased from 280$\pm$10 ppm before the industrial revolution to 373 ppm in 2002, the highest CO_2 level recorded in the past 20 million years. This spectacular increase is similar to the change that occured in going from glacial to interglacial periods. However, the present increase has happened within less than 200 years, while glacial-interglacial variations occured on a time scale of thousands of years.

M. Follows and T. Oguz (eds.), The Ocean Carbon Cycle and Climate, 251–295.

Most of the increase in radiative trapping since preindustrial times has been attributed to anthropogenic CO_2, with a smaller amount due to methane, nitrous oxide, chlorofluorocarbons, and other anthropogenic gases and aerosols. Since only the anthropogenic CO_2 which stays in the atmosphere adds to the greenhouse effect, it is critical that we quantify the amount of CO_2 taken up by ocean and land. Section 2 discusses the mechanisms behind the uptake of anthropogenic CO_2 by the oceans and illustrates them with results from ocean general circulation models (GCMs) and observations.

Anthropogenically induced increases in carbon dioxide and other atmospheric gases result in an increase in direct trapping of long wave radiation emitted from Earth's surface and have most likely led to climate change and global warming (IPCC Report, 2001). The resulting changes in the ocean physics and biology will modify the biological and solubility pumps, which in turn will affect the oceanic carbon uptake. The complex feedback mechanisms between climate change and ocean carbon uptake are discussed in Section 3.

1. BASIC MECHANISMS AT WORK: THE PRE-ANTHROPOGENIC CARBON CYCLE

CO_2 not only dissolves in water but also reacts with it to form bicarbonate ($HCO_3{}^-$) and carbonate ($CO_3{}^{2-}$) ions. The total dissolved inorganic carbon (DIC) is thus

$$DIC = [\mathrm{CO_2}] + [\mathrm{HCO_3^-}] + [\mathrm{CO_3}^{2-}], \qquad (1)$$

where the fractional contributions of the three species for average surface ocean water are 0.5%, 88.9% and 10.9%, respectively. Total alkalinity (ALK) can be defined from a statement of charge balance of the ocean and is well approximated by:

$$ALK \approx [\mathrm{HCO_3^-}] + 2\cdot[\mathrm{CO_3}^{2-}] + [B(OH)_4^{-1}] + [OH^-] - [H^+] + others, \qquad (2)$$

where $[B(OH)_4^{-1}]$ is the concentration of the borate ion and the fractional contributions of the first three species are 76.8%, 18.8%, and 4.2%, respectively. Observations show that DIC varies in the ocean from a surface value of about 1970 umol/kg to deep values of 2280 umol/kg, while ALK increases from about 2280 umol/kg at the surface to 2350 umol/kg at depth. What are the processes that maintain these gradients against the homogenizing tendencies of oceanic transport and mixing ? What determines the complex features of pre-anthropogenic air-sea CO_2 fluxes ? The mechanisms responsible for all of the above are the biological and solubility pumps.

Dissolved inorganic chemicals such as DIC and phosphate (PO_4) are consumed by photosynthesis at the ocean surface, and transported as sinking particles or as dissolved organic matter to the deep ocean. Remineralization converts organic matter back into dissolved inorganic matter deep in the ocean.

This biological process, called the **soft-tissue pump** [*Volk and Hoffert*, 1985], leads to a net transport of inorganic carbon from surface to depth. Another mechanism important in the ocean is the formation and dissolution of calcite and aragonite associated with coccolithophorids and foraminifera:

$$Ca^{2+} + CO_3^{2-} \Leftrightarrow CaCO_3. \tag{3}$$

Equations 1 and 2 show that the formation of mineral calcium carbonate shells at the sunlit surface is associated with a decrease in both DIC and ALK, while dissolution of $CaCO_3$ shells deeper in the water column results in an increase in DIC and ALK. Due to the double charge of the carbonate ion, the formation and dissolution of calcium carbonate changes ALK twice as much as it changes DIC. The net effect is therefore a transfer of ALK and DIC to depth in a 2:1 ratio, a process called the **carbonate pump** [*Volk and Hoffert*, 1985]. The soft-tissue and carbonate pumps add up to the full **biological pump**.

Carbon dioxide is more soluble in cold water than in warm water. Colder deep waters are therefore expected to hold more DIC than warmer surface waters. Both the North Atlantic and the Southern Ocean are areas of deep water formation. Here, cold high latitude surface waters rich in DIC are submerged to the deep creating the Antarctic bottom water (AABW) and the North Atlantic Deep water (NADW). The NADW penetrates southward at depths between 1500 m and 2500 m, as an essential part of the oceanic thermohaline circulation (THC). The AABW penetrates to the north along the bottom in all ocean basins. This circulation pattern results also in a net increase in DIC at depth relative to the surface. The net transport of carbon from the surface of the ocean to the deep which is due to temperature and salinity effects only is called the **solubility pump**. We will show that the DIC gradient between the surface and deep ocean in the absence of biology is indicative of the strength of the **solubility pump**. Simple box models show that without the biological and solubility pumps, $pCO_{2\,atm}$ would increase to $\sim$ 420 ppm, 50% higher than its pre-industrial value [*Marinov et al.*, in prep.].

1.1 THE REPRESENTATION OF OCEANIC CARBON PUMPS IN BOX MODELS

Using the simplest representation of the biological pump in the context of a two box model of the ocean with a surface and a deep box (Figure 1), the equation for the deep ocean balance of PO_4 can be written as:

$$V_d \frac{\partial PO_{4\,d}}{\partial t} = \nu \cdot (PO_{4\,s} - PO_{4\,d}) + \Pi_{PO_4} \tag{4}$$

where ν is a vertical exchange rate in m^3/s, $PO_{4\,s}$ and $PO_{4\,d}$ are the surface and deep phosphate concentrations in mol/m^3, and V_d is the volume of the

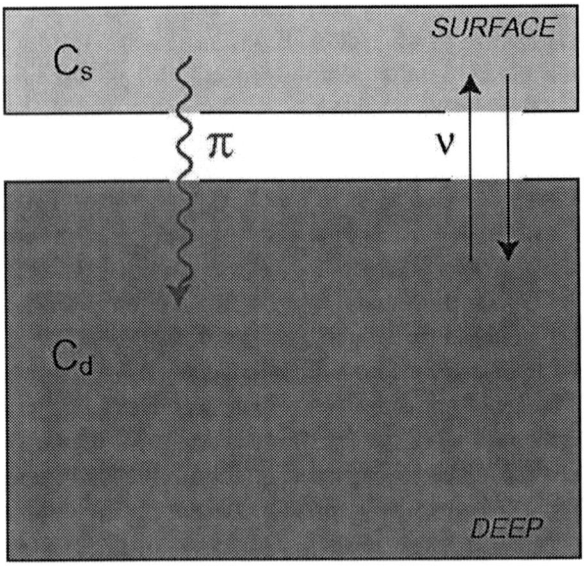

Figure 1. The two-box model representation of the oceanic biogeochemical loop. C_s and C_d are surface and deep concentrations, respectively. Π (mol/yr) is the biological pump, while ν (m^3/yr) is the vertical exchange rate between the reservoirs.

deep ocean in m^3. The net biological export production is represented by the sinking particle and dissolved organic matter flux Π_{PO_4} measured in mol/s of PO_4. Here, the time rate of change of deep PO_4 depends on the surface to deep concentration difference, which increases as more organic matter falls to the deep and gets remineralized back into inorganic PO_4. Biological production consumes PO_4 and DIC, and produces O_2 at the surface. In the interior of the ocean, organic matter remineralization consumes O_2, and produces inorganic PO_4 and DIC. All these processes tend to occur in constant "stoichiometric" ratios of C:P:$O_2 = 117 : 1 : -170$ [*Anderson and Sarmiento*, 1994]. Writing Eq. 4 at steady state for O_2, PO_4 and DIC yields

$$\begin{cases} \nu \cdot (\mathrm{PO_4}_d - \mathrm{PO_4}_s) &= \Pi_{\mathrm{PO_4}} \\ \nu \cdot (\mathrm{O_2}_d - \mathrm{O_2}_s) &= \Pi_{\mathrm{O_2}} \\ \nu \cdot (DIC_d - DIC_s) &= \Pi_{DIC}, \end{cases}$$

such that

$$DIC_d - DIC_s = r_{\mathrm{C:P}} \cdot (\mathrm{PO_4}_d - \mathrm{PO_4}_s). \quad (5)$$

$$\mathrm{O_2}_d - \mathrm{O_2}_s = r_{\mathrm{O_2:P}} \cdot (\mathrm{PO_4}_d - \mathrm{PO_4}_s), \quad (6)$$

where $r_{\mathrm{C:P}}$ is the stoichiometric ratio of total carbon to phosphate, which also takes into account the carbon in $CaCO_3$. Plugging in mean ocean values of 234

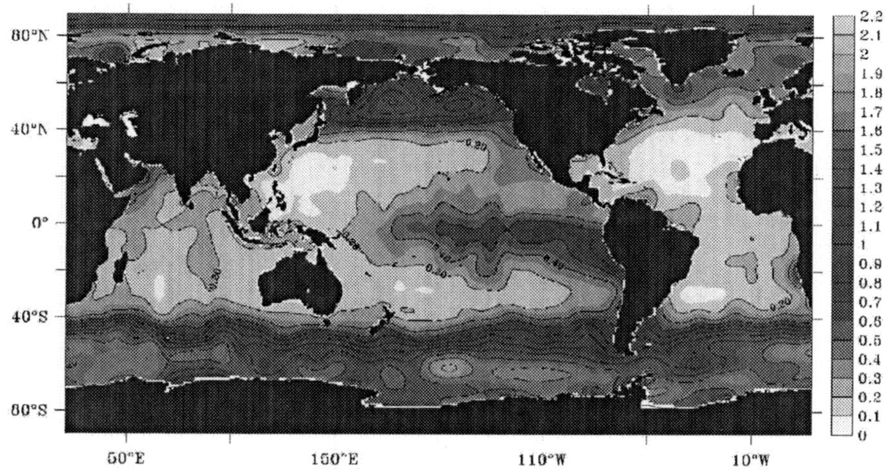

Figure 2. Observed annual mean surface phosphate at the ocean surface, in umol/kg. From NOAA NESDIS [1994].

mmol/m^3 for $O_{2\,s}$, 2.1 mmol/m^3 for $PO_{4\,d}$ and 0 mmol/m^3 for $PO_{4\,s}$, and using $r_{O_2:P}$= - 170 [*Anderson and Sarmiento*, 1994] yields anoxic bottom waters with oxygen concentrations of -123 mmol/m^3. This result is clearly wrong. Ocean data show deep oxygen concentrations around 170 mmol/m^3. Where does the two box model go wrong ?

The limitation of the two box model is related to the fact that most water that participates in the formation of deep water originates in small surface regions of the high latitudes and therefore has properties different from the global surface mean. In particular, high latitude waters are characterized by cold temperatures resulting in higher than average O_2 and inefficient biology resulting in large quantities of unutilized nutrients (1.3 mmol/m^3 of PO_4), also called **preformed nutrients**. Figure 2 shows that nutrients in high latitudes are much higher than average low latitude nutrients. Because of the large scale ocean thermohaline circulation (THC), average nutrient concentrations of the high latitude surface waters feeding the deep ocean provide more appropriate values for the above problem. We modify our model to account for polar surface waters by including a third, high latitude box in our calculation, as shown in Figure 3. The atmosphere is represented by a single box with uniform $pCO_{2\,atm}$ of 280 ppm. The steady-state balance for O_2, PO_4, and DIC in the deep box becomes:

$$(\mathrm{PO}_{4\,d} - \mathrm{PO}_{4\,h}) \cdot (T + f) = (\Pi_l + \Pi_h) \tag{7}$$

$$(\mathrm{O}_{2\,d} - \mathrm{O}_{2\,h}) \cdot (T + f) = (\Pi_l + \Pi_h) \cdot r_{\mathrm{O}_2:\mathrm{P}} \tag{8}$$

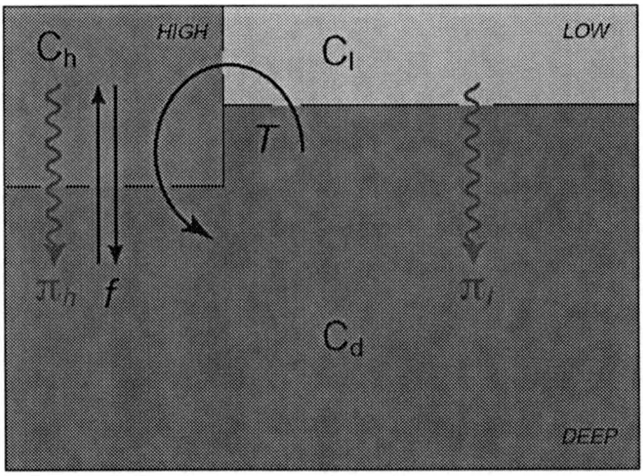

Figure 3. The three-box model of *Sarmiento and Toggweiler* [1984]. Box h represents the surface polar ocean of both hemispheres, box l the low and mid latitude surface ocean, box d the deep ocean. Photosynthesis converts phosphate into sinking particles, with Π_h and Π_l the measure of this export production. T is the unidirectional thermohaline overturning circulation represented as an advection process, while f is a diffusive mixing term. Both high and low latitudes exchange CO_2 with the atmosphere (not shown). Similar models were proposed independently by *Knox and Mc Elroy* [1984] and *Siegenthaler and Wenk* [1984].

$$(DIC_d - DIC_h) \cdot (T + f) = (\Pi_l + \Pi_h) \cdot r_{\mathrm{C:P}} \tag{9}$$

where h and l represent the high and low latitude boxes, respectively. Note that in this model, export production includes both low (Π_l) and high latitude (Π_h) production. Here, T (m^3/s) represents the conveyor belt circulation, an advection process which moves water up from the deep box into the low latitude surface, poleward to high latitudes and back to the deep box. The diffusion term f (m^3/s) mixes deep and high latitude surface waters. Equations 7 and 8 give:

$$\mathrm{O_2}_d = \mathrm{O_2}_h + r_{\mathrm{O_2:P}} \cdot (\mathrm{PO_4}_d - \mathrm{PO_4}_h) = 316 - 170 \cdot (2.1 - 1.3) = 180\,\mathrm{mmol/m^3}, \tag{10}$$

close to the observed value of deep oxygen. This result confirms the intuition gained from our analysis of the failure of the two box model.

The low and high latitude $p\mathrm{CO_2}$ concentrations are connected through gas exchange with the atmosphere. The steady state balance in the low latitude surface box takes into account this gas exchange:

$$(\mathrm{PO_4}_d - \mathrm{PO_4}_l) \cdot T = \Pi_l \tag{11}$$

$$(DIC_d - DIC_l) \cdot T + k_l \beta_l A_l \cdot (p\mathrm{CO_2}_a - p\mathrm{CO_2}_l) = r_{\mathrm{C:P}} \cdot \Pi_l, \tag{12}$$

$$k_l \beta_l A_l \cdot (p\mathrm{CO_2}_a - p\mathrm{CO_2}_l) = k_h \beta_h A_h \cdot (p\mathrm{CO_2}_h - p\mathrm{CO_2}_a) \tag{13}$$

Here, k_l and k_h are gas exchange rates (m/s) representative of the low and high latitudes, respectively; A_l, A_h are ocean areas (m^2) effectively exposed to gas exchange; and β_l, β_h are solubilities of CO_2 (mol/($m^3 \cdot \mu$atm)), which depend on local temperature and salinity.

In the course of this chapter we will explore different ways of measuring the strength of the biological pump. Perhaps the simplest way of thinking about the strength of the biological pump is as the DIC gradient between deep ocean and the high latitude box. Combining Eqs. 7 and 9 gives:

$$DIC_d - DIC_h = r_{\text{C:P}} \cdot (\text{PO}_{4\,d} - \text{PO}_{4\,h}). \tag{14}$$

The total strength of the carbon pump has a significant impact on $p\text{CO}_{2\,atm}$, by determining how much carbon is stored in the deep ocean relative to the surface ocean and the atmosphere. An increase in the carbon pump means that more carbon is sequestered in the deep ocean which will decrease $p\text{CO}_{2\,atm}$. Concretely, we see that a decrease in $\text{PO}_{4\,h}$, an increase in $\text{PO}_{4\,d}$ or an increase in $r_{\text{C:P}}$ in Eq. 14 increase the biological pump strength $DIC_d - DIC_h$. Since the deep ocean is a huge reservoir, DIC_d is difficult to change. An increase in $DIC_d - DIC_h$ therefore implies a reduction in DIC_h and thus a decrease in $p\text{CO}_{2\,h}$, which allows more atmospheric CO_2 drawdown.

Since $\text{PO}_{4\,d}$ is basically set by the total amount of PO_4 in the system, which is assumed constant for our purpose, Eq. 14 tells us that the strength of the biological pump is directly related to the amount of PO_4 that goes unutilized in the high latitudes, i.e., the **preformed phosphate** ($\text{PO}_{4\,h}$). A decrease in preformed nutrients strengthens the biological pump; the maximum pump strength is achieved when preformed PO_4 is zero as in the 2 BM. It is clear now that the anoxic bottom waters in the 2 BM are due to an unrealistically strong biological pump. Using a preformed value of 1.3 mmol/m^3 in a 3 BM yields a lower biological pump and a realistic bottom oxygen (Eq. 10). Equations 7 and 11 yield an expression for $\text{PO}_{4\,h}$:

$$\text{PO}_{4\,h} = \frac{\text{PO}_{4\,d} \cdot f - \Pi_h + \text{PO}_{4\,l} \cdot T}{f + T}. \tag{15}$$

In comparison to the two box model where surface nutrients are fixed (Eq. 5), changes in T, f and Π_h can change the preformed nutrient concentration, the strength of the biological pump and therefore $p\text{CO}_{2\,atm}$. These parameters confer upon the three box model additional degrees of freedom compared to the two box model,allowing the carbon cycle to respond to glacial-interglacial changes.

1.1.1 The biological pump: glacial-interglacial changes. One of the most interesting problems in the study of climate is the attempt to understand the glacial-interglacial cycles, characterized by significant - about 80 ppm- and fairly regular variations of atmospheric CO_2, as seen in the Vostok ice core [*Petit et al.*, 1999].

Sarmiento and Toggweiler [1984], *Knox and Mc Elroy* [1984], and *Siegenthaler and Wenk* [1984] used simple box models in an attempt to explain the decrease in $pCO_{2\,atm}$ in glacial times compared to interglacials. Since changes in the solubility pump due to colder temperatures account for only 25% of the 80 ppm drop during each of the glacial periods [*Bacastow*, 1996] and much of this is compensated by a corresponding salinity increase [*Broecker*, 1982], it has been suggested that an increased biological pump strength played the principal role in lowering $pCO_{2\,atm}$ in glacial times [e.g., *Sigman and Boyle*, 2000]. Equations 14 and 15 allow many alternatives for achieving an increase in biological pump strength during glacial times, as listed below:

1.) *Sarmiento and Toggweiler* [1984], *Knox and Mc Elroy* [1984] and *Siegenthaler and Wenk* [1984] proposed that a decrease in preformed high latitude phosphate occured during glacial times. According to Eq. 14 this would result in a stronger biological pump which sequesters more carbon in the deep ocean and draws more atmospheric CO_2 into the ocean. But why should we expect a decrease in high latitude surface nutrients during glacial times ? Two separate mechanisms could be responsible for increased utilization of available nutrients during glacials.

1a.) An increase in the downward organic flux Π_h would increase $PO_{4\,h}$ according to Eq. 15. *Toggweiler et al.* [Part 2, 2003] illustrate this mechanism in Figure 4 where a ten fold increase in Π_h more than doubles the surface to deep DIC gradient. The maximum strength of the biological pump is achieved when $PO_{4\,h} = PO_{4\,l} = 0$ and $f = 0$, i.e., when the three box model becomes a two box model. An explanation offered for the increases in high latitude productivity during glacial times is the *Martin et al.* [1990] iron fertilization mechanism. Iron concentration is relatively low in the ocean, and particularly scarce in certain nutrient rich areas (so-called high nutrient low chlorophyll or HNLC regions) such as the Southern Ocean. *Martin et al.* proposed that the growth of Southern Ocean phytoplankton is physiologically limited by the lack of iron in the present climate. Due to colder, drier, and windier climate, the atmosphere transported much more iron-bearing dust to the oceans during the last glacial maximum. This addition of soluble iron to the Southern Ocean might have increased photosynthetic uptake of CO_2 and nutrients from the surface ocean, thus increasing CO_2 uptake during glacials.

1b.). Another possible explanation for the low $PO_{4\,h}$ stemming from the original BM analyses is an increase in polar stratification and the resulting reduction in Southern Ocean convective mixing f during glacial times. Consider a model in which solubility effects due to varying surface temperature and salinity are neglected. In such a model the gas exchange and distribution of CO_2 is due only to biology and f mixes high latitude surface waters with deep waters rich in DIC and nutrients. Due to insufficient biological uptake relative to the vertical supply of carbon in the high latitudes, carbon stored in the deep ocean is thus outgassed to the atmosphere through the high latitude surface box. A decrease in f during glacial times would therefore act as a cap on the biological outgassing of CO_2 to the atmosphere in high latitudes. More CO_2 would stay in the ocean and atmospheric CO_2 concentration would decrease. This mechanism was systematically analyzed by *Toggweiler* [1999] in the context of 3, 4, 6, and 7 box models.

2.) Another possible mechanism for lowering $pCO_{2\,atm}$ stemming from Eq. 14 is to change the total oceanic inventory of DIC and therefore DIC_d, which could be feasible by changing the total carbon inventory in land plants. *Shackellton* [1977] shows, however, that this mechanism works in the opposite direction. *Broecker* [1982] suggested that a change in the deep ocean inventory of PO_4 due to PO_4 dissolution from exposed shelves was responsible for the observed glacial-interglacial difference in $pCO_{2\,atm}$. This mechanism was however shown to result in a delay of the CO_2 change well beyond what is observed in trapped air-bubbles in ice cores [*Sigman and Boyle*, 2000].

Others contend that fixed nitrogen rather than phosphorus limits the net effect of the biological CO_2 pump on geologic timescales and that the ratio of N fixation/ denitrification plays a key role in establishing atmospheric CO_2 [*McElroy*, 1983; *Falkowski*, 1997]. Nitrogen is involved in photosynthesis and remineralization just like phosphorus. Additionally, bio-available oceanic nitrogen is supplied through river input, rain, and in situ fixation, and is lost from the ocean through denitrification. If nitrogen limits biological production, an increase in its oceanic inventory during glacial times would have stimulated the production of phytoplankton, contributing to the observed decrease in $pCO_{2\,atm}$. This could have been brought about by a decline in denitrification during glacials [*Ganeshram et al.*, 2002], increased continental weathering, or enhancement of nitrogen fixation by increased aeolian fluxes of iron [*Falkowski*, 1997].

3.) It is possible that physical processes such as stratification and mixing dictate ecosystem structure. Since some species do better in certain environments than others, the bulk stoichiometric ratio of carbon to phosphate $r_{C:P}$ might change if we alter the physics of the environment, thus changing the biological pump (Eq. 14). At the same time we note that the stoichiometric coefficient $r_{C:P}$ depends on the fraction of biological production going into the formation of $CaCO_3$, which might have changed between glacials and interglacials. Mechanisms based on changes in the carbonate cycle might have played some part in the glacial-integlacial changes in biological pump; these are discussed by *Archer and Maier-Reimer* [1994].

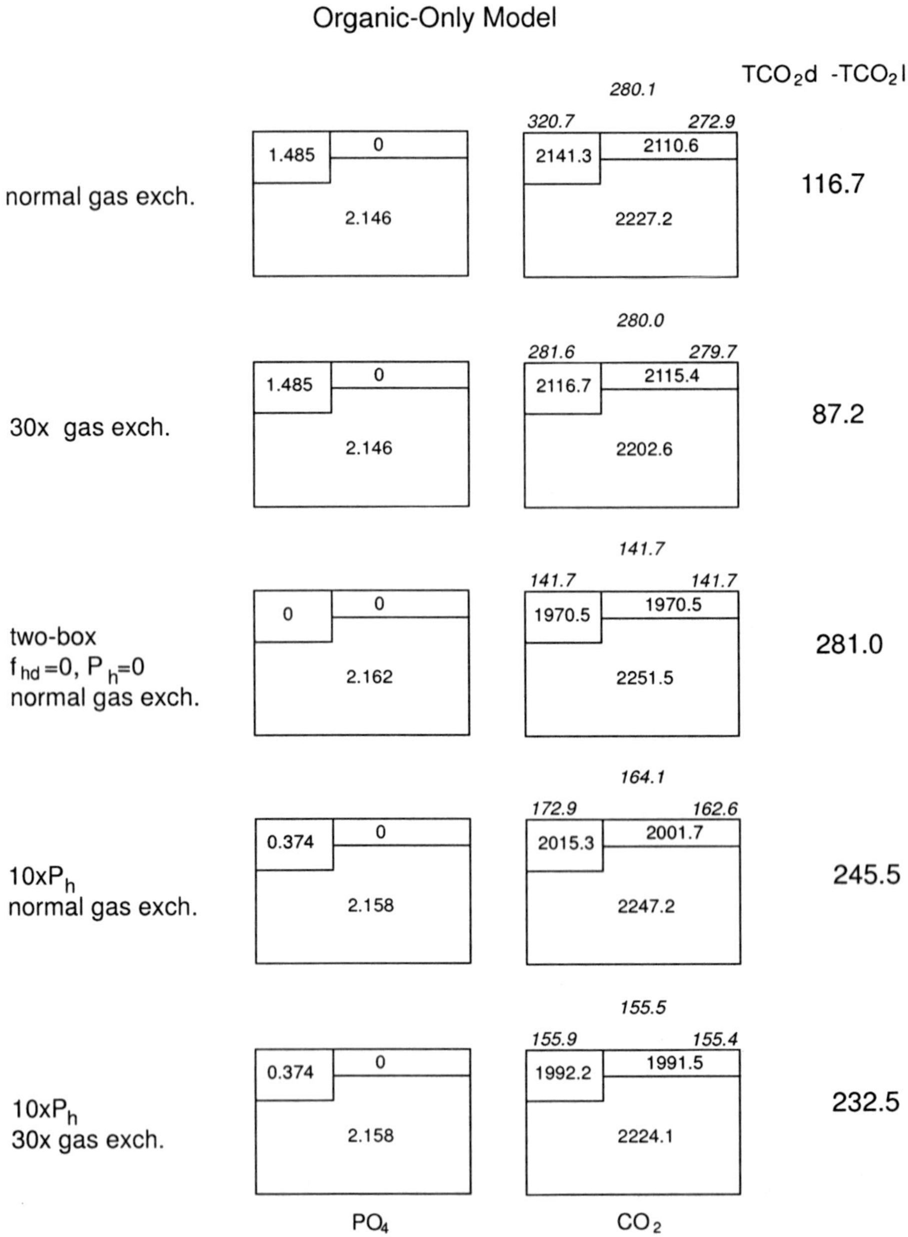

Figure 4. Figure 2 of *Togweiler et al.* [Part 2, 2003]. PO $_4$ DIC, pCO $_2$concentrations, and $DIC_d - DIC_l$ (here TCO_2d - TCO_2l) for five solutions of a soft-tissue only model. Standard simulations have T=20 Sv and f_{hd}=60 Sv (f in this paper). $PO_{4\,l}$ is set to zero. The top two panels give results from the 3BM with normal and fast gas exchange; the bottom two panels give results from a ten fold increase in export production, P_h (Π_h in this text). Total amount of CO_2 in normal gas exchange models (panels 1, 3, 4) is kept constant to the amount given in panel 1; total amount of CO_2 in fast gas exchange models (panels 2, 5) is kept constant to the amount given in panel 2. An increase in export production or a decrease in gas exchange increase the biological pump strength, as measured by the DIC gradients in italics, and as shown in our analysis in Section 2.1.

1.1.2 The biological pump: the effect of surface disequilibrium. Recently, a series of papers have been written proposing that gas exchange disequilibrium in the high latitudes of the Southern Ocean may play an important role in the glacial-interglacial cycle of atmospheric CO_2. We discuss here how one can use the box models to gain insight into how this functions.

Volk and Hoffert [1985] first proposed the average low latitude surface-to-deep gradient of DIC in the ocean, $DIC_d - DIC_l$, as the relevant metric for the strength of the carbon pumps. We rewrite Eqs. 11- 13 as:

$$DIC_d - DIC_l = r_{\mathrm{C:P}} \cdot (\mathrm{PO_4}_d - \mathrm{PO_4}_l) + dCgasex, \quad \text{where} \tag{16}$$

$$dCgasex = -k_h \beta_h A_h \cdot (p\mathrm{CO_2}_h - p\mathrm{CO_2}_a)/T. \tag{17}$$

Equation 16 suggests that another mechanism potentially responsible for the increase in strength of the biological pump during glacial times is a change in the degree of equilibration of surface waters with atmospheric CO_2, $dCgasex$. Consider a scenario in which either the thermohaline circulation (T) is strong, the gas exchange rate (k_h) is low, or the effective gas exchange area (A_h) is small. In such a scenario, water sinking into the Southern Ocean has less time to equilibrate with the atmosphere by outgassing carbon dioxide, and as a consequence deep water ends up trapping more DIC than the equivalent watermass in a scenario where these parameters are reversed in magnitude. This increases the total storage of CO_2 in the deep ocean. In terms of our equations, note that in a biology only scenario $p\mathrm{CO_2}_h > p\mathrm{CO_2}_a$ and $dCgasex < 0$ (Eq. 17). Decreasing k_h or A_h or increasing T thus increase the strength of the biological pump as given by Eq. 16 and as illustrated graphically in the top two panels of Figure 4 from *Toggweiler et al.* [Part 2, 2003].

This mechanism is equivalent to the sea-ice mechanism proposed by *Stephens and Keeling* [2000], and the gas exchange mechanism of *Toggweiler et al.* [Part 2, 2003]. Interestingly, this mechanism occurs also as a result of self-sustained climate oscillations in *Gildor et al.* [2002]. Increased sea ice coverage decreases A_h and results in a decrease in gas exchange efficiency in high latitudes, and therefore a stronger biological pump. *Stephens and Keeling* [2000] conjecture that complete coverage with ice of the Southern Ocean during glacials would have decreased atmospheric pCO_2 by 80 ppm relative to the present CO_2 level. For more on this and other glacial-interglacial mechanisms see also the Chapter 10, "Glacial-Interglacial Cycles" in this book.

1.2 THE REPRESENTATION OF CARBON PUMPS IN GCMS AND THE EFFECT OF THE PUMPS ON AIR-SEA CARBON FLUXES

The distribution of carbon in the ocean is affected by both the biological and the solubility pumps, which depend in turn on the detailed circulation characteristics and gas exchange at the surface. General circulation models (GCMs) offer us a closer, more detailed look at what sets the strengths of these pumps in the ocean. GCMs also enable us to examine another important aspect of the global carbon cycle, namely, what sets the pattern of CO_2 flux at the surface of the ocean.

The ocean circulation model we will use for illustration in this chapter is the Modular Ocean Model, Version 3 in the exact setup described by *Gnanadesikan et al.* [2002]. The biogeochemical component of the model follows the OCMIP2 specifications [*Najjar and Orr*, 1999]. The model includes surface production and sub-surface consumption of organic phosphorus, carbon and oxygen in a ratio of P:C:O_2 = 1:117:-170 [*Anderson and Sarmiento*, 1994], carbon chemistry and physical transport of nutrients and DIC. Dissolved organic phosphate and calcium remineralization follow prescribed functions of depth. Biological production of organic phosphorus J_{Prod} (mol/m^3/s) occurs only in the euphotic layer (top 75 m) and is given by:

$$J_{Prod} = 1/\tau \cdot (PO_4 - PO_4{}^*) \;\; if \;\; PO_4 > PO_4{}^* \qquad (18)$$

$$J_{Prod} = 0 \;\; if \;\; PO_4 < PO_4{}^*. \qquad (19)$$

In our standard run, PO_4 is restored to the observed (Levitus) surface nutrients, represented by $PO_4{}^*$, with $\tau = 30$ days. The air-sea gas transfer of CO_2 in the model follows the standard formulation of *Wanninkhof* [1992] which can be written in a simplified manner as:

$$Flux = k \cdot \beta \cdot (pCO_{2\,ocean} - pCO_{2\,atm}), \qquad (20)$$

where k is a gas transfer coefficient (m/s) which is a function of wind speed, Schmidt number, and sea ice fraction and β is the CO_2 solubility (mol/m$^3\cdot\mu$atm). $pCO_{2\,ocean}$ is the partial pressure at the surface of the ocean computed at each time step from prognostic variables of the model: temperature (T), salinity (S), DIC, and ALK.

Four different simulations will be presented in this section. Three simulations illustrate the impact of changing gas exchange or sea ice coverage on the air-sea flux, by analogy with the box model mechanisms presented in Section 1.1.1. In these simulations the model is run to equilibrium with a preindustrial $pCO_{2\,atm}$ of 278 ppm. A fourth simulation show the impact of driving $PO_{4\,h}$ to zero in the Southern Ocean, by analogy with mechanism 1, Section 1.1.1. In this case the model is run to equilibrium while keeping the total ocean-atmosphere carbon budget fixed.

The air-sea flux of CO_2 can be separated into a solubility component due to the solubility pump and a biotic component due to the biological pump. The solubility component is mainly a consequence of the heat and water fluxes, which change the solubility of carbon and lead to locally large air-sea CO_2

fluxes. The biological formation of dissolved and particulate organic carbon and calcium carbonate and the transport of these materials to other regions of the world where they are remineralized or dissolved leave their imprint on the biological component of the carbon flux. We separate the two effects by running two separate models.

In a first **biology only model** we turn the solubility pump off by setting the ocean temperature and salinity constant everywhere at the surface in the calculation of the air-sea gas exchange of CO_2 [*Toggweiler et al.*, Part 2, 2003; *Marinov et al.*, in prep.]. The values chosen for T and S are observed average surface values of 10°C and 34.7 psu, respectively. Our model includes both the carbonate and soft tissue pumps.

The biological air-sea CO_2 flux is intrinsically related to the nutrient distribution at the ocean surface, $PO_4{}^*$, shown in Figure 2. In high surface nutrient concentration regions such as the Southern Ocean, North Atlantic, and North Pacific, biological uptake J_{Prod} is inefficient relative to the upward supply of excess carbon and nutrients. Consider the Southern Ocean, a region of strong upwelling of deep waters characterized by the largest amount of preformed nutrients. Here J_{Prod} is not strong enough to take up all of the PO_4 from the surface water. As remineralized nutrients and CO_2 upwell, the pCO_2 at the ocean surface builds up, resulting in the escape of CO_2 to the atmosphere.

In a steady state, large uptake in the subtropics is required in order to counterbalance high latitude degassing. While the subtropics are overall a CO_2 uptake region,waters high in nutrients and DIC are brought up by upwelling along the coast of NW South America, in the sub-Arctic Western Pacific, Equatorial Eastern Pacific and by the upward entrainment of the equatorial undercurrent water, contributing locally to large biological CO_2 fluxes out of the ocean.

In a second **solubility model** biology is turned off [*Murnane et al.*, 1999]. The air-sea flux of CO_2 is the result of changes in CO_2 solubility due primarily to heat fluxes and to a smaller extent to water fluxes. Warming of relatively cold upwelled water in equatorial regions decreases gas solubility and results in loss of CO_2 to the atmosphere. This loss is offset by the gain of CO_2 in high latitudes due to cooling of the surface waters. Solubility effects generally oppose the effect of the biology on the air-sea flux. Figure 5 shows the strong **compensation** between the biological (starred line) and solubility air-sea flux (dashed line).

Finally, a **full model** which includes both biological and solubility effects is run. The full air-sea CO_2 flux represented by the full line in Figure 5 is very well approximated by the sum of the the biological and the solubility air-sea flux fluxes (dotted line). Carbon cycle models do a reasonable job at reproducing the observed spatial distribution of the pre-industrial air-sea CO_2 flux. Indeed, our full CO_2 flux is very similar to the flux estimated from observa-

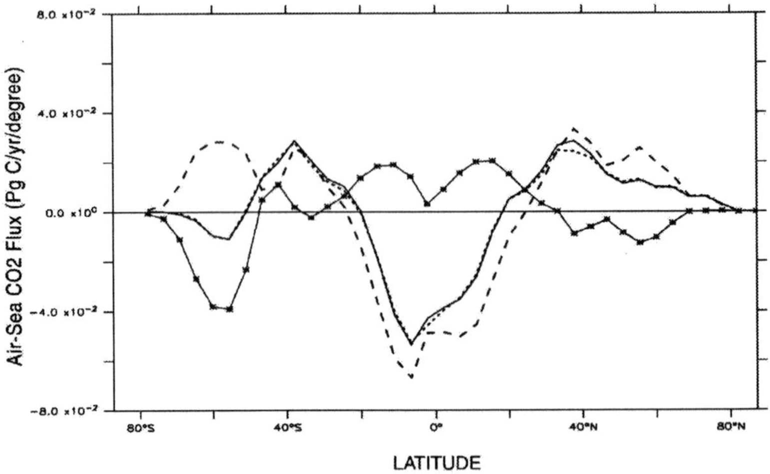

Figure 5. Equilibrium air-sea DIC flux components from a set of Princeton MOM3 simulations (PgC/yr/degree). The starred and dashed lines are air-sea fluxes from biology only and solubility only models, respectively. The dotted line is the sum of these two, while the full line is the total flux from a standard pre-industrial run which includes both biology and solubility effects. All models were run to equilibrium with a fixed atmosphere of 280 ppm. The standard full run (full line) is the so-called "LL" run of *Gnanadesikan et al.* [2002].

tions by *Takahashi* [1999] shown in Figure 6. This is to be expected since in our model nutrients, temperature, and salinity are restored towards observations at the surface of the ocean.

In order to explore analogies with the box model we calculate the surface to deep DIC gradients in the solubility, biology only, and full models. Since the DIC of the waters feeding the deep ocean is hard to determine exactly in GCMs, we express our carbon pumps by analogy with the $DIC_d - DIC_l$ metric proposed in Section 1.1.2. DIC gradients here are calculated as deviations from the corresponding mean surface DIC concentration, such that all surface values are in each case zero. Figure 7a shows that the sum of the solubility only and biology only DIC gradients (dotted line) is an excellent approximation to the full model DIC gradient (full line). The difference between the dotted and full lines is due to nonlinearities in the carbon chemistry. For example, the addition of biology to a solubility model changes the surface chemistry, modifying the reponse of the system to addition or removal of heat.

The biological pump and the corresponding air-sea flux depend strongly on the preformed nutrients and the efficiency of gas exchange, as suggested by our discussion in Section 1.1.2. Figure 7b shows results from the biology only model in which we have either increased the gas exchange, or we have depleted surface nutrients in the Southern Ocean. We discuss each of these in turn.

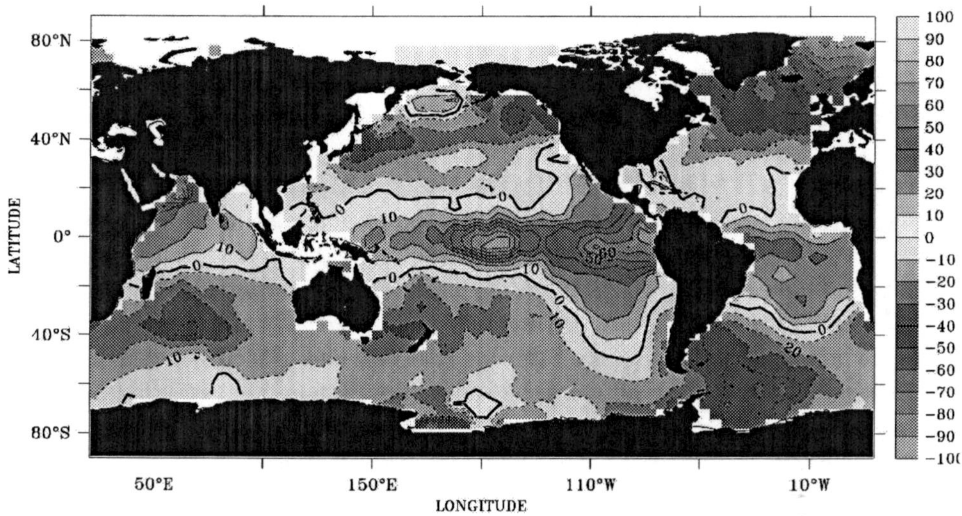

Figure 6. Annually averaged CO_2 flux at the ocean surface in $mol/m^2/yr$ for year 1995. Estimated from measurements of sea-air pCO_2 difference by *Takahashi et al.*, [1999].

Figure 7. a. Gradients in DIC from our set of Princeton GCM simulations: a solubility only model (dashed line), a biology only model (starred line) and a full model with biology and solubility effects (full line). The sum of the latter two (dotted line) is a good approximation to the DIC gradient in the full model (full line). b. DIC gradient from a biology only model with normal gas exchange (starred line), infinitely fast gas exchange (squares), and normal gas exchange after Southern Ocean nutrient depletion south of 30°S (triangles). The normal gas exchange model is the same one used for Figure 5.

In a first simulation with the biology only model we set the gas exchange to be as fast as possible by requiring surface DIC values at the ocean surface to be in equilibrium with the overlying atmosphere at each time step. Increasing gas exchange means that more of the deep CO_2 escapes to the atmosphere through the high latitude surface waters. The result is a net decrease in the surface-to-deep DIC gradient and therefore in the strength of the biological pump compared to the standard biological run. This mechanism is analogous with those observed in box models and discussed in Section 1.1.2 and is also shown in the top two panels of Figure 4. Incomplete equilibration of the Southern Ocean due to slower gas exchange could have contributed to a stronger biological pump and a lower atmospheric pCO_2 during glacials [*Toggweiler et al.*, 2003; *Stephens and Keeling*, 2000].

Traditionally, GCM based studies of the ice-age CO_2 drawdown focused on mechanisms that enhance the fertility of the ocean or on mechanisms that increase the ratio of organic matter relative to $CaCO_3$ in the sinking flux [*Archer and Maier-Reimer*, 1994; *McElroy*, 1983]. An increase in nutrient utilization in high latitudes was one of the first explanations offered for the observed decrease in atmospheric CO_2 going from interglacials into glacials (mechanism 1a., Section 1.1.1). Fertilizing the ocean with iron so as to increase biological productivity has also been proposed as a potential mechanism for offsetting anthropogenic emissions of carbon dioxide [*Martin*, 1991].

Assuming that the total amount of phosphate in the ocean is fixed, Equation 14 implies that the maximum gradient in DIC is achieved when high latitude nutrients are depleted, i.e., when $PO_{4\,h}$ is set to zero. In this case the organic pump operates at maximum efficiency, retaining the largest amount of CO_2 in the deep ocean. In the GCM we can approximate a more efficient biological pump by performing nutrient depletion simulations. We restore surface phosphate to zero in three different areas of the ocean: Southern Ocean south of 30°S, North Atlantic (30°N-80°N), North Pacific (30°N-67°N), and the tropical band (18°S-18°N). According to our definition of biological productivity (Eq. 18), setting PO_4^* to zero maximizes J_{Prod} and strengthens the biological pump, which transfers CO_2 from the atmosphere to the ocean. After 500 years of simulation, the decrease of $pCO_{2\,atm}$ is 63 ppm for nutrient depletion in the Southern Ocean, 15 ppm for depletion in the North Atlantic, 3.7 ppm for depletion in the North Pacific and 4.4 ppm in the tropical depletion simulation.

As areas of deep water formation, high latitude waters influence strongly the characteristics of the deep ocean. As expected from our 3 box model analysis, forcing nutrients to zero in high latitudes has a much stronger impact on the biological pump than forcing them to zero in low latitudes. The North Atlantic is a region where surface water is fed directly in the abyss. The Southern Ocean is a region where abyssal waters come to the surface with very high nutrients and DIC; here these waters can lose much of the DIC that is added to the deep ocean by the biological pump. Because of these circulation characteristics,the potential change in the biological pump following depletion is much stronger in the Southern Ocean than in the North Atlantic. Because it is not an area of strong exchange with the deep ocean, the North Pacific is a weak sink of CO_2 under depletion.

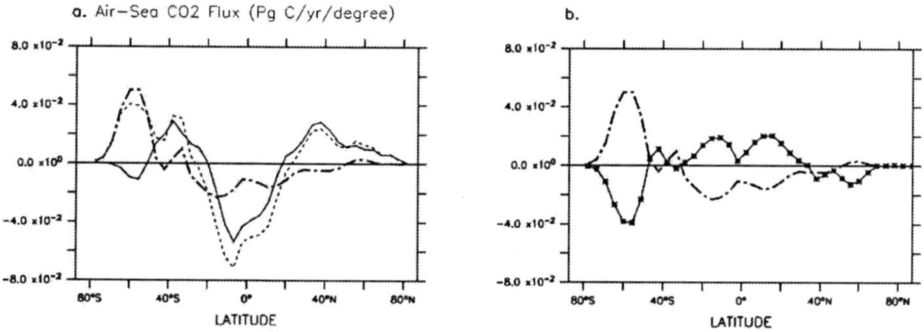

Figure 8. (a) Equilibrium air-sea fluxes of CO_2 in a full model before depletion (pre-industrial, full line) and after depletion of nutrients everywhere south of 30°S (dotted line). The difference between these two gives the dash-dotted line. Note that the air-sea flux after depletion is similar to the solubility only flux in Figure 5. (b) The contribution due to depletion (dash-dotted line) nearly cancels out the biological air-sea flux (starred line, see also Figure 5). Depletion puts the ocean in a state where there is no net air-sea flux due to biological cycling.

Figures 7b and 8 show the impact of Southern Ocean nutrient depletion on the biological pump and on the total air-sea fluxes of CO_2, respectively. Figure 8a shows the standard air-sea flux before depletion (full line, includes both solubility and biological effects), and the air-sea flux after performing nutrient depletion (dotted line). The dash-dotted line is the difference between the above two lines, and represents the change in air-sea flux due to the depletion.

Under this depletion scenario and for this particular circulation and gas exchange, the biological pump in the Southern Ocean operates at its maximum, as shown by the high DIC gradient in Figure 7b. Biology is highly efficient such that all the nutrients brought from the deep and the CO_2 associated with them are consumed during photosynthesis, increasing significantly the net export production. Consequently CO_2 is not allowed to escape to the atmosphere and the total biological CO_2 flux is close to zero. This is exactly the meaning conveyed by Figure 8b,which shows that the air-sea flux component due to depletion (dash-dotted line) acts in such a way as to cancel out the initial (pre-depletion) biological air-sea flux (starred line). The significant cancellation north of 30°S suggests that the Southern Ocean plays a critical role in setting up the biological pump in the rest of the ocean. Note that the total air-sea flux after depletion (dotted line in Figure 8a) is roughly equal to the pre-depletion solubility flux (dashed line in Figure 5).

Archer et al. [2000] made the point that different models respond differently to nutrient depletion simulations. In particular, they found that the decrease in atmospheric pCO_2 with nutrient depletion is considerably larger in simple box models than in GCMs.

To gain some insight into why this might be so, let us consider again Eqs. 14 and 16. While in the 3 box model there is only one type of preformed nutrient ($PO_{4\,h}$), the net preformed nutrient "seen" by the deep ocean is in reality a combination of unused PO_4 from different regions of the ocean, most notably from the North Atlantic and Southern Ocean. The actual value of $PO_{4\,h}$ therefore depends not only on preformed PO_4 in the North Atlantic and Southern Ocean but also on the fractional composition of deep water, which is a function of advection T and mixing coefficients f characteristic to each region. Thus, different ways of representing large scale circulation, deep water formation, and mixing processes in the box models relative to the GCMs could result in different net values of $PO_{4\,h}$ and therefore in different responses of the biological pump to nutrient depletion. At the same time, $dCgasex$ is a combination of surface disequilibrium contributions from the deep water formation regions. Recently, Toggweiler conjectured that the Southern Ocean surface water equilibrates to a different degree in box models and GCMs because of different mechanisms for deep water ventilation [*Toggweiler et al.*, Part 2, 2003]. In box models deep water is ventilated through high latitude boxes of 10% to 20% of the ocean area while in GCMs ventilation occurs in very small areas where deep convection reaches the surface. GCM surface waters are therefore far less equilibrated with the atmosphere than corresponding water masses in box models and respond less to surface nutrient depletion. Differences in surface disequilibrium can therefore be critical in explaining differences in biological pumps between box models and GCMs.

2. ANTHROPOGENIC CO_2 UPTAKE BY THE OCEANS

In order to monitor and eventually moderate climate change due to human interference on the natural system,we must understand the mechanisms driving the carbon sinks in the ocean and land, and accurately project their future evolution. The following two sections review some of the mechanisms behind the present ocean carbon sink without climate change (Section 2) and in the presence of climate change (Section 3).

The growth rate of atmospheric CO_2, as shown in Figure 9 [*Sarmiento and Gruber*, 2002], was about half the rate of fossil CO_2 emissions in the 1990s. The implication is that the ocean and the terrestrial biosphere are taking out about half of the anthropogenic fossil CO_2 released to the atmosphere. Figure 9 also points to the considerable interannual variability in the total rate of uptake, which has been linked to processes such as El Nino.

Partitioning the anthropogenic carbon sink between ocean and land is a non-trivial task which has been boosted by the development of the atmospheric oxygen method of Ralph Keeling [*Keeling et al.*, 1992; *Bender et al.*, 1994;

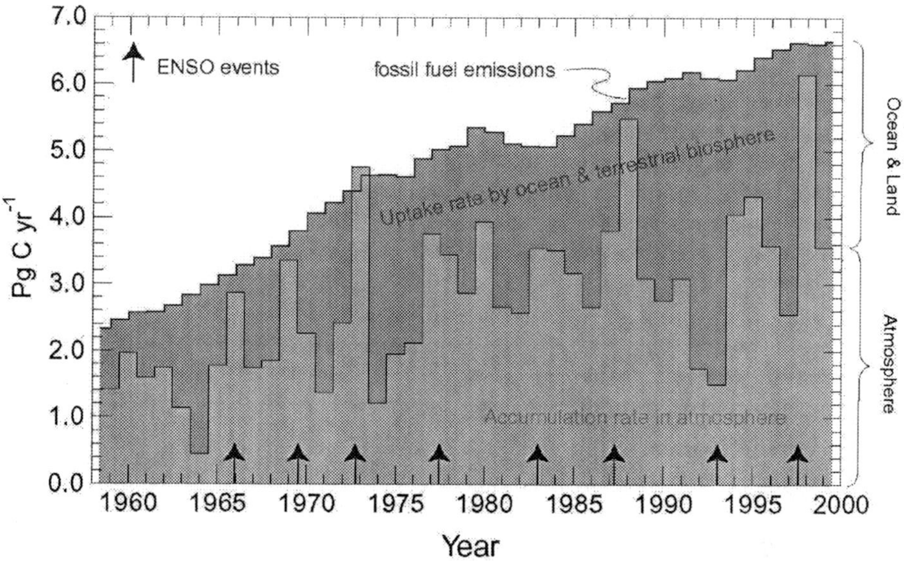

Figure 9. Growth rate of carbon reservoirs. Since 1958, the yearly accumulation rate of atmospheric carbon dioxide has grown, on average, from 1 to 3.0 Pg C/yr (light blue area). Over the same period, fossil-fuel emissions (red line) have grown from about 2.5 Pg C/yr to about 6.5 Pg C/yr. Net uptake by the ocean or terrestrial biosphere (green region) must account for the difference. Note the large interannual variation in the atmospheric CO_2 growth rate. Higher growth rates generally appear to be associated with El Nino episodes (black arrows), the exception being the period following the Mt. Pinatubo eruption in the early 1990s. Figure 3 from *Sarmiento and Gruber*, [2002].

Keeling et al., 1996; *Battle et al.*, 2000; *Manning*, 2001]. More recent breakthroughs improve the global CO_2 budget by correcting for the oxygen trend due to warming of the ocean in the 1990s [*Keeling et al.*, 2002; *Plattner et al.*, 2001; *Plattner et al.*, 2002]. The resulting breakdown of carbon sources and sinks is presented in Table 1.

Uptake of CO_2 by land implies an excess of primary production over respiration and other oxidation processes such as organic matter decomposition. Land-use changes in the form of deforestation lead to a net loss of carbon from plants and soil, whereas reforestation on land that was previosuly farmed or logged can lead to a net gain of carbon. The exact breakdown of land carbon sources and sinks is subject to intense debate. *Goodale et al.* [2002] suggest that northern forests and woodlands provided a total sink for 0.6-0.7 Pg of C per year during the early 1990s, with most of the estimated sink in temperate forests affected by fire suppression, agricultural abandonment, and plantation forestry. Carbon may accumulate also in agricultural soils, nonforested areas responding to fire supression, wood products both in use and in landfills, sediments of reservoirs and rivers. *Pacala et al.* [2001] suggest that the addition of

Table 1. An updated compilation of carbon sources and sinks. (a) Fossil fuel emissions and atmospheric CO_2 increase are based on the 2001 IPCC report [*Houghton et al.*, 2001]; these are our most strongly constrained estimates. (b) The land and ocean sink estimates are updated from the 2001 IPCC report to take into account changes in the O_2 method due to ocean warming [*Keeling and Garcia*, 2002].These particular results are confirmed by new ocean estimates by *Gloor et al.* [2003] and *McNeil et al.* [2003]. (c) The land breakdown is very different from the 2001 IPCC report. The tropical deforestation component we show is much smaller than in the IPCC report, and is based on a recent analysis by *De Fries et al.* [2002]. The magnitude of this term continues to be controversial. The total land sink is calculated as a residual; its value is strongly debated.

Carbon Sources and Sinks ($PgC\ yr^{-1}$)[a]		
	1980's	1990's
Emissions (fossil fuel, cement)	+5.4±0.3	+6.3±0.4
Atmospheric Increase	-3.3±0.1	-3.2±0.1
Net Oceans/Land	-2.1±0.3	-3.1±0.4
Land and Ocean Breakdown using Oxygen and Carbon Dioxide Observations[b]		
Ocean Sink	-1.8± 0.6	-2.2±0.5
Net Land Sink	-0.3± 0.7	-0.9 ±0.7
Land Breakdown[c]		
Tropical deforestation	0.6 (0.3 to 0.8)	0.9 (0.5 to 1.4)
Total Land Sink	-0.9 (-0.6 to -1.1±0.7)	-1.8 (-1.4 to -2.3±0.7)

these terms to the forest air-to-land flux doubles the total land sink of CO_2 for the 1980-1990 decade in the US. If these non-forest sink terms were equally important everywhere else in the world, they could potentially account for the rest of the CO_2 land sink.

The net oceanic sink accounts for about one-third of the total carbon emissions, and increases between the 1980s (1.8±0.6 PgC/yr) and 1990s (2.2±0.5 PgC/yr) (Table 1).We know that the pre-anthropogenic ocean accounted for 98.5% of the total atmosphere-ocean CO_2 inventory. Why then is the oceanic uptake of anthropogenic CO_2 so small ? What are the basic processes governing the uptake of excess carbon ?

2.1 THE CARBONATE SYSTEM

Only one in about 20 moles of CO_2 added to the ocean stays as CO_2, while the remaining ones react with carbonate ion to form bicarbonate via the "buffering" reaction:

$$CO_2 + CO_3{}^{2-} + H_2O \Longleftrightarrow 2\,HCO_3{}^- \qquad (21)$$

Following a CO_2 addition of 20 moles to the ocean, DIC increases by 20 moles while the dissolved CO_2 concentration only increases by 1 mol, such that $\partial DIC/\partial CO_2 \simeq 20$ in a pre-industrial setting. As we add anthropogenic CO_2 to the ocean, the carbonate ion concentration decreases, thus increasing the proportion of CO_2 added that stays in its dissolved form and decreasing $\partial DIC/\partial CO_2$. This results in a decrease in the overall capacity of the ocean to take up additional carbon. Carbonate buffering ensures that only 85% of the anthropogenic carbon dioxide we are adding to the atmosphere today will eventually dissolve into the ocean; this uptake capacity decreases as we put even more CO_2 in the atmosphere [*Sarmiento et al.*, 1992].

In the extreme scenario in which all of the $CO_3{}^{2-}$ available today in the ocean (around 4000 GtC eq) is depleted, every mole of CO_2 entering the ocean would stay as dissolved CO_2. In this limit at equilibrium about 33% of the added CO_2 would go into the ocean and 66% would remain in the atmosphere. On a time scale of about 10,000 years, the reaction of anthropogenic CO_2 in the deep ocean with calcium carbonate sediments accounts for a 9-15% increase in the uptake capacity of the ocean [*Archer et al.* 1998]. On even longer time scales CO_2 is neutralized by reactions with $CaCO_3$ and silicates on land. Because of their long time scales, these reactions have little relevance for the immediate anthropogenic uptake of CO_2.

2.2 THE ROLE OF GAS EXCHANGE AND CIRCULATION: THE GREEN'S FUNCTION SOLUTION

In this section we assume that the natural carbon cycle is in steady state and is unaffected by the anthropogenic CO_2 addition, which we treat as a perturbation on the system. For the purpose of this section we ignore the land carbon sources and sinks. Let us consider the carbon balance for atmospheric CO_2:

$$\frac{\partial C(t)}{\partial t} = S(t) - U(t), \tag{22}$$

where $C(t)$ is the atmospheric CO_2 concentration in ppm, $S(t)$ is the carbon input to the atmosphere at time t and $U(t)$ is the oceanic uptake at time t.

Following *Maier-Reimer and Hasselmann* [1987] we use a Green's Function approach to gain insight into the rate of oceanic uptake. We consider a simplified problem in which C undergoes a jump at $t = 0$, that is, $C(t) = C_o$ for $t < 0$ and $C(t) = S_o + C_o$ at $t = 0$. This corresponds to $S(t) = S_o \cdot \delta(t)$. For $t > 0$ we then seek a solution to

$$\frac{\partial C}{\partial t} = -U(t), \tag{23}$$

$$\text{with initial condition} \quad C(0) = S_o + C_o. \tag{24}$$

In other words, we ask what is the atmospheric concentration and coresponding oceanic uptake at time t following an instantaneous addition of S_o ppm of CO_2 at time $t = 0$? In order to solve this problem we need a functional form for the oceanic uptake $U(t)$. Assuming that the ocean responds linearly to the CO_2 addition to the atmosphere, a well motivated choice is:

$$U(t) = \frac{1}{\tau} \cdot (C(t) - (C_o + A_o \cdot S_o)), \tag{25}$$

where τ is a characteristic time scale for the ocean uptake. If A_o is the fraction of the initial source pulse that stays in the atmosphere at equilibrium, $C_o + A_o \cdot S_o$ is the amount of carbon left in the atmosphere once the system reaches equilibrium and $U(t)$ decreases exponentially to 0 at equilibrium. We substitute Eq. 25 into Eq. 23 and solve for $C(t)$ subject to the initial condition (24):

$$C(t) = C_o + S_o \cdot G(t), \tag{26}$$

$$\text{where} \quad G(t) = A_o + (1 - A_o)e^{-t/\tau}. \tag{27}$$

If there is more than one characteristic time scale for oceanic uptake τ_j, one can write the function $G(t)$ as a superposition of a number of exponentials of different amplitude A_j and relaxation time τ_j [*Maier-Reimer and Hasselmann*, 1987]:

$$G(t) = A_o + \sum_j A_j e^{-t/\tau_j}, \tag{28}$$

$$\text{where } A_o + \sum_j A_j = 1. \tag{29}$$

Note that A_o is associated with an infinite relaxation time scale. The Green's function $G(t)$ is the fraction of the initial source pulse still in the atmosphere at time t; $G(0) = 1$ and $lim_{t \to \infty} G(t) = A_0$. For an arbitrary emission function $S(t)$ the solution to Eqs. 23 - 24 can be formally written as:

$$C(t) = \int_0^t S(t')G(t - t')dt' + C_o, \tag{30}$$

where the Green's function $G(t)$ is the linear response to a pulse at time t'. Note that substituting $S(t) = S_o \cdot \delta(t)$ reduces Eq. 30 to Eq. 26, as expected.

Maier-Reimer and Hasselmann [1987] and *Sarmiento et al. [1992]* perturbed a full GCM with a step function initial change representing an instantaneous 25% increase, doubling and quadrupling of the background pCO_2 = 280 ppm: i.e., S_o = 70 ppm, 280 ppm, 840 ppm, respectively. A response of the form (26), where

$$G(t) = A_o + A_1 e^{-t/\tau_1} + A_2 e^{-t/\tau_2} + A_3 e^{-t/\tau_3} + A_4 e^{-t/\tau_4} \tag{31}$$

Table 2. Coefficients for exponential fits to the atmospheric response to a pulse input. A_j and τ_j are the amplitude and exponential relaxation time scales from Eq. 28. The numbers in the initial input column are the fraction by which the atmospheric CO_2 was increased above its pre-industrial value of 280 ppm. Table A1 in *Sarmiento et al.* [1992].

Initial Input	A_o	A_1	τ_1	A_2	τ_2	A_3	τ_3	A_4	τ_4
0.25	0.164	0.245	358.8	0.302	60.8	0.229	10.3	0.059	1.0
1.00	0.174	0.275	376.6	0.307	67.7	0.189	10.7	0.054	0.9
3.00	0.208	0.358	433.3	0.261	83.9	0.131	11.2	0.042	0.8

was least-square fitted to the atmospheric CO_2 response to a pulse input in the GCM, in order to determine A_j and τ_j. The results of one such perturbation experiment are shown in Table 2. Once $G(t)$ is determined, one can in principle use Eq. 30 to determine the atmospheric concentration given any source of excess CO_2.

In the quadrupling experiment the atmospheric concentration increase by 840 ppm represents an addition of 1780 GtC to the atmosphere, of which about 1424 GtC eventually end up in the ocean. An addition of the order 1400 GtC eq can change significantly the $CO_3{}^{2-}$ reservoir, which is presently about 4000 GtC. This experiment therefore falls into a nonlinear regime in which $CO_3{}^{2-}$ changes strongly and the asymptotic oceanic uptake drops accordingly to 79.2%, as contrasted with 83.6% for the 25% addition case. Though our representation of Green's functions is strictly correct only for an ocean that responds linearly to the atmosphere as in Eq. 25, the quadrupling experiment is considered in order to estimate the changes in the relaxation time scales and amplitudes due to chemical nonlinearities in the system [*Sarmiento et al.*, 1992].

The representation of the atmospheric concentration in terms of exponentials allows us to assess the importance of different physical and chemical processes on the storage of CO_2 in the ocean. The equations show that $C(t)$ is the sum of terms of the form $S_o \cdot A_j \cdot e^{-t/\tau_j}$. Each of these terms decreases exponentially as oceanic processes with time scales τ_j gradually absorb CO_2 from the atmosphere. The bigger the τ_j, the longer it takes for the particular process to absorb carbon. The time scales given in Table 2 suggest that τ_1 (350 years) is the relaxation time scale associated with deep water thermohaline circulation, τ_2 (60 years) is representative of intermediate water ventilation, τ_3 (10 years) is associated with the ventilated thermocline or gyre circulation, and τ_4 (1 year) is the time scale for the equilibration of mixed layer CO_2 with atmospheric CO_2 through gas-exchange. Thus, while on short time scales the penetration of CO_2 into the mixed layer is an important process, as t increases the absorption of CO_2 is gradually limited by larger scale circulation processes.

The coefficients A_j represent the fraction of S_o that is involved in processes with the relaxation time scale τ_j; the sum of all the coefficients is one. Alternatively, we can think of the A_j coefficients as the relative capacities of the reservoirs j which are filled up at rates given by τ_j. Consider an initial input of 0.25×280 ppm. A_o=16.4% is the fraction of the source that stays permenently in the atmosphere. Thus, the **potential uptake** of the ocean is $\int_o^\infty U(t)dt = S_o(1 - A_o)$, i.e. 83.6% of the added carbon dioxide. According to Table 2 on a time scale of 1 year the anthropogenic signal penetrates roughly $A_4/(1 - A_o) = 5.9/83.6 = 7.06\%$ of the total volume of the ocean; on a time scale of 10 years the signal penetrates into an additional $A_3/(1 - A_o) = 22.9/83.6 = 27.39\%$ of the ocean volume, and so forth.

Note that it is not straightforward to account for the nonlinearity of the system with the Green's function approach described above. *Maier-Reimer and Hasselmann* [1987] and *Sarmiento et al.* [1992] show that the magnitude of the A_j coefficients and the corresponding time scales τ_j depend on the size of the pulse, the initial atmospheric CO_2 concentration and the time history of the background atmospheric CO_2. In particular, as the initial input increases, the capacity of the ocean to hold CO_2 at equilibrium goes down and more of the initial CO_2 will stay in the atmosphere, i.e., A_o increases. This general conclusion holds for more realistic scenarios in which the source function increases linearly or exponentially in time.

One can apply the Green's function method to simple box models. In a 1 BM of the ocean we can still write $G(t) = A_o + A_1 e^{-t/\tau_1}$ where τ_1 is the time scale associated with the surface gas exchange. $A_1 = 1 - A_o$ is the potential uptake of the ocean as determined by carbon buffering, i.e., approximately 85%. For the 2 BM of Figure 1 we can write $G(t) = A_o + A_1 e^{-t/\tau_1} + A_2 e^{-t/\tau_2}$, where τ_2 is the gas exchange time scale and τ_1 is the deep water ventilation time scale. τ_1 depends on the mixing between deep and surface waters ν; large ν results in fast mixing and smaller τ_1. The ratio of A_1 to A_2 is roughly the ratio of the bottom to the surface box volumes and $A_1 + A_2 = 1 - A_o \simeq 85\%$. *Joos et al.* [1997] explored the sensitivity of the oceanic uptake of CO_2 to different vertical diffusivity values in a box model. For a careful choice of depth-dependent diffusivity with higher values near the surface and lower values at depth, their 12 layer BM was able to reasonably predict both the uptake of anthropogenic CO_2 and natural ^{14}C observations.

We will next try to understand the role of the gas exchange in the uptake of anthropogenic CO_2. Consider the time rate of change in the mixed layer of a gas that exchanges only with the atmosphere, such as surface oxygen $O_{2\,sfc}$:

$$\frac{\partial O_{2\,sfc}}{\partial t} = \frac{1}{\tau} \cdot (O_{2\,atm} - O_{2\,sfc}). \tag{32}$$

A typical magnitude for τ can be estimated as follows:

$$\tau = \frac{z_{ml}}{k_w} = \frac{75\,\mathrm{m}}{4\,\mathrm{m/day}} = 18\,\mathrm{days}. \tag{33}$$

where k_w is a gas exchange coefficient (in m/day) and z_{ml} is the mixed layer depth (in m). As the gas exchange time scale τ is very short relative to the circulation time scale, surface O_2 in water is well equilibrated with atmospheric O_2. This is the reason why in GCMs, box models and in the real ocean the surface concentration of O_2 is very close to the O_2 saturation concentration.

The surface exchange of CO_2 is different from that of O_2 because CO_2 is a weak acid in seawater that reacts with water and carbonate (Eq. 21). The time rate of change of DIC in the surface mixed layer is well approximated by:

$$\frac{\partial DIC}{\partial t} = \frac{\partial DIC}{\partial CO_{2\,sfc}} \cdot \frac{\partial CO_{2\,sfc}}{\partial t} = \frac{k_w}{z_{ml}} \cdot (CO_{2\,atm} - CO_{2\,sfc}), \tag{34}$$

such that

$$\frac{\partial CO_{2\,sfc}}{\partial t} = \frac{1}{\tau} \cdot (CO_{2\,atm} - CO_{2\,sfc}) \text{ and} \tag{35}$$

$$\tau = \frac{\partial DIC}{\partial CO_{2\,sfc}} \cdot \frac{z_{ml}}{k_w} \simeq 20 \cdot \frac{75\,\mathrm{m}}{4\,\mathrm{m/day}} \simeq 1\,\mathrm{year}. \tag{36}$$

The time scale τ corresponds to τ_4 in Table 2. The characteristic time scale on which gas exchange occurs is therefore longer for CO_2 than for O_2 by the factor $\partial DIC/\partial CO_2$. Note that as more CO_2 is added to the atmosphere this ratio decreases, thus moderately decreasing τ as shown in Table 2.

The characteristic time scale of 1 year for a 75 m top layer to reach equilibrium with the atmosphere is smaller than the time-scale of the present anthropogenic perturbation, which is roughly an exponential with an e-folding time of several decades. Surface waters thus track the atmospheric perturbation closely. *Sarmiento et al.* [1992] showed that a doubling of the gas exchange coefficient k_w increases the overall excess CO_2 uptake only moderately (11% in the Princeton GCM). This experiment confirms previous intuition gained from box models [*Siegenthaler*, 1983; *Siegenthaler and Joos*, 1992] that vertical exchange is the dominant process in limiting perturbation CO_2 flux into the ocean. Thus, intermodel differences in the rate at which deep waters are exposed to the surface and then submerged into the deep or in the convective parameterization can cause large differences between the resulting carbon uptake.

The difference between the present oceanic uptake of 40% of the total atmosphere-ocean increase (Table 1) and the potential uptake of 85% of the total excess atmosphere-ocean CO_2 is explained by the slowness of the oceanic circulation, which does not mix the new anthropogenic CO_2 signal into the deep sufficiently fast to allow for extensive surface uptake. Slow mixing between surface and deep should continue to produce ocean uptake for several centuries after an input of anthropogenic CO_2 into the atmosphere [*Siegenthaler and Hasselman*, 1978; *Sarmiento et al.*, 1992]. This also means that CO_2 will continue to enter the ocean for decades or centuries even after a potential stabilization of atmospheric CO_2; the equilibrium time scale is of the order of hundreds of years to a thousand years. The role of circulation in the uptake of anthropogenic CO_2 is further discussed in the following section.

2.3 THE ROLE OF CIRCULATION AND MIXING: MODEL RESULTS AND OBSERVATIONS

The anthropogenic signal, which can be defined as the increase in oceanic carbon dioxide relative to the pre-industrial times, is small compared to the natural signal and requires very accurate measurements in the ocean [*Sabine et al.*, 1997]. Few time-series that allow long-term, precise measurements exist. Given the present limitations to directly estimating anthropogenic CO_2 from observations, ocean models have been one of the principal tools for quantifying air-sea fluxes and oceanic uptake of anthropogenic carbon dioxide.

Figure 10 shows typical air-sea CO_2 flux results from a 3D simulation of anthropogenic carbon uptake with the same Princeton model described in Section 1.2 and Figures 5 and 8. The anthropogenic contribution is calculated directly in models by subtracting the pre-anthropogenic air-sea flux (full line) from the greenhouse-forced air-sea flux (dotted line). By comparing Figure 10 with Figure 5 we see that the increase of atmospheric pCO_2 above its pre-industrial level tends to increase oceanic uptake in natural sink regions such as high latitudes and decrease the release of CO_2 to the atmosphere in natural CO_2 outgassing regions. The anthropogenic uptake by the ocean (dash-dotted line in Figure 10) is calculated as the difference between the dotted and full lines. One should bear in mind that when applied to CO_2 source (outgassing) regions, uptake of anthropogenic carbon represents an actual decrease in outgassing (of the CO_2 already present in the ocean) relative to the pre-industrial state. Reduced outgassing implies that more carbon stays in the ocean; this excess carbon dioxide is what is refered to as the **anthropogenic** CO_2 contribution.

According to models, high uptake of anthropogenic CO_2 occurs as a consequence of strong exchange with uncontaminated interior waters [*Sarmiento et al.*, 1992]. Such exchange occurs in upwelling and convective overturning areas, such as the upwelling Equatorial region, the North Atlantic, and the Southern Ocean, as reflected in the air-sea flux patterns in Figure 10.

The Southern Ocean is particularly important in this regard. As upper circumpolar deep waters with low, pre-anthropogenic CO_2 upwell to the surface

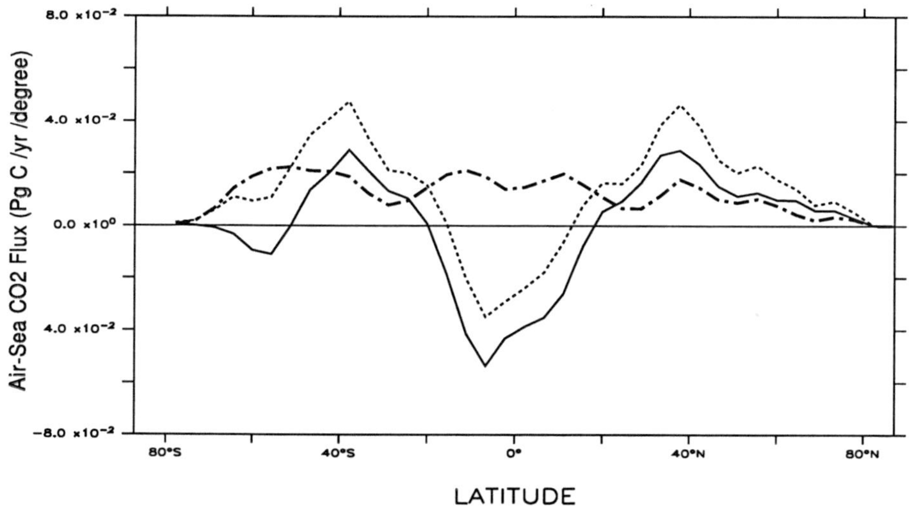

Figure 10. Anthropogenic air-sea flux of CO_2 from a standard Princeton biogeochemistry model. The model was first run to equilibrium with an atmosphere of 280 ppm while keeping thc total amount of *DIC* in the system constant; pre-anthropogenic air-sea flux (full line) is the same as in Figure 5.CO_2 was then released into the atmosphere folowing the OCMIP protocol; the dotted line is the resulting air-sea flux at the present time. The difference between these two fluxes is the excess (anthropogenic) flux (dash-dotted line).

in the Southern Ocean, they interact with an atmosphere which now contains higher, anthropogenic CO_2. Anthropogenic CO_2 penetrates this water, which is then transported northward before sinking as Subantarctic Mode Water (SAMW) and Antarctic Intermediate Water (AAIW) somewhere between 40° and 50°S [*Caldeira and Duffy*, 2000]. Excess carbon is thus carried to the southern subtropics, where it is stored along with CO_2 that enters the main thermocline directly in lower latitudes. This circulation pattern accounts in models for the large cumulative flux South of 40°S and for the northward shift in the total column inventory compared to the cumulative flux of anthropogenic CO_2 [Sarmiento et al., 1992].

The Ocean Carbon Model Intercomparison Project (OCMIP) is a recent effort to systematically compare different models. Participating models use identical ocean biogeochemistry and gas exchange parametrizations but different model physics (i.e, different grids, resolution, eddy mixing parametrizations). Models were forced by a spline fit to historical observations from 1765 until 1990 under prescribed, constant surface conditions (temperature, salinity, wind speed, alkalinity).

Figure 11 from *Orr et al.* [2001] shows the zonal mean and zonal integral of anthropogenic air-sea CO_2 fluxes in 1990 in four different OCMIP models. Significant differences in the uptake of anthropogenic CO_2 are due to different

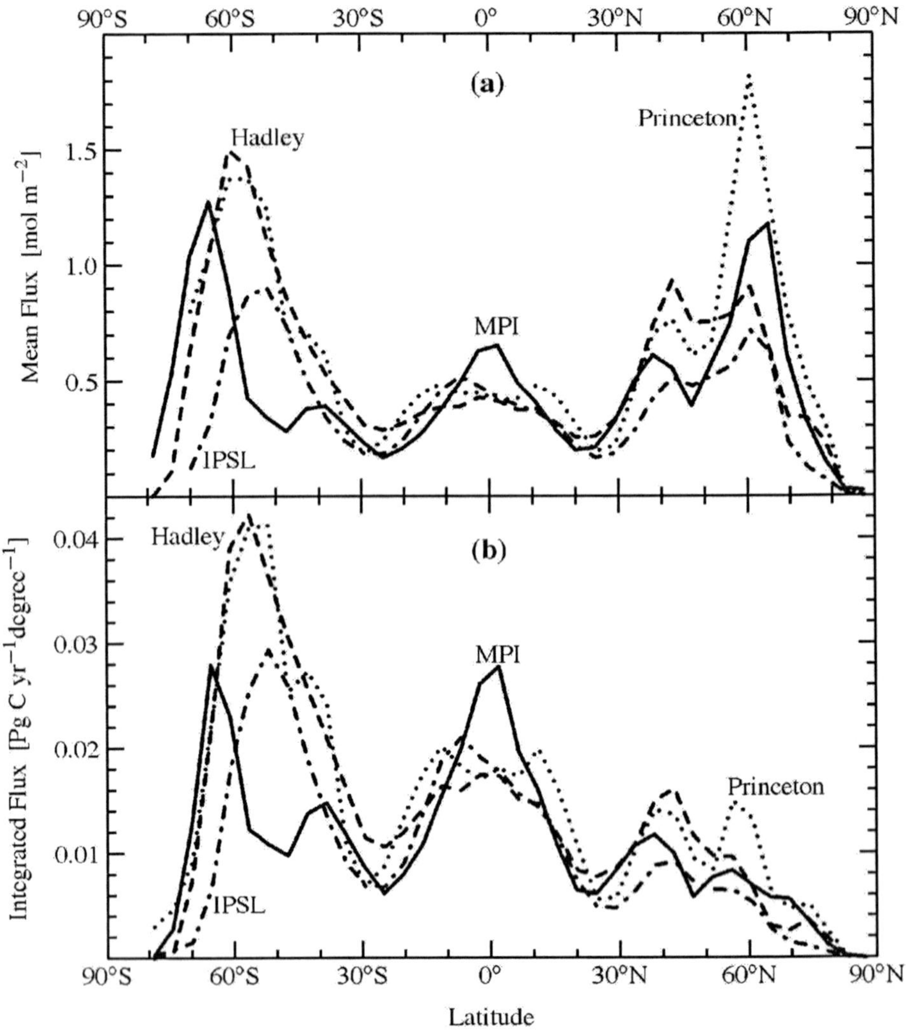

Figure 11. Zonal air-to-sea flux of anthropogenic CO_2 for the global ocean in 1990, given as (a) the zonal mean (mol/m^2) and (b) the zonal integral (PgC/yr/°lat). The models compared are Hadley, MPI, IPSL, Princeton/GFDL. Figure 1 from *Orr et al.* [2001]. Note the large uptake and inter-model differences in the Southern Ocean.

physics of these models. For the MPI, IPSL, Hadley and Princeton models, flux estimates south of 30°S range from 35 to 48% of each model's global uptake. Thus, the Southern Ocean south of 30°S, with its deep wintertime mixing and strong upwelling, has higher excess carbon uptake and inventory than expected from its area of 31% of the global ocean area [*Orr et al.*, 2001]. The Southern Ocean is also found to account for about 50% of the inter-model variance, consistent with findings from sensitivity studies that this is the region with highest variability of CO_2 uptake in response to changes in physical parametrizations [*Matear*, 2001; *Marinov et al.*, in prep.]. This suggests that tracer uptake in the Southern Ocean is poorly determined by models and highly uncertain.

A critical component for global carbon cycle studies is the evaluation of modeled anthropogenic carbon uptake and inventory through comparisons with tracer observations in the ocean. The tracers most used for such analyses are estimates of anthropogenic CO_2 itself, estimates of bomb ^{14}C, and measurements of chlorofluorocarbons (CFCs). Because of the lack of pre-industrial CO_2 observations, estimating the oceanic inventory of anthropogenic CO_2 from observations has been done indirectly. Starting from the analyses of *Brewer* [1978] and *Chen and Millero* [1979], *Gruber et al.* [1996] proposed that the total anthropogenic CO_2 concentration in the ocean is given by

$$C_{anth} = C - C_{bio} - C_{eq-280} - C_{diseq}, \tag{37}$$

where C is the presently observed DIC in the ocean, C_{bio} is the contribution due to the soft tissue and carbonate pumps, C_{eq-280} is the calculated DIC equilibrium concentration with a pre-industrial pCO_2 of 280 ppm, and C_{diseq} is the DIC concentration that results from the air-sea CO_2 disequilibrium at the time the water parcel left the surface ocean. Figure 12 shows the anthropogenic DIC calculated with this method from observations of present DIC.

Anthropogenic CO_2 inventories since 1800 from models are globally in rough agreement with observational estimates. However, differences between models and observations can be large on a regional scale. Models [*Orr et al.*, 2001; *Matear*, 2001] seem to over-predict at high latitudes the inventory of anthropogenic CO_2 estimated from observations with the *Gruber et al.* [1996] method or with other methods [i.e., *McNeil et al.*, 2002]. Figure 12 [*Gruber and Sarmiento*, 2002] shows high penetration of anthropogenic CO_2 in the subtropics and low penetration into the deep Southern Ocean. Note that anthropogenic carbon has penetrated significantly below about 2000 meters of the water column only in the North Atlantic, where surface waters sink directly into the abyss. Between 50°S and 50°N CO_2 penetration is generally shallower and excess CO_2 storage is lower in models compared to data estimates. This difference between models and observations could be explained by inadequate ocean physics in the models, such as a problematic parametrization of Southern Ocean physical processes. Too much surface water ventilation in the Southern Ocean south of 50°S, for example, could lead to overestimating anthropogenic storage in the region. A northward transport out of the Southern Ocean that is too weak (due to inadequate isopycnal mixing, too low advection, or inadequate formation of SAMW or AAIW) could explain the reduced penetration and low storage in the southern subtropics in the models.

While this intercomparison points to the incomplete representation of physical processes in the GCMs, one has to also take into account the random

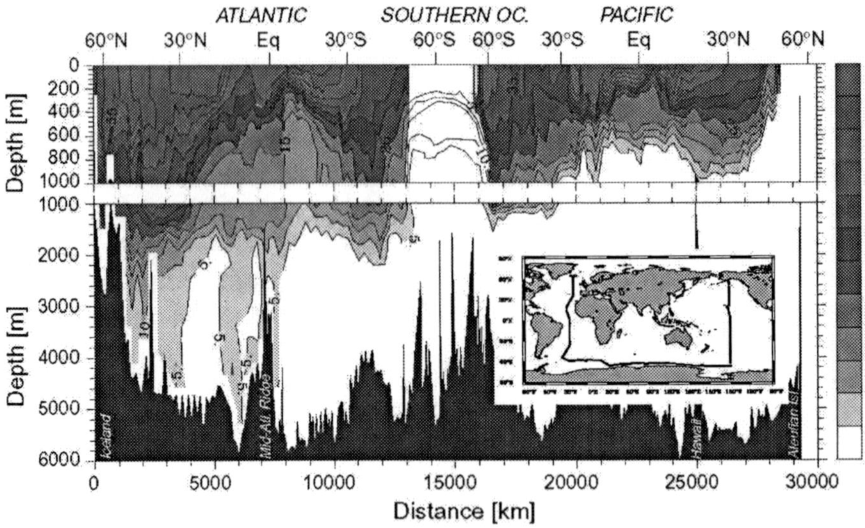

Figure 12. Anthropogenic carbon concentrations in the ocean along the track shown in the inset. The black area represents the ocean floor. The anthropogenic carbon is separated from the natural background by using the C^* method of *Gruber et al.* [1996] applied to high-precision measurements of dissolved inorganic carbon. Uncertainties in this separation technique are so large south of 60°S that they are shown there as contour lines only. Anthropogenic carbon has penetrated significantly below about 2000 meters of the water column only in the North Atlantic, where surface waters sink directly into the abyss; note the low Southern Ocean uptake. Figure 3 from *Gruber and Sarmiento* [2002].

and systematic errors associated with data estimation methods. For example, *Coatanoan et al.* [2002] show that two studies of the same Indian Ocean data set [*Sabine et al.*, 1998 and *Goyet et al.*, 1999] using two different methods (the *Gruber et al.* [1996] method and a regression technique) produced quite different estimates for anthropogenic CO_2 storage and uptake. Therefore understanding uncertainties associated with these techniques is a much needed step, and model calibration must be principally done with tracers such as CFCs, where these uncertainties do not exist.

Since they enter the ocean via gas exchange and are transported in the ocean as passive and conservative tracers, CFCs have been proposed as "ventilation clocks" for the ocean, i.e., as indicators of water age and deep ocean ventilation. CFC-11 and CFC-12 have been proposed as analogs of anthropogenic CO_2 because they grew exponentially from 1950 to 1995 just as anthropogenic CO_2 has since the Industrial Revolution. CFCs have the advantage over anthropogenic CO_2 and bomb ^{14}C that they have no natural background and they can be measured with high precision, with extended coverage from the recent World Ocean Circulation Experiment.

A number of studies have used CFCs to evaluate models of the ocean circulation [e.g., *England*, 1995; *England and Hirst*, 1997, *Dutay*, 1998, etc.], decadal scale ventilation characteristics [*Dutay et al*, 2002], and to validate model parametrization schemes [*Robitaille et al.*, 1995]. Model sensitivity studies in which one varies the gas exchange formulation or the circulation characteristics should help us understand better the effect of these processes on relations between bomb ^{14}C, CFCs, and anthropogenic CO_2 inventories. Observed tracer-tracer relations could guide our choice of appropriate gas exchange or mixing parametrizations, and lead to significant model improvement.

3. THE OCEAN CARBON CYCLE AND CLIMATE CHANGE

3.1 CLIMATE RESPONSE TO CO_2 INCREASE

Simulations of the anthropogenic CO_2 increase in atmosphere-ocean models coupled with ocean biogeochemistry have traditionally been done in models with constant oceanic circulation, sea temperature, and biology such as those analyzed in Section 2. However, all coupled climate models to date predict that continous emissions of anthropogenic CO_2 lead to increased radiative forcing resulting in higher sea surface temperature (SST) and a stronger hydrological cycle, accompanied by significant changes in ocean circulation and mixing [IPCC report, 2001].

At present, the high latitude ocean loses heat and gains fresh water from precipitation and runoff. In simulations of greenhouse gas forcing, increased poleward transport of moisture in the atmosphere [*Schiller et al.*, 1997] results in increased precipitation in high latitudes and especially in the formation region of the NADW. In some models, this causes a freshening of high latitude waters [*Manabe and Stouffer*, 1993] and a strong decline in surface density, particularly in the formation regions of NADW and AABW. At the same time, increased SST further decreases surface density, stabilizing the upper column throughout the ocean (Fig. 13e). This surface stabilization reduces vertical mixing by flattening isopycnals, and diminishing convective mixing and thermohaline overturning. The relative roles of the temperature increase and of the surface freshening on surface stabilization remains controversial [*Dixon et al., 1999*; *Mikolajewicz and Voss*, 2000], with some models showing that freshening in the North Atlantic is counteracted by advection of high salinity waters from the tropics [IPCC report, 2001].

All of the above changes are likely to result in a modification of the thermohaline circulation, with some models indicating a reduction in the North Atlantic maximum meridional overturning circulation (a measure of the THC) of 10%-50% following a CO_2 increase in the atmosphere up to the year 2100 [*Manabe and Stouffer*, 1993; *Stocker and Schmittner*, 1997]. Most models so far project significant changes in the marine carbon cycle and a reduction in the oceanic carbon uptake due to all of these processes, which may in turn accelerate atmospheric CO_2 growth [*Sarmiento and Le Quere*, 1996; *Matear and Hirst*, 1999; *Cox et al.*, 2000].

Of some concern is a scenario in which the entire thermohaline circulation collapses after passing a critical threshold, when the global atmospheric temperature has increased by $3.7 - 7.4°C$ [*Manabe and Stouffer*, 1993; *Schmittner and Stocker*, 1999; *Dixon et al.* 1999]. The reduction or shut-down of the THC leads to a reduction in the meridional transport of heat in the Atlantic, and a regional cooling which counteracts the temperature increase. The meridional distributions of precipitation and warming [*Schiller et al.*, 1997], sea-level rise [*Knutti and Stocker*, 2000], and biogeochemistry of the ocean [*Joos et al.*, 1999; *Sarmiento and Le Quere*, 1996] would all be profoundly affected by this scenario [*Stouffer and Manabe*, 1999].

Physical processes such as convection, diffusion and sub-grid scale (eddy) motions, while fundamental for calculating the oceanic uptake of natural and anthropogenic CO_2 (Section 2), are not resolved and are instead parameterized in climate models. Climate change predictions depend strongly on the response of the hydrological cycle to warming, the parameterization of mixing processes and of deep water formation, the vertical transport of heat, and the rate of temperature increase [*Stocker and Schmittner*, 1997]. Differences in the representation of these processes and our crude rendition of sea-ice processes account for large uncertainties and inter-model differences in climate warming simulations.

3.2 CO_2 RESPONSE TO GLOBAL WARMING

3.2.1 Oceanic feedbacks. The continuous release of anthropogenic CO_2 into the atmosphere changes the physical and biogeochemical properties of the oceans. A change in these properties could influence the future uptake of anthropogenic CO_2 by the oceans and therefore $pCO_{2\,atm}$. The present section discusses some of these property changes, which can be either direct changes in the anthropogenic uptake itself (point 1 below), or indirect changes through alterations of the natural solubility and biological pumps in the ocean (points 2-4). For the purpose of this paper we define positive feedback as the interaction betwen climate change and the carbon cycle that results in an increase in $pCO_{2\,atm}$ and a decrease in the ocean carbon uptake.

1. As anthropogenic CO_2 accumulates in the surface ocean, the carbonate chemistry of the ocean continuously changes. This process modifies the *carbonate buffering* and is a positive feedback acting to decrease the uptake capacity for CO_2 as more anthropogenic carbon is added to the ocean (see Section 2.1).
2. The continuous emission of CO_2 increases radiative forcing and results in a net heat flux into the ocean. Gases are less soluble in warmer waters. Warming of the surface waters therefore increases surface pCO_2 and drives CO_2 out of the ocean. This is a positive feedback; the chemical capacity of the ocean to dissolve anthropogenic CO_2 decreases as SST rises.
3. Coupled atmosphere-ocean models predict increases in vertical stratification and a general weakening of the vertical component of isopycnal mixing and convective overturning. As a consequence, the amount of water submerged from or upwelled to the surface changes. This slowing of vertical exchange will alter both the natural carbon cycle and the uptake of anthropogenic CO_2. Models so far predict an overall decrease in cumulative oceanic uptake of anthropogenic carbon due to circulation changes, i.e., a positive feedback (see Table 3 and discussion in this chapter).
4. Perhaps the largest unknown and the hardest to systematize are changes in the carbon uptake due to changes in ocean biology. The ocean biological pump can be affected by numerous climate-related factors such as: temperature, cloudiness or light availability, nutrient availability, ocean physics, sea-ice coverage. In a recent study, *Falkowski et al.* [1998] pointed to three factors, listed below, that could affect the strength of the biological pump and therefore the oceanic uptake of CO_2. For a more comprehensive review of these factors see also the 2001 IPCC report.

- a. Changes in surface nutrient utilization in nutrient rich areas. Changes in the supply of iron to high nutrient low chlorophyl [HNLC] regions might affect the utilization of surface nutrients, thereby changing export production and oceanic carbon storage. The amount of iron-bearing dust in the atmosphere depends on climate (precipitation, wind strength, moisture) and on the extent of dust sources, which could vary due to anthropogenically or climate induced changes in vegetation cover. Increases in erosion (due to enhanced agriculture, urbanization, conversion of land for agricultural use) and the hydrological cycle will likely affect transport of iron-bearing dust to the ocean [*Tegen and Fung*, 1995; *Harrison et al.*, 2001]. This could change the biological pump strength and alter atmosphere-ocean carbon partitioning. A similar mechanism might have worked in the past (see mechanism 1a, Section 1.1.1; *Martin*, [1990]). Note that surface nutrient availability depends on stratification and vertical mixing, which might change with climate warming as well [*Sarmiento et al.*, 1998].

- b. Changes in total oceanic inventories of nutrients could be triggered by changes in the delivery of critical nutrients (N, P, Fe, Si) from external sources (rivers, atmosphere, sediments) or by changes in nitrogen fixation and denitrification rates. These processes might have occured on glacial-interglacial time scales and could occur in the future as indirect effects of climate change [*Falkowski*, 1998]. If Falkowski's hypothesized link between nitrogen fixation and external iron supply exists [*Falkowski*, 1997], the fixed nitrogen inventory depends on the changing hydrological cycle, as it affects the availability of iron to the ocean. Changes in stratification and mixing with climate warming might also affect nitrogen fixation rates and therefore the total nitrogen inventory.

- c. Changes in the average elemental composition of organic material. Climate-related factors such as temperature, cloudiness (light availability), sea-ice extent, stratification and mixing might affect the structure of marine ecosystems. A significant shift in ecosystem structure and particularly in the composition of phytoplankton species could change the average strength of the biological pump and result in a modified $pCO_{2\,atm}$ [IPCC report, 2001].

Note that mechanisms (a to c) above are analogous to mechanisms (1 to 3) proposed in Section 1.1.1 as explanations for glacial-interglacial CO_2 changes. The simple intuition gained from our three box model analysis and Eqs. 14 and 16 can help us understand how the biological pump responds to climate change.

Modeling studies of future climate have so far investigated only some of the feedback mechanisms described above. Recent studies that looked at the feedback between climate change and ocean uptake include *Maier-Reimer et al.* [1996], *Sarmiento and Le Quere* [1996], *Sarmiento et al.* [1998], *Matear and Hirst* [1999], *Joos et al.* [1999], *Cox et al.* [2000]. All of these studies performed and inter-compared two types of runs: climate control runs and global warming runs. A constant climate control run fixes CO_2 and other greenhouse gases at their pre-industrial values. Global warming simulations are based on prescribed increases of CO_2 and other greenhouse gases into the future (so-called "emission scenarios").

Sarmiento and Le Quere [1996] and *Sarmiento et al.* [1998] coupled their ocean-carbon model to the Princeton/GFDL atmosphere-ocean climate model. Their diagnostic marine biological model includes biological pump parametrizations and the full carbon system equations, and is coupled with fixed biological export fluxes. The *Joos et al.* [1999] model is highly idealized in that it uses a zonally averaged ocean model coupled to an atmospheric energy balance model, and a simplified marine biological model. The model accounts for the potential fertilization of land by elevated CO_2. *Matear and Hirst* [1999] performed simulations in a coupled atmosphere-ocean climate model similar to *Sarmiento et al.* [1998]. However, this study uses the Gent-McWilliams eddy mixing parametrization in the ocean and a prognostic parameterization of export production. These two factors have the potential to significantly alter the feedbacks between climate change and oceanic CO_2 uptake. Results from these three studies are summarized in Table 3 and are described below.

Table 3. The response of oceanic CO_2 uptake to climate-induced feedbacks in three studies discussed in the text. Climate baseline refers to simulations with anthropogenic CO_2 emissions but preindustrial ocean temperature and circulation. Percent changes refer to uptake changes for the climate simulation -based on the IPCC IS92a scenario- relative to the climate baseline. Warming and circulation effects are changes due to warming and variation in transport processes in the absence of the biological pump. "Total effect" is the uptake change when all climate feedbacks are present; "total uptake" is the cumulative ocean uptake for a model which includes all feedbacks. Note that decreases in ocean carbon uptake (i.e., negative numbers in the table) correspond to positive feedbacks relative to $pCO_{2\,atm}$. *Sarmiento et al.* fix biological export fluxes while the other two studies use variable export fluxes. *Joos et al.* use a simplified, 2D ocean model coupled to an atmospheric energy balance model; the other two studies use 3D coupled atmosphere-ocean GCMs. Warming and circulation tend to decrease the oceanic uptake of CO_2, with circulation having a larger effect later on in the century, while biology increases the uptake of CO_2.

Response of Oceanic Uptake to climate-induced ocean feedbacks			
	Sarmiento et al. ['98]	*Matear and Hirst* ['99]	*Joos et al.* ['99]
Time Span	1990-2065	1850-2100	1765-2100
Climate Baseline (PgC)	289	402	530
Warming Effect	-11.4%	-11.9%	-12.8%
Circulation Effect	-21.8%	-10.2%	-2.8%
Biological Effect	+23.9%	+8.2%	+6.2%
Total Effect	-9.3%	-13.9%	-9.4%
Total Uptake (PgC)	262	346	480

In the context of modelling studies, the **Solubility feedback** is defined as the interaction between climate change and the ocean carbon uptake in a solubility only climate change scenario. In the absence of biology, a varying climate affects oceanic CO_2 uptake through (a) temperature or heat flux changes (SST feedback) and (b) changes in stratification and convection (circulation feedback). Note that these are precisely mechanisms 2 and 3 from the beginning of this section. The temperature and circulation effects on CO_2 uptake can be separated from each other by assuming that they are linearly independent, i.e. by neglecting nonlinearities of the carbonate chemistry. The climate simulations are therefore run in two separate abiotic models in which either the circulation or the sea surface temperature are kept constant.

Under global warming conditions, increasing SSTs cause a reduction in solubility, which drives CO_2 out of the ocean and decreases the net capacity of the ocean for taking up CO_2. Models also predict an increase in upper water column stratification and a decrease in high latitude convective overturning with climate warming (Fig 13 e and d, respectively). Overall decreases in convection and THC result in less communication between deep and surface waters. This results in a weaker solubility pump in high latitudes, i.e., less loss of heat to the atmosphere and less CO_2 flux into the high latitude ocean.

Climate warming simulations performed with abiotic models confirm that the SST feedback and the circulation feedback are both positive feedbacks on $pCO_{2\,atm}$ which contribute to a decrease in the net uptake of excess CO_2 by the ocean (Table 3). In the *Sarmiento et al.* [1998] abiotic simulations, the SST feedback accounts for 82% of the decrease in cumulative uptake in CO_2 early in the simulation (years 1765-1990), while transport processes take over later on in the simulations. From 1990 to 2065, the slow-down of ocean circulation processes accounts for about 66% of the decreases in cumulative ocean uptake, with temperature increase accounting for the remaining 34%. The inclusion of the Gent-McWilliams eddy-advection scheme is hypothesized to reduce the importance of convection in the Southern Ocean and seems to account for the smaller effect of the circulation feedback of 10.2% in the *Matear and Hirst* study (Table 3). The small circulation feedback of *Joos et al.* is difficult to evaluate, since this study uses a highly simplified, two dimensional ocean model.

Biological feedback is the interaction between climate change and the uptake of CO_2 due to biological mechanisms, as described by mechanisms 4 a-c in this chapter.

The biological models used in the global climate change studies presented above are highly simplified; most of them only allow for simple changes in surface nutrient availability with climate change (mechanism 4a). The resulting biological feedbacks are necessarily tentative parameterizations of the real processes occuring in nature. In the context of our models the biological feedback is dictated only by changes in vertical stratification and the thermohaline circulation. These modify the amount of DIC and nutrients brought from depth by the overturning circulation, altering the strength of the natural biological pump and the CO_2 partitioning between ocean and atmosphere. Note that current thinking holds that at current CO_2 levels, increasing CO_2 in the ocean by anthropogenic addition does not directly affect biological processes in the ocean; modeling studies so far have been based on this premise.

Sarmiento et al. [1998] use a constant-biota model which fixes the export and production at their initial values. The exception are areas where phosphate becomes depleted at the surface; here, production is set to zero. In this simple model, increased stratification and slower circulation (Fig. 13e and d) lower the upward supply of deep carbon and nutrients to the euphotic zone. The Southern Ocean is a region where deep mixing ensures communication between surface and deep waters by outcropping isopycnals. A flattening of isopycnals and a decrease in vertical mixing with climate warming reduces the communication between deep waters (rich in DIC and PO_4) and surface waters (poor in DIC and PO_4) and, in a model where biological uptake is fixed at its pre-industrial value, results in a decrease in surface-water PO_4 (Fig 13f) and accumulation of excess DIC in the deep. The biological pump becomes more effective especially in the Southern Ocean, as reflected by the large flux of CO_2 into the region. The net change due to the biological pump is a 24% increase in oceanic uptake of CO_2 (Table 3). Figure 13b provides an interesting counterpart to Figure 8 and model studies which show that the Southern Ocean biological pump reacts more strongly to nutrient depletion experiments than any other region of the ocean (see Section 1.2).

The above mechanism is equivalent to mechanism 1b in Section 1.1.1. The net effect of increased stratification and decreased Southern Ocean convective mixing in a 3 box model is a decrease in f. When biological export fluxes Π_h are fixed, a decrease in f results in a decrease in preformed nutrients PO_{4h}, in agreement with Eq. 15. The resulting depletion of nutrients acts to increase the biological pump strength and the oceanic uptake of CO_2 as shown by Eq. 14 and Figures 4 and 8.

Note that the strength of the biological feedback depends on both the physical and the biological models being used, i.e. on the specification of f and Π_h in our model. *Joos et al.* [1999] compare their standard prognostic biological model, where export fluxes are allowed to change in response to climate, to a constant export production model. Table 3 presents results from their prognostic biological model runs. In this model increased stratification and the resulting surface nutrient depletion decreases export production. This effect works against the decrease in surface nutrient supply and tends to make the biological pump less efficient. While the overall effect in the *Joos et al.* prognostic model is still an increase in the oceanic uptake of CO_2, this increase is smaller compared to a constant export production model such as that employed by *Sarmiento et al.* [1998].

Finally, the inclusion of the Gent-McWilliams parametrization of the eddy effects [*Gent et al.*, 1995] by *Matear and Hirst* [1999] results in a smaller increase in stratification with climate change. This then results in less nutrient depletion and therefore a smaller increase in the oceanic uptake of CO_2 compared to *Sarmiento et al.* [1998]. For each of the above model studies, the negative biological feedbacks partially cancel out the positive solubility feedbacks on the oceanic uptake. The net effect of including climate change is a decrease in the oceanic uptake of CO_2.

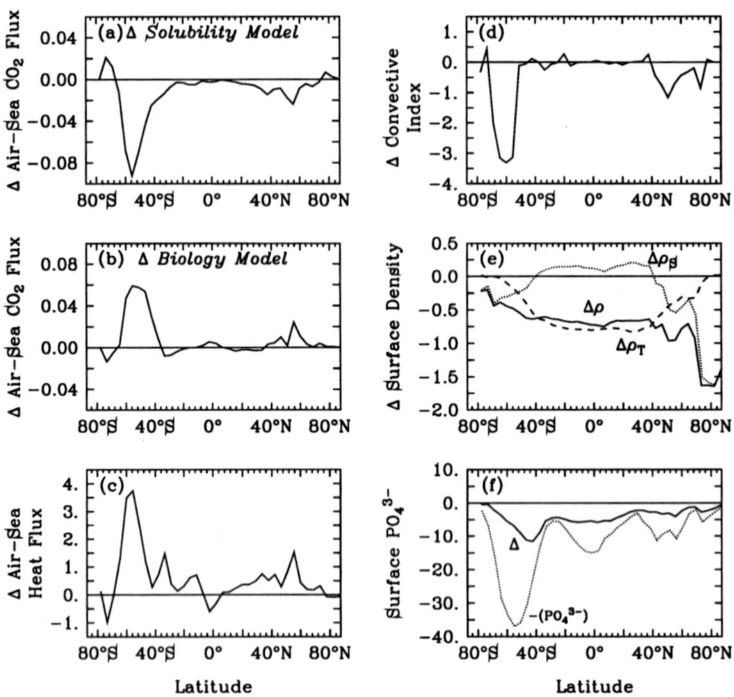

Figure 13. Figure 2 from *Sarmiento et al.* [1998] showing the Princeton GCM response to anthropogenic climate warming. Zonal integrals for the decade 2056-65 of: a. the change in air-sea CO_2 fluxes (positive for ocean uptake) due to warming and changes in ocean transport estimated by the difference between the solubility Global warming (GW) and solubility baseline (Pg/degree/yr); b. the change in air-sea CO_2 fluxes due to the biological pump calculated as the difference between the constant-biota GW and the constant biota baseline air-sea CO_2 fluxes, after removing the changes due to warming and transport (Pg/degree/yr); c. GW minus baseline heat fluxes in W/m; d. GW minus baseline of the convection index; e. Solid line: change in total surface density (GW minus control) in kg/m^3; dashed and dotted lines show density changes due to temperature and salinity, respectively. f. Constant biota GW minus ct. biota baseline surface PO_4 in mmol/m^2. Dotted line is the negative of baseline PO_4.

More recently, *Bopp et al.* [2001] investigated the response of marine productivity to climate change using two different ocean biogeochemical schemes. The first links marine export production to phosphate utilization at the surface, temperature, and light intensity but, just like the schemes above, does not include explicit biology. The second is an NPZD (nutrient-phytoplankton-zooplankton-detritus) type model in which phytoplankton growth depends on the local conditions of light, temperature, and vertical eddy diffusion.

Interestingly, increased stratification associated with climate warming leads to both reduced nutrient supply, as in previous studies, and to increased light supply in the now thinner mixed layer. Reduced nutrient supply mainly in wintertime causes a decrease in the biological production. Increased light supply in the now more strongly stratified high latitudes gives longer growing seasons in these areas and an overall poleward shift in oceanic biological production. *Bopp et al.* find that the effect of increased light supply prevails in high latitudes, while the effect of reduced upwelling of nutrients is most important in low latitudes. As a result export production increases by 30% in high latitudes and decreases by 20% in low latitudes for a doubling of atmospheric CO_2. The overall effect is a modest decrease in export production of 6%, similar to the decrease in export production found by *Joos et al.* [1999]. One would expect that the high latitude increase in biological production will dominate, resulting in a net increase in oceanic CO_2 uptake.

The response of biology to climate change and its feedback on the system is highly dependent on the model being used (Table 3). The *Bopp et al.* study makes the point that production and surface nutrient concentration depend on a variety of mechanisms such as light availability and trophic interactions, which need to be included in future climate feedback studies. A natural next step in model development would then be to replace the nutrient based biological models with more complex, ecosystem dynamics (NPZD) type models as in the studies of *Cox et al.* [2000] and *Friedlingstein et al.* [2001].

4. SUMMARY

This chapter focuses on understanding the mechanisms responsible for the oceanic uptake of carbon. The oceanic uptake of natural CO_2 is determined by the carbon pump, which is the sum of the biological and solubility carbon pumps (Section 1). The DIC gradient between the surface and the deep ocean proves to be a good metric for the strength of the carbon pumps. Simple box models can help us understand the basic mechanisms underlying the variation of atmospheric pCO_2 between glacials and interglacials and the relationship between the biological pump strength and preformed nutrients, oceanic circulation, and gas exchange (Section 1.1).

One of the most important problems of carbon cycle research is to determine where all the anthropogenic carbon added to the oceans is going. While Section 1 analyzes the steady state oceanic uptake, Section 2 studies the transient problem, i.e., the behavior of the oceanic uptake and the corresponding atmospheric response to anthropogenic CO_2 perturbations. The uptake of excess carbon can be characterized as a sum of decaying exponential functions with time scales τ_j characteristic for different oceanic processes and with coefficients related to the relative capacities of the reservoirs j filled up with CO_2 at rates τ_j. Mixing between surface and deep waters determines the rate at which these reservoirs are being filled up. Differences in convective mixing or isocpycnal mixing are thus partly responsible for differences in carbon uptake between different models.

Model studies suggest that the Southern Ocean plays a critical role in the uptake of anthropogenic CO_2 from the atmosphere (Figures 11 and 12) and in the feedback processes between the oceanic carbon cycle and climate change (Figure 13). Changes in oceanic SST, circulation, mixing, gas exchange following climate warming will result in changes of the natural biological and solubility pumps in the ocean, which will influence in their turn the further uptake of carbon by the ocean. In particular, GCM studies suggest that climate warming results in increased stratification, a decrease in the vertical component of isopycnal mixing and a decrease in convective overturning primarily in the Southern Ocean.

The resulting decrease in preformed nutrients suggests a strengthening of the biological pump and therefore an increase in the biological uptake of CO_2. At the same time, decreasing high latitude convection results in reduced uptake by the solubility pump in the high latitude ocean. Finally, increasing SSTs cause an additional reduction in the solubility pump, driving CO_2 out of the ocean. The solubility effects win over the biological effects. According to the most recent models available, in the net, climate warming decreases the capacity of the ocean for further carbon uptake and thus represents a positive feedback on the atmospheric carbon dioxide. This result, as well as the large uncertainty associated with it, should raise a red flag and stimulate scientists to improve both models and theory in order to better understand natural and anthropogenic CO_2 uptake by the ocean.

ACKNOWLEDGMENTS

The authors are grateful for supportfrom DOE grant DE-FG02-00ER63009.

REFERENCES

Anderson, L.A. and J.L.Sarmiento, Redfield ratios of remineralization determined by nutrient data analysis, *Global Biogeochem. Cycles*, 8, 65-80, 1994.

Archer, D.E.,H. Kheshgi, and E. Maier-Reimer, Dynamics of fossil fuel CO_2 neutralization by marine $CaCO_3$, Global Biogeochem. Cycles, 12, 259-276, 1998.

Archer, D.E., G. Eshel, A. Winguth, W. Broecker, R. Pierrehumbert, M. Tobis, and R. Jacob, Atmospheric pCO_2 sensitivity to the biological pump in the ocean, *Global Biogeochem. Cycles,* 14, 1219-1230, 2000.

Bacastow, R.B., The effect of temperature change of the warm surface waters of the oceans on atmospheric CO_2, *Global Biogeochem. Cycles*, 10, 319-333, 1996.

Battle, M., M. Bender, T. Sowers, P.P. Tans, J.H. Butler, J.W. Elkins, J.T. Ellis, T. Conway, N. Zhang, P. Lang, and A.D. Clarke, Atmospheric gas concentrations over the past century measured in air from firn at the South Pole, *Nature,* 383, 231-235.

Bender, M.L., P.P. Tans, J.T. Ellis, J. Orchardo, and K. Habfast, High precision isotope ratio mass spectroscopy method for measuring the O_2/N_2 ratio of air, *Geochim. Cosmochim. Acta* 58, 4751-4758, 1994.

Bender, M.L. and M.O. Battle, Carbon cycle studies based on the distribution of O_2 in air, *Tellus* 51B, 165-169, 1999.

Bopp, L., P. Monfray, O. Aumont, J.-L Dufresne, H. LeTreut, G. Madec, L. Terray, and J. Orr, Potential impact of climate change on marine export production, *Global Biogeochem. Cycles*, 15(1), 81-100, 2001.

Brewer, P.G., Direct observation of the oceanic CO_2 increase, *Geophys. Res. Lett.*, 5(12), 997-1000, 1978.

Broecker, W.S., Glacial to Interglacial Changes in Ocean Chemistry, *Progress in Oceanography*, 2, 151-197, 1982.

Broecker, W.S., Ocean Chemistry during glacial time, *Geochimica et Cosmochimica Acta* 46, 1689-1705, 1982.

Broecker, W.S., S. Sutherland, W. Smethie, TH. Peng, G. Ostlund, Oceanic radiocarbon: separation of the natural and bomb component, *Global Biogeochem. Cycles* 9(2), 263-288, 1995.

Caldeira, K. and P.B. Duffy, The role of the Southern Ocean in Uptake and storage of anthropogenic carbon dioxide, *Science*, 287, 620-622, 2000.

Chen, C.T.A., and F.J. Millero, Gradual increase of oceanic CO_2, *Nature*, 277, 205-206, 1979.

Coatanoan, C., C. Goyet, N. Gruber, C.L. Sabine, and M. Warner, Comparison of two approaches to quantify anthropogenic CO_2 in the ocean: Results from the northern Indian Ocean, *Global Biogeochem. Cycles,* 15(1), 11-25, 2001.

Cox, P.M., R.A. Betts, C.D. Jones, S.A. Spall, I.J. Totterdell, Acceleration of global warming due to carbon-cycle feedbacks in a coupled climate model, *Nature,* 408, p.184, 2000.

Cramer, W., A. Bondeau, F.I. Woodward, I.C. Prentice, R.A. Betts, V. Brovkin, P.M. Cox, V. Fisher, J.A. Foley, A.D. Friend, C. Kucharik, M.R. Lomas, N. Ramankutty, S. Sitch, B. Smith, A. White, and C. Young-Molling, Global response of terrestrial ecosystem structure and function to CO_2 and climate change: results from six dynamic global vegetation models, *Global Change Biology,*2001.

DeFries R.S., R.A. Houghton, M.C. Hansen, C.B. Field, D. Skole, J. Townshend, Carbon emissions from tropical deforestation and regrowth based on satellite observations for the 1980s and 1990s, *Proceedings of the National Academy of Sciences of the United States of America* 99(22), 14256-14261, 2002.

Dixon, K.W., T.L. Delworth, M.J. Spelman and R.J. Stouffer, The influence of transient surface fluxes on North Atlantic overturning in a coupled GCM climate change experiment, *Geophys. Res. Lett*, 26,2749-2752, 1999.

Dutay, J.-C., Influence du melange vertical et de la couche melangee sur la ventilation de l'ocean. Simulations numeriques des traceurs transitoires tritium-helium-3 et CFCs aves le modele OPA. Ph.D. Thesis, Univ. Pierre et Marie Curie.

Dutay, J.-C. et al., Evaluation of ocean model ventilation with CFC-11: comparison of 13 global ocean models, *Ocean Modeling*, 4, 89-120, 2002.

England, M.H., Using chlorofluorocarbons to assess ocean climate models, *Geophys. Res. Lett.* 22, 3051-3054, 1995.

England, M.H., Hirst, A.C., Chlorofluorocarbon uptake in a world ocean model. Sensitivity to surface thermohaline forcing and subsurface mixing parametrizations, *J. Geophys. Res.* 102, 15709-15731, 1997.

Falkowski, P.G., R.T. Barber, and V. Smetacek, Biogeochemical controls and feedbacks on ocean primary production, *Science,* 281, 200-206, 1998.

Falkowski, P.G, Evolution of the nitrogen cycle and its influence on the biological sequestration of CO2 in the ocean, *Nature*, 387(6630), 272-275, 1997.

Feely, R.A., C.L. Sabine, R.M. Key, and T.H. Peng, CO_2 synthesis results: estimating the anthropogenic carbon dioxide sink in the Pacific ocean, US. JGOFS News, 9, 1-5, 1999.

Friedlingstein, P., L. Bopp, P. Ciais, J.-L. Dufresne, L. Fairhead, H. LeTreut, P. Monfray, and J. Orr, Positive feedback between future climate change and the carbon cycle. Note du Pole de Modelisation, *Geophysical Research Letters,* 28, 1543-1546, 2001.

Ganeshram, R.S., T.F. Pedersen, S.E. Calvert, and R. Francois, Reduced nitrogen fixation in the glacial ocean inferred from changes in marine nitrogen and phosphorus inventories, *Nature*, 415, 156-159, 2002.

Gent, P.R., J. Willebrand, T.J. Mc Dougall and J.C. McWilliams, Parametrizing eddy-induced transports in ocean circulation models, *J. Phys. Oceanogr.*, 25, 463-474, 1995.

Gildor, H., E. Tziperman and R.J. Toggweiler, The sea-ice switch mechanism and glacial-interglacial CO_2 variations, *Global Biogeochemical Cycles*, 16, 10.1029/2001GB001446, 2002.

Gloor, M., N. Gruber, J.L. Sarmiento, C.L. Sabine, R.A. Feely, C. RÃűdenbeck, A first estimate of present and preindustrial air-sea CO_2 flux patterns based on ocean interior carbon measurements and models, *Geophys. Res. Lett.* Vol. 30 No. 1, 2003.

Gnanadesikan, A., R.D. Slater, N. Gruber, and J.L. Sarmiento, Oceanic vertical exchange and new production: a comparison between models and observations, *Deep Sea Res. Part II,* 49, 363-401, 2002.

Goodale C.L., M.J. Apps, R.A. Birdsey, C.B. Field, L.S. Heath, R.A. Houghton, J.C. Jenkins, G.H. Kohlmaier, W. Kurz, S.R. Liu, G.J. Nabuurs, S. Nilsson, A.Z. Shvidenko,Forest carbon sinks in the Northern Hemisphere, *Ecological Applications,* 12(3), 891-899, 2002.

Gruber, N. and J.L. Sarmiento, Biogeochem./Physical Interactions in Elemental Cycles,in *THE SEA: Biological-Physical Interactions in the Oceans*, edited by A. R. Robinson, J. J. McCarthy, and B. J. Rothschild, John Wiley and Sons, Volume 12, 337-399, 2002.

Gruber, N., J.L. Sarmiento, and T.F. Stocker: An improved method for detecting anthropogenic CO_2 in the oceans. *Global Biogeochem. Cycles,* 10(4), 809-837, 1996.

Gruber, N., Anthropogenic CO_2 in the Atlantic Ocean, *Global Biogeochem. Cycles,* 12, 165-191, 1998.

Harrison S.P., K.E. Kohfeld, C. Roelandt, T. Claquin, The role of dust in climate changes today, at the last glacial maximum and in the future, *Earth-Science review*, 54(1-3), 43-80 Sp.Iss SI, 2001.

Heimann, M. and Maier-Reimer, E., On the relations between the oceanic uptake of CO_2 and its carbon isotopes,*Global Biogeochem. Cycles*, 10, 89-110, 1996

Houghton, J.T. et al., eds., Climate Change 2001: The Scientific Basis, Cambridge U. Press, New York, 2001. See http://www.ipcc.ch.

Joos, F., G.K. Plattner, T.F. Stocker, O. Marchal, A. Schmittner, Global Warming and Marine Carbon Cycle Feedbacks on Future Atmospheric CO_2, *Science,* 284, 464-467, 1999.

Joos, F., J.C. Orr, and U. Siegenthaler, Ocean carbon transport in a box-diffusion versus a general circulation model, *J. Geophys. Res.,* 102 (C6), 12367-12388, 1997.

Keeling, R.F. and S.R. Shertz, Seasonal and interannual variations in atmospheric oxygen and implications for the global carbon-cycle, *Nature,* 358(6389), 723-727, 1992.

Keeling, R.F., S.C. Piper, and M. Heimann, Global and hemispheric CO_2 sinks deduced from changes in atmospheric O_2 concentrations, *Nature*, 381, 218-221, 1996.

Keeling, R.F. and H.E. Garcia, The change in oceanic O-2 inventory associated with recent global warming, *Proceedings of the National Academy of Sciences of the United States of America*, 99(12), 7848-7853, 2002.

Knorr, W., and M. Heimann, Uncertainties in global terrestrial biosphere modeling, Part I: a comprehensive sensitivity analysis with a new photosynthesis and energy balance scheme, *Global Biogeoche. Cycles*, 15(1), 227-246, 2001.

Knox, F., and M.B. McElroy, Changes in atmospheric CO_2: Influence of the marine biota at high latitude, *J. Geophys. Res.*, 89, 4629-4637, 1984.

Knutti, R. and T.F. Stocker, Influence of the thermohaline circulation on projected sea level rise, *J. Clim.*, 13, 1997-2001, 2000.

Maier-Reimer, E., U. Mikolajewicz, and A. Winguth, Future ocean uptake of CO_2 - Interaction between ocean circulation and biology, *Climate Dynamics,* 12, 711-721, 1996.

Maier-Reimer, E. and K. Hasselmann, Transport and storage of CO_2 in the ocean - an inorganic ocean-circulation carbon cycle model, *Climate Dynamics,* 2, 63-90, 1987.

Manabe, S. and R.J. Stouffer, Century-scale effects of increased atmospheric CO_2 on the ocean-atmosphere system, *Nature,* 364, 215-218, 1993.

Manning, A.C., Temporal variability of atmospheric oxygen from both continuous measurements and a flask sampling network: Tools for studying the global carbon cycle. Ph. D. thesis, U.C. San Diego, La Jolla, California, U.S.A., 190p.

Marinov, I., A. Gnanadesikan, J.L. Sarmiento, and R.D. Slater, The biological carbon pump: influence of nutrient depletion and circulation, in preparation.

Martin, J.H., Glacial-Interglacial CO_2 change: the iron hypothesis, *Paleoceanography,* Vol. 5, no. 1, 1-13, 1990.

Matear, R.J. and A.C. Hirst, Climate change feedback on the future oceanic CO_2 uptake, *Tellus,* 51B(3), 722-733, 1999.

Matear, R.J., Effects of numerical advection schemes and eddy parametrizations on ocean ventilation and oceanic anthropogenic CO_2 uptake, *Ocean Modelling,* 3: 217-248, 2001.

McElroy, M.B., Marine biological controls on atmospheric CO_2 and climate, *Nature*, 302, 328-329, 1983.

McNeil, B.I., B. Tilbrook, and R.J. Matear, Accumulation and uptake of anthropogenic CO_2 in the Southern Ocean, south of Australia between 1968 and 1996, *JGR-Oceans,* 106, 31431-31445, 2001.

McNeil, B.I., R.J. Matear, R.M. Key, Bullister, J.L. Sarmiento, Anthropogenic CO_2 uptake using the global CFC data-set, *Science,* 299, 235-239, 2003.

Mikolajewicz, U. and R. Voss, The role of the individual air-sea flux components in CO_2-induced changes of the ocean's circulation and climate, *Clim. Dyn.*, 16, 627-642, 2000.

Murnane, R.J., J.L. Sarmiento, C. LeQuere, Spatial distribution of air-sea fluxes and the interhemispheric transport of carbon by the oceans, *Global Biogeochem. Cycles*, v.13, no 2, 287-305, 1999.

Najjar, R. and J. Orr, Design of OCMIP-2 simulations of chlorofluorocarbons, the solubility pump and common biogeochemistry, http://www.ipsl.jussieu.fr/OCMIP.

Orr, J.C., Ocean Carbon-cycle Intercomparison Project (OCMIP). Phase 1 (1995-1997). IGBP/GAIM Report Series, Report no. 7.

Orr, J.C., E. Maier-Reimer, U. Mikolajewicz, P. Monfray, J.L. Sarmiento, J.R. Toggweiler, N.K. Taylor, J. Palmer, N. Gruber, C.L. Sabine, C.L. Le Quere, R.M. Key, and J. Boutin, Estimates of anthropogenic carbon uptake from four three-dimensional global ocean models, *Global Biogeochem. Cycles,* 15(1), 43-60, 2001.

Pacala, S.W., et al., Consistent land- and atmosphere-based US carbon sink estimates, *Science*, 292(5525), 2316-2320, 2001.

Peng, T.H., R. Wanninkhof, J.L. Bullister, R.A. Feely, and T. Takahashi, Quantification of decadal anthropogenic CO_2 uptake in the ocean based on dissolved inorganic carbon measurements, *Nature*, 396, 560-563, 1998.

Petit, J.R. et al, Climate and atmospheric history of the past 420,000 years from the Vostok ice core, Antarctica, *Nature*, 399, 429-436, 1999.

Plattner, G.K., F. Joos, T.F. Stocker, O. Marchal, Feedback mechanisms and sensitivities of ocean carbon uptake under global warming, *Tellus Series-B- chemical and physical meteorology* 53(5), 564-592, 2001.

Plattner, G.K., F. Joos, T.F. Stocker, Revision of the global carbon budget due to changing air-sea oxygen fluxes, *Global Biogeochem. Cycles,* 16(4), art. 1096, 2002.

Robitaille, D.Y. and A.J. Weaver, Validation of sub-grid-scale mixing schemes using CFCs in a global ocean model. *Geophys. Res. Lett.* 22, 2917-2920, 1995.

Sabine, C.L., R.M. Key, K.M. Johnson, F.J. Millero, A. Poisson, J.L. Sarmiento, D.W.R. Wallace, and C.D. Winn, Anthropogenic CO_2 inventory of the Indian Ocean, *Global Biogeochem. Cycles,* 13, 179-198, 1999.

Sabine, C.L., Feely RA, Key RM, Bullister JL, Millero FJ, Lee K, Peng TH, Tilbrook B, Ono T, and Wong CS, Distribution of anthropogenic CO_2 in the Pacific Ocean, *Global Biogeochem. Cycles,* 16(4), 1083, 2002.

Sarmiento, J.L. and J.R. Toggweiler, A new model for the role of the oceans in determining atmospheric pCO_2, *Nature,* 308, 620-624, 1984.

Sarmiento, J.L., J.C. Orr, and U. Siegenthaler, A perturbation simulation of CO_2 uptake in an ocean general- circulation model, *J. Geophys. Res.-Oceans,* 97, 3621-3645, 1992.

Sarmiento, J.L., C. Le Quere, S.W. Pacala, Limiting future atmospheric carbon-dioxide, *Global Biogeochem. Cycles,* 9(1): 121-137, 1995.

Sarmiento, J.L. and C. Le Quere, Oceanic carbon dioxide uptake in a model of century scale global warming, *Science,* 274, 1346-1350, 1996.

Sarmiento, J.L., T.M.C. Hughes, R.J. Stouffer, and S. Manabe, Simulated response of the ocean carbon cycle to anthropogenic climate warming, *Nature,* 393, 245-249, 1998.

Sarmiento, J.L. and N. Gruber, Sinks for anthropogenic carbon, *Physics Today,* 55(8), 30-36, 2002.

Schiller, A., U. Mikolajewicz and R. Voss, The stability of the North Atlantic thermohaline circulation in a coupled ocean-atmosphere general circulation model, *Clim. Dyn.*, 13, 325-347, 1997.

Schmittner, A. and T.F. Stocker, The stability of the thermohaline circulation in global warming experiments, *J.Clim.*, 12, 1117-1133, 1999.

Shackleton, N.J., Carbon-13 in Uvigerina: tropical rainforest history and the equatorial Pacific carbonate dissolution cycles, *The fate of fossil fuel CO_2 in the Oceans*, Plenum Press, p. 401-427, 1977.

Shackleton, N.J., The 100,000-year ice-age cycle identified and found to lag temperature, carbon dioxide, and orbital eccentricity, Science, 289, 1897-1902, 2000.

Siegenthaler, U. and H. Oeschger, Predicting future atmospheric carbon-dioxide levels, *Science,* 199, 388-395, 1978.

Siegenthaler, U. and T.H. Wenk, Rapid atmospheric CO_2 variations and ocean circulation, *Nature,* 308, 624-626, 1984.

Siegenthaler, U., Uptake of excess CO_2 by an outcrop-diffusion model of the ocean, *J. Geophys. Res.,* 88, 3599-3608, 1983.

Sigman, D.M. and E.A. Boyle, Glacial/interglacial variations in atmospheric carbon dioxide, *Nature*, 407, 859-869, 2000.

Stephens B.B. and R.F. Keeling, The influence of Antarctic sea ice on glacial-interglacial CO_2 variations, *Nature*, 404, 171-174, 2000.

Stocker, T.F. and A. Schmittner, Influence of CO_2 emission rates on the stability of the thermohaline circulation, *Nature*, 388, 862-865, 1997.

Stouffer, R.J. and S. Manabe, Response of a coupled ocean-atmosphere to increasing atmospheric carbon dioxide: sensitivity to the rate of increase, *J. Clim.*, 12, 2224-2237, 1999.

Takahashi, T., R.H. Wanninkhof, R.A. Feely, R.F. Weiss, D.W. Chipman, N. Bates, J. Olafsson, C. Sabine, and S.C. Sutherland, Net sea-air CO_2 flux over the global oceans: An improved estimate based on the sea-air pCO_2 difference, in *Proceedings of the 2nd International Symposium, CO_2 in the Oceans,extended abstracts*, ed. by Yukihiro Nojiri, Center for Global Env Research, Tsukuba, Japan, 1999.

Tegen, I., and I. Fung, Contribution to the atmospheric mineral aerosol load from land surface modification, *J. Geophys. Res.*, 100, 18707-18726, 1995.

Toggweiler, J.R., Variation of atmsopheric CO_2 by ventilation of the ocean's deepest water, *Paleoceanography*, 14, 571-588, 1999.

Toggweiler, J.R., R. Murnane, S. Carson, A. Gnanadesikan, and J.L. Sarmiento, Representation of the carbon cycle in box models and GCMs - 1. Solubility pump, *Global Biogeochem. Cycles*, 17(1), 1026, doi:10.1029/2001GB001401, 2003.

Toggweiler, J.R., R. Murnane, S. Carson, A. Gnanadesikan, and J.L. Sarmiento, Representation of the carbon cycle in box models and GCMs - 2. Organic pump, *Global Biogeochem. Cycles*, 17(1), 1027, doi:10.1029/2001GB001841, 2003.

Volk, T. and M.I. Hoffert, Ocean carbon pumps: Analysis of relative strengths and efficiencies in ocean driven atmospheric CO_2 changes, in *The Carbon Cycle and Atmospheric CO_2: Natural variations Archaean to present,* ed. by E.T. Sundquist and W.S. Broecker, Geophys. Monogr.Ser., 32, 99-110, AGU, Washington, D.C., 1985.

Wallace, Monitoring global ocean carbon inventories. *Ocean Observing System Development Panel, OOSDP Background Report, 5*, Texas AM University, Texas, 1995.

Wanninkhof, R., Relationship between gas exchange and wind speed over the ocean, *J. Geophys. Res.*, 97, 7373-7382, 1992.

OCEAN CARBON CYCLE IN A CHANGING CLIMATE: CLIMATE CHANGE DETECTION

RICHARD J. MATEAR
CSIRO Marine Research and the Antarctic Cooperative Research Center, GPO Box 1538, Hobart Tasmania, Australia.
richard.matear@marine.csiro.au

INTRODUCTION

The atmospheric concentration of CO_2 and other greenhouse gases are increasing and changing the radiative properties of the atmosphere. To study the response of the ocean-climate system to this climate perturbation, investigators have integrated ocean-climate models with an atmospheric CO_2 concentration that rises over time to double, triple or quadruple the control CO_2 level [Bi et al., 2001; Hirst, 1999; Manabe and Stouffer, 1993; Goosse and Renssen, 2001]. The oceanic response in these experiments typically includes widespread surface warming, retreat of sea ice and a general increase in upper ocean stratification, all of which may impact upon the ocean biogeochemistry. The substantial reductions in formation rate and/or density of intermediate and deep water masses that occur in these simulations would likely affect biogeochemical cycling in the ocean and the air-sea exchange of oxygen and CO_2. This chapter uses a biogeochemical model to examine the effect of global warming on biogeochemical cycling in the ocean focusing on the oceanic oxygen distribution. The chapter discusses the simulated oxygen changes in the ocean and demonstrates the potential of using oxygen changes to track global warming and provide observations to assess climate model simulations.

M. Follows and T. Oguz (eds.), The Ocean Carbon Cycle and Climate, 297–315.

1. MODELS AND EXPERIMENTS

We utilize the CSIRO climate model [*Gordon and O'Farrell*, 1997; *Hirst et al.*, 2000] together with a low-order oceanic biogeochemical model [*Matear and Hirst*, 1999] to investigate the term impact of climate change on the ocean biogeochemical cycles. The CSIRO climate model is a coupled general circulation model with atmospheric, oceanic, sea ice and land surface components [*Farrell et al.*, 1995; *Gordon and O'Farrell*, 1997]. The atmospheric component is spectral with rhomboidal truncation at wavenumber 21, and in the vertical has nine levels on a sigma coordinate system. The model has full diurnal and annual cycles, gravity wave drag, a mass flux convection scheme, semi-Lagrangian water vapour transport, and a relative humidity based cloud parameterization. The sea ice component [*O'Farrell*, 1998] includes the cavitating fluid rheology of *Flato and Hibler*, [1990] and the three-layer ice thermodynamics of *Semtner*, [1976]. The coupling of the atmospheric and oceanic model components is done using flux adjustments (fixed for the particular model version) to mitigate spurious climate drift.

The oceanic component is based on the code of *Cox*, [1984], and has a resolution of 5.6 degrees of longitude and 3.2 degrees of latitude, with 21 levels in the vertical. Oceanic tracers are advected by the resolved large-scale velocity field together with an eddy-induced transport velocity parameterized as per *Gent et al.*, [1995]. Internal diffusion of tracers includes a predominant component along the neutral density surfaces (with isoneutral diffusivity K_I of 1000 $m^2\ s^{-1}$) and a weak vertical component with vertical diffusivity K_V as parameterised by *Gargett*, [1984] (with a default minimum of 0.3×10^{-4} $m^2\ s^{-1}$, and a typical range in the stratified interior ocean of $0.3\text{-}1\times10^{-4}$ $m^2\ s^{-1}$). Convective mixing is simulated by setting the vertical diffusivity to 100 $m^2\ s^{-1}$ in statically unstable regions. Further details of the ocean component are found in *Hirst et al.*, [2000].

To represent the atmospheric concentrations of greenhouse gas (i.e. CO_2, methane, ozone, etc) we convert their radiative affect into an equivalent CO_2 concentration. Then we use this equivalent CO_2 concentration to simulate impact of changing levels of greenhouse gas on the climate system. We perform two simulations with the CSIRO climate model: 1) a control experiment with a constant atmospheric level of equivalent CO_2 (330μatm) and 2) a greenhouse gas forcing experiment where the level of equivalent CO_2 follows the IS92a radiative forcing scenario [*Houghton et al.*, 1995] from the year 1880 until the year 2083, and thereafter is held constant (Fig. 1). The final elevated level of equivalent CO_2 corresponds to a value an atmospheric CO_2 level that is three times 1880 value (990μatm).

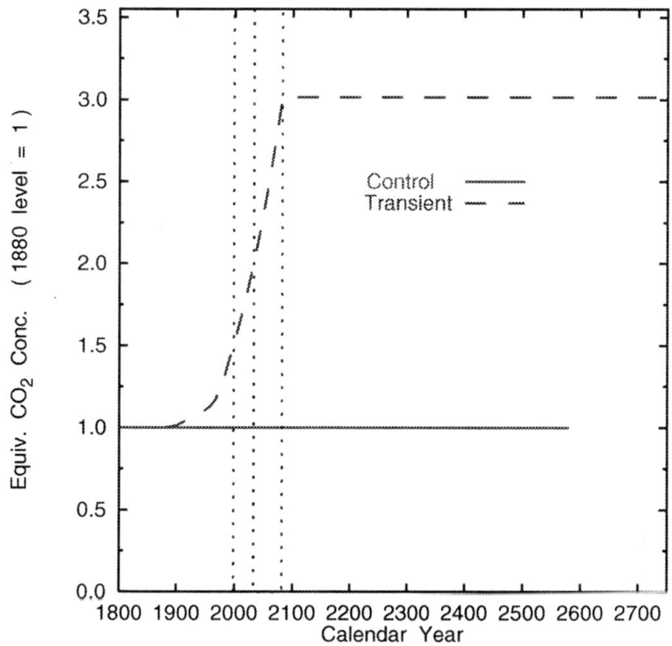

Figure 1. For the control and transient experiment the prescribed atmospheric equivalent CO_2 concentration (relative to 1880 level). Dotted vertical lines indicate the present time (year 2000), and the times of equivalent CO_2 doubling (2033) and tripling (2082) from the 1880 level

The biogeochemical model is described by *Matear and Hirst*, [1999] and is run using archived climate model output following a technique described by *Aumont et al.*, [1998]. The archived data used are the three-dimensional fields of ocean temperature, salinity and currents, and the surface fields of wind speed and sea-ice cover. Archived monthly mean data provided the forcing fields for our biogeochemical model, which determines the oceanic distributions of dissolved inorganic carbon (DIC), alkalinity, phosphate and dissolved oxygen.

The biogeochemical model includes a simple representation of the surface export production of biological matter as a function of the temperature, mixed-layer depth and nutrient concentration (phosphate) in the euphotic zone, with the sinking particulate organic matter (POM) and calcium carbonate remineralization according to prescribed functions of depth. The euphotic zone production of organic matter consumes dissolved phosphorous, nitrogen and carbon and releases oxygen according to the *Redfield et al.*, (1963) ratio P:N:C:O_2 of 1:16:106:-138, while the subsurface remineralization conversely releases/consumes these elements in like ratio.

The conservation equation describing the evolution of phosphate, dissolved oxygen, DIC and alkalinity are similar and only the conservation equation for dissolved oxygen concentrations will be presented. The conservation for dissolved oxygen (C_O) is given by

$$\frac{\partial C_o}{\partial t} = -\nabla_3 \bullet (\mathbf{u}C_o) + \nabla_I \bullet (K_I \nabla_I C_o) + \frac{\partial}{\partial z}(K_z \frac{\partial C_o}{\partial z}) + Q_F^o - Q_R^o \tag{1}$$

The first term on the right represents the usual three-dimensional advection, the second term represents diffusion along the neutral density surfaces and the third term represents vertical diffusion. $Q_R{}^o$ denotes the biological production and consumption due photosynthesis and remineralization, respectively. The oxygen flux into the surface ocean across the air-sea interface ($Q_F{}^o$) is given by

$$Q_F^O = K(pO_2{}^{atmosphere} - pO_2{}^{ocean}) \tag{2}$$

where K is the gas exchange coefficient, which is a function of wind speed and temperature, $pO_2{}^{atmosphere}$ is the atmospheric partial pressure of oxygen (fixed at 0.2048 atm), and $pO_2{}^{ocean}$ is the effective partial pressure oxygen in the surface ocean water. For our simulations, the K fields from the Ocean Carbon Model Inter-Comparison Project are used [*Dutay et al.*, 2002] with an adjustment for the changes in wind speed simulated by our climate model under global warming. $pO_2{}^{ocean}$ is related to surface oxygen concentration, C_O, by:

$$pO_2{}^{ocean} = pO_2{}^{atmosphere} \frac{C_o}{C_o^{sat}(T)} \tag{3}$$

where $C_o{}^{sat}(T)$ is the oxygen concentration in equilibrium with the atmosphere, which is a monotonic decreasing function of surface ocean temperature, T. From equations (2) and (3), an increase in surface temperature leads to an outgasing of oxygen from the ocean (negative Q_F) and a reduction in surface water oxygen concentration, if all other factors remain unchanged.

In the euphotic zone the export production of POM is computed in terms of its particulate organic carbon (POC) content by:

$$E_{POC} = S(\frac{I}{I_o})(\frac{h_e}{h})(\frac{P}{P + P_o})(1.066)^T \tag{4}$$

where S is the scaling factor in mol C m^{-2} s^{-1} , I is the daily averaged incident solar radiation at the surface, I_o is the solar constant, h_e is the depth of the euphotic layer (50 m), h is the depth of the mixed layer, P is the surface phosphate concentration (mol/kg), P_o is the half-saturation constant (0.1μmol/kg) and T is the sea-surface temperature (in °C). Scaling factor S is set empirically to obtain a surface phosphate field that appears consistent with observations. The last term in equation 4 reflects the affect on temperature on phytoplankton growth and it is based on *Eppley*, 1972 [1972]. In the euphotic layer, oxygen production due to the formation of POM is

$$Q_R^O = aE_{POC}h_e \tag{5}$$

The coefficient a represents the molar ratio between the consumption of dissolved oxygen and production of DIC by the remineralized organic carbon and is set at -138/106. Below the euphotic zone, the consumption rate of oxygen by remineralization of POM is given by

$$Q_R^O = -aE_{POC}\frac{\partial}{\partial z}[R(z)] \tag{6}.$$

The operand $R(z) = (z_R/z)^{0.9}$ is a standard profile for the downward flux of particulate organic matter (e.g., *Yamanaka and Tajika*, 1996), where z_R is the scale depth of remineralization (set at 100 m). In the model, any organic matter reaching the ocean bottom is remineralized in the bottom layer. Further, remineralization according to (5) also occurs in anoxic regions, as a result of denitrification. In these denitrification regions POM is converted back into its inorganic constituents without consuming oxygen.

Equations similar to equation (1) can be written equation for phosphate, DIC and alkalinity. For the phosphate equation there is no air-sea exchange term and the source term due to the formation and remineralization of POM identical to equations (5) and (6) except a is set to 1/106. For alkalinity, there is no air-sea exchange term.

For the DIC equation, the CO_2 air-sea flux is given by

$$Q_F^C = K(pCO_2^{atmosphere} - pCO_2^{ocean}) \tag{7}$$

where K is the gas exchange coefficient, which is a function of wind speed and temperature[*Dutay et al.*, 2002], $pCO_2^{atmosphere}$ is the atmospheric partial pressure of carbon dioxide and pCO_2^{ocean} is the partial pressure of carbon dioxide in the surface ocean. The partial $pCO_2^{atmosphere}$ was specified (Figure 1) while the surface ocean pCO_2^{ocean} was determined using the full carbon chemistry equations which required modeled temperature, salinity, alkalinity and DIC from the surface ocean [*Peng and Broecker*, 1987].

For the DIC and Alkalinity, source terms are included for the formation and remineralization of POM, equations (5) and (6) are used with *a* set to 1 and -16/106, for DIC and alkalinity respectively. The alkalinity source term reflects the impact of the production and remineralization of particulate organic carbon on alkalinity through changing the nitrate concentration [*Broecker and Peng*, 1982]. For DIC and alkalinity, an additional source term was included for the formation and remineralization of $CaCO_3$. Following *Yamanaka and Tajika*, [1996], the $CaCO_3$ formation and export was a constant 8% fraction of the POM export. The remineralization profile, R(z), of $CaCO_3$ was set to $e^{(-z/3500)}$ where *z* is depth in meters. For every 1 mole of $CaCO_3$ formed DIC is reduced by 1 mole and alkalinity by 2 moles, respectively. The remineralization of $CaCO_3$ occurs in the same ratio.

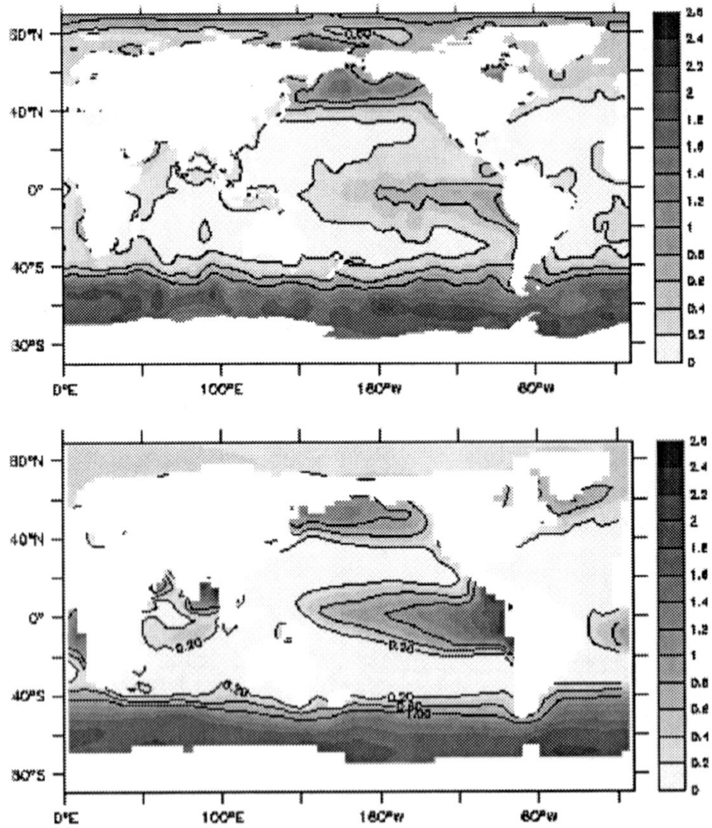

Figure 2. Annual mean phosphate concentrations in μmol/kg at 25m from a) the observations [Levitus et al., 1993] and b) the control run for the year 1980.

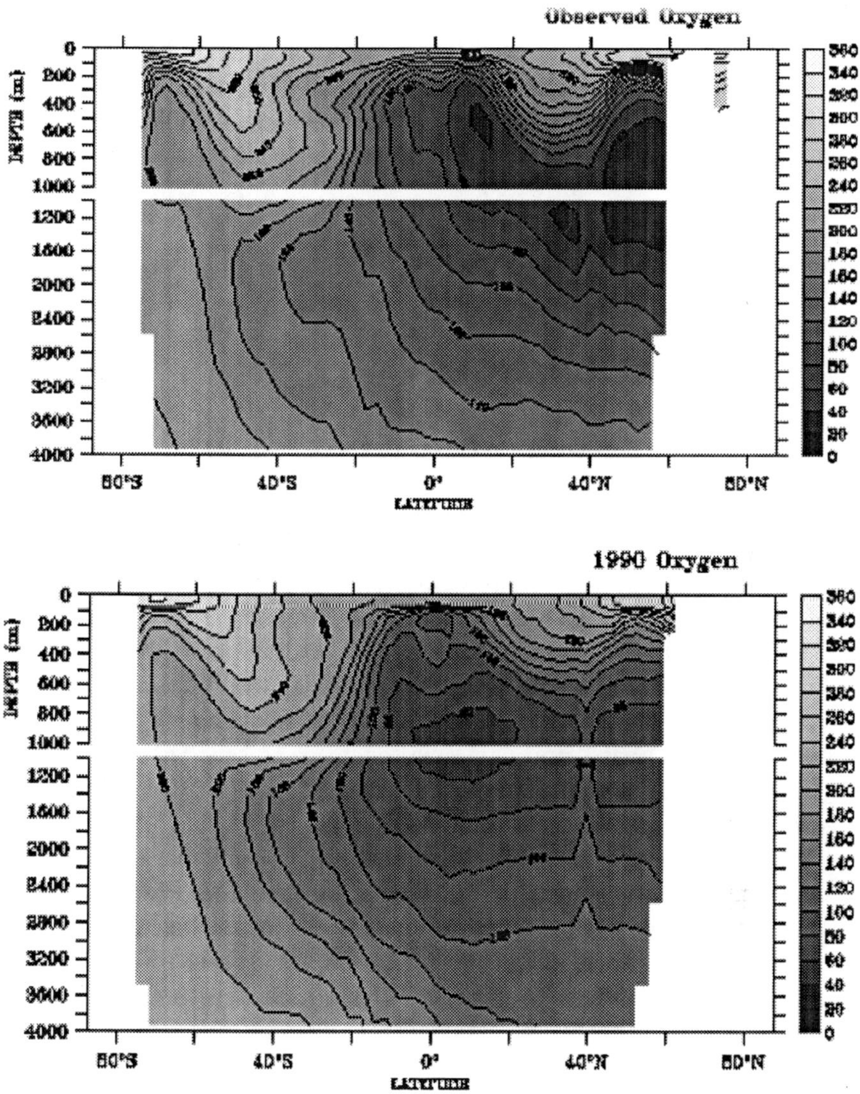

Figure 3. Zonally-averaged annual mean field of dissolved oxygen (μmol kg^{-1}) from (a) the observations [Levitus 1993], and (b) the climate change integration for the year 1980.

The biogeochemical model is then integrated as described above over the duration of the climate model integrations. As expected the distributions of the biogeochemical variables are generally stable during the control integration. The observed phosphate concentrations at 25m [*Levitus et al.*, 1993] are compared to annual mean values simulated by the model. The model reproduces the large scale observed phosphate fields with high concentrations in the Southern Ocean, eastern Equatorial Pacific and North Pacific (Fig. 2). Figures 3 shows that the zonally averaged annual mean oxygen distribution in the ocean computed for 1980 of the climate change integration, compares well, in broad terms, to that obtained from the observations .

2. RESULTS

2.1 PHYSICAL CHANGES

Figure 4a shows the time series of globally-averaged sea surface temperature (SST) for the two experiments. Similar to other climate model experiments (e.g., *Sarmiento et al.*, 1998), the SST shows a rapid increase during the period of increasing equivalent CO_2 concentration in the atmosphere, followed by a gradual increase during the subsequent period of stable elevated equivalent CO_2 concentration. The change in surface water density follows a similar pattern. The density reduction in the surface is mainly caused by surface warming in the tropics and mid-latitudes, and by surface freshening in polar regions. The simulated freshening of the surface polar oceans with global warming is caused by increased in high latitude precipitation and, in some regions, by reduced rate of brine rejection associated with reduced sea-ice formation.

The upper ocean responds rapidly to changes in buoyancy forcing, but the deep ocean density adjusts to the altered surface conditions on much longer, millennial, time scale. The consequence of these differences in time-scale is increased upper ocean stratification as shown by increased stratification in the Southern Ocean (Fig. 4b), which is responsible for the reduction and/or shoaling of the oceanic overturning circulation and the reduced oceanic ventilation of the deep ocean under global warming. The increased stratification of the upper ocean with global warming is also evident in the reductions in maximum winter mixed layer depth projected under global warming. For the calculation of the mixed layer depth we use the monthly mean temperature and salinities fields to determine potential density then set the mixed layer depth as the depth where the sub-surface density exceeds the surface density by 0.05 kg/m^3. The observations show that the maximum winter mixed layers exceed 1000m in the North Atlantic near Iceland and Greenland and in the Southern Ocean in the Ross and Weddell Seas (Fig. 5). In these regions, the deep winter mixing of the upper ocean provides an efficient mechanism for ventilating the upper 1000 m of the ocean. In the Southern Ocean, the observations show a latitudinal zone of deep winter mixed layers centered at approximately 40°S. This zone of deep mixed layers broadly defines the sub-Antarctic zone, which is the source region for Sub-Antarctic Mode Water. Such a band of deep winter mixed layers is evident in the control simulation but in the model the extent of the deep mixed layer regions is more widespread and shallower than the observations (Fig. 5). In the Southern Ocean, the control simulations reproduces the deep winter mixed layers in the Weddell and Ross Seas, a region of shallower mixed layers between the Antarctic continent and the sub-Antarctic zone and deep winter mixed layers in the sub-Antarctic zone.

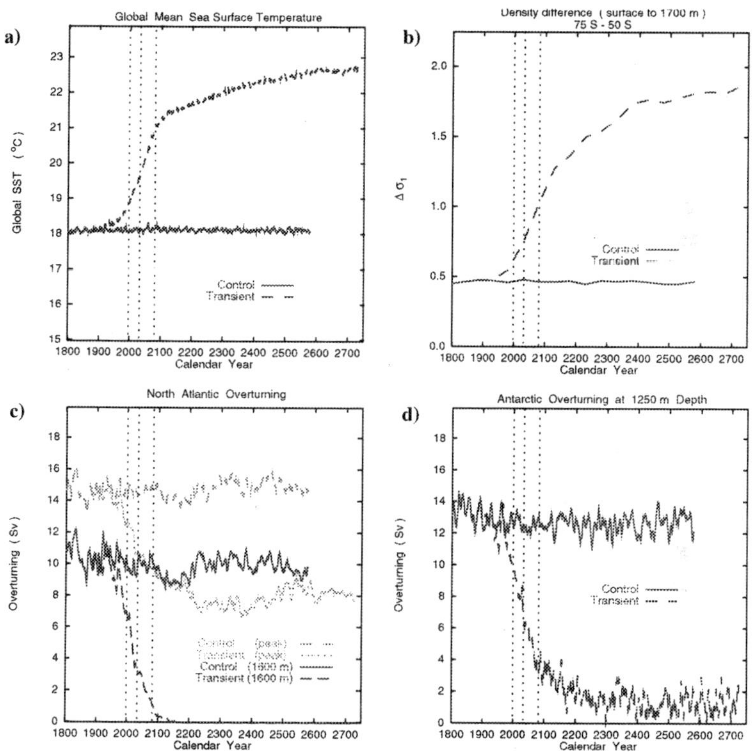

Figure 4. Time series of model indices for the control and climate change ("transient") integrations: a) Global average sea surface temperature; b) Difference in Aug.-Sept. potential density (referenced to 1000 m) between the ocean surface and 1,750-m depth, averaged over the ocean region south of 50°S; c) Strength of overturning in the North Atlantic Deep Water formation circulation, and strength of this overturning which penetrates downward through the 1600-m depth level (106 m^3 s^{-1}); d) Strength of overturning of the Antarctic Bottom Water formation circulation which penetrates downward through the 1250-m depth level (106 m^3 s^{-1}). Dotted vertical lines indicate the present time (calendar year 2000), and the times of equivalent CO_2 doubling (2033) and tripling (2082) from the 1880 level.

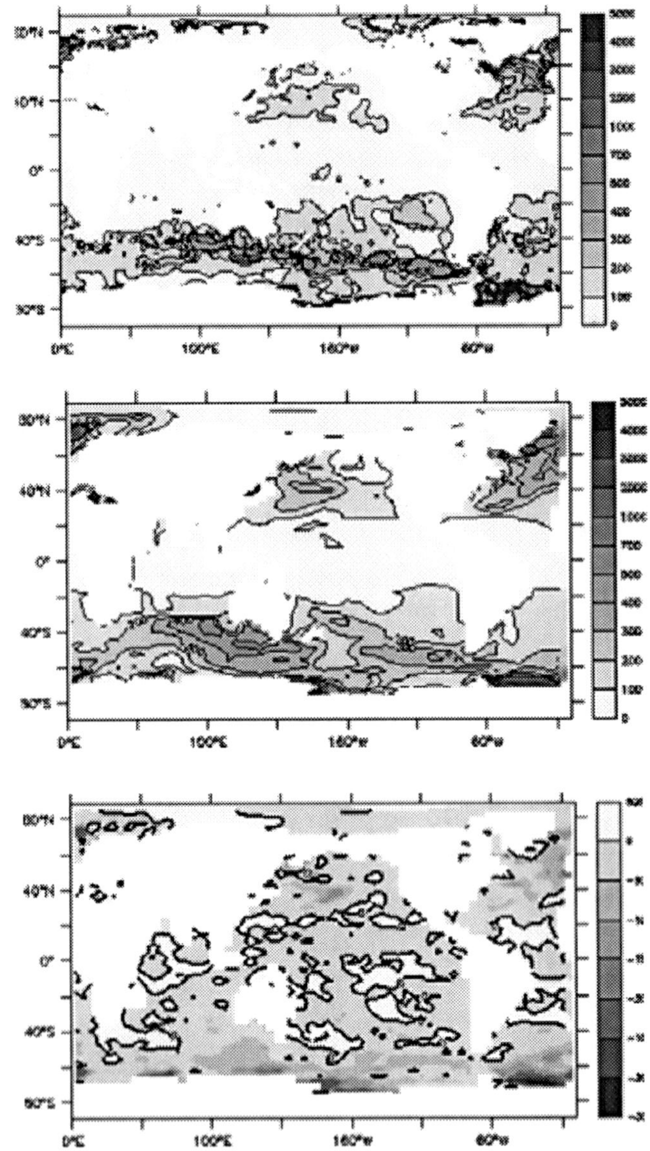

Figure 5. Maximum winter mixed layer depth determined from monthly temperature and salinity fields for the a) observations [Conkright et al., 2002], b) 1980-2000 of the control run and c) 2000-2010 from the transient run.

However, the models lacks the well define spatial structure evident in the observations which reflects both the coarse resolution and the parameterization of vertical mixing in the upper ocean in our OGCM. The change in maximum mixed layer depths for years 2000-2010 between the transient and control experiments shows clear shoaling the maximum winter mixed layer depth in the Southern Ocean. The shoaling of the mixed layer is a general feature of the entire Southern Ocean with the greatest shoaling occurring in the deep winter layer areas of the Ross and Weddell Seas. Reduced winter mixed layers in the Southern Ocean further increases the density stratification in the upper ocean by reducing the density of the surface water.

The time series of our simulated North Atlantic and Antarctic deep water overturning is shown in Figures 4c and 4d. The simulated strength of the North Atlantic overturning prior to the year 2000 of 14 to 16 Sv agrees with the estimated transport at 24°N of 17±4 Sv [*Roemmich and Wunsch*, 1985]. Under global warming, the strength of the North Atlantic overturning declines by about 50% and by about the 23rd century, North Atlantic Deep Water no longer penetrates below about 1.5 km depth. The simulated strength of Antarctic overturning prior to the year 2000 varies between 8 and 12 Sv. Geochemical based estimates of Antarctic deepwater overturning give 8.1 to 9.4 Sv using CFCs [*Orsi et al.*, 1999] and 15 Sv using PO_4^* and ^{14}C [*Broecker et al.*, 1998]. Our model simulations is consistent with these observational estimates. Under global warming, we project the cessation of the penetration of Antarctic Deep Water below 1250m depth in the Southern Ocean.

The simulated reduction in deepwater formation combined with the general reductions in Southern Ocean winter mixed layers and high-latitude outcropping of density surfaces markedly reduces the ventilation of ocean interior under global warming.

2.2 OXYGEN CHANGES

The three primary factors affecting oceanic dissolved oxygen concentrations are changes in (1) surface temperature, (2) export production, and (3) ventilation. An increase in surface ocean temperature decreases oxygen solubility; hence decreases surface oxygen concentration (equation (3)). The signature of the reduced surface concentration may then be advected or mixed into the ocean interior. An increase in export production increases the remineralization of sinking particulate organic matter causing more rapid depletion of subsurface oxygen (via equation (5)), hence reduces interior oxygen concentration. Finally, a decrease in the ventilation rate increases the residence time of water in the subsurface ocean allowing for more POM remineralization to occur within a given water mass and reduces the supply of oxygen rich surface water both process reduce the oxygen concentrations in the ocean interior.

With climate change, our model projects a reduction in the oceanic uptake of oxygen (Fig. 6). The reduced oxygen uptake by the ocean with climate change is dominated by reduction in the oxygen uptake by the ocean south of 40°S (Fig. 6) and we focus on this region. The zonal averaged change in dissolved oxygen with climate change for the 2000 to 2010 period shows a maximum decrease of 35 umol kg^{-1} (Fig. 7a).

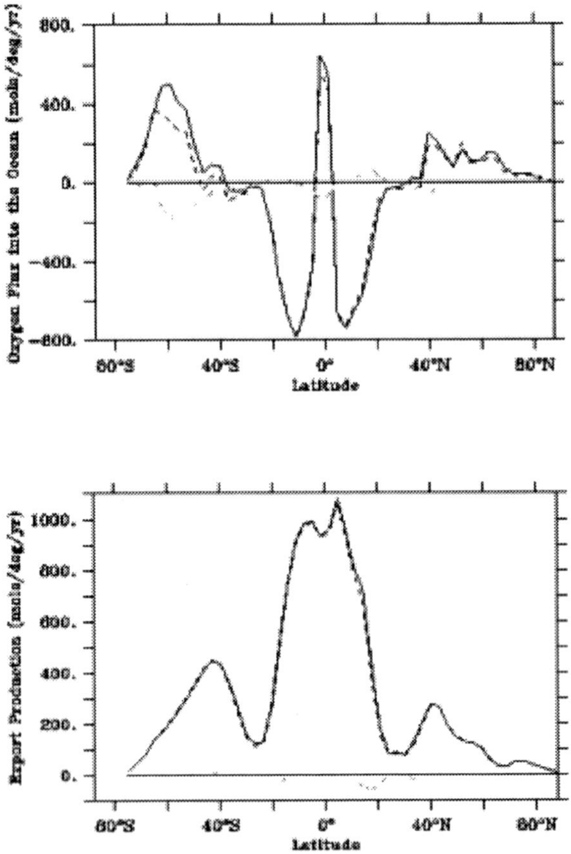

Figure 6. For the 1990-2000 period the annual mean zonal averaged a) oceanic uptake of oxygen (O_2 mol/deg/yr) for the control (solid line) and the transient (dashed line) experiments and for the difference between the two experiments (dotted line); and b) export production (C mol/ deg/yr) of the control (solid line), transient (dashed line) experiments and difference between these two experiments (dotted line).

The maximum decrease occurs south of 60°S at a depth of about 350m. In contrast, changes in the surface concentrations are much smaller, with an actual increase in dissolved oxygen in the surface waters south of 60°S due to a regional cooling (but freshening) of the surface water (Fig 7c and d). At the surface, the oxygen anomaly is controlled only by changes in solubility because the equilibration of surface water with atmospheric (≈7 days) is much faster than the influence of ocean circulation. This prevents the changes in ocean circulation from impacting the surface concentration of dissolved oxygen. Below the surface, dissolved oxygen levels are controlled by the supply of oxygen rich surface water through water mass subduction and winter deepening of the mixed layer, and by the consumption of oxygen by the remineralization of organic matter. To quantify how changes in oxygen solubility at the surface under global warming impact dissolved oxygen concentrations in the ocean interior the projected changes in temperature and salinity are converting into an oxygen solubility change, ΔO_2^{sol}, as follows.

$$\Delta O_2^{sol} = O_2^{sat}(T^{transient}, S^{transient}) - O_2^{sat}(T^{control}, S^{control})$$

where T and S represent the simulated in situ temperature and salinities of the transient and control experiment and the function O_2^{sat} gives the saturated oxygen concentration at a given temperature and salinity. For the 2000-2010 period, the solubility changes in dissolved oxygen are less than 5μmol/kg (Fig 7b).

Figure 6b showed that the change in export production in the Southern Ocean with climate change was small. By integrating the change in export production from 1880 to 2010 we determined that dissolved oxygen concentrations in the upper 1000 m of the Southern would be reduced by less than 2 $\mu mol\ kg^{-1}$. Our calculations of the oxygen changes associated with solubility and export production changes are small compared to the simulated change in dissolved oxygen. This demonstrates that simulated reduction in dissolved oxygen concentrations in the Southern Ocean are dominated by changes in ocean circulation.

The simulate reductions in the rate of subduction of AABW and the maximum winter mixed layer depth with global warming reduce the oceanic uptake of oxygen by the Southern Ocean and the supply of oxygen to the interior. These changes dominate the reduction in dissolved oxygen levels in the Southern Ocean interior. The maximum decrease in dissolved oxygen south 60°S at 350 m reflects the reduction of the maximum mixed layer depth with global warming.

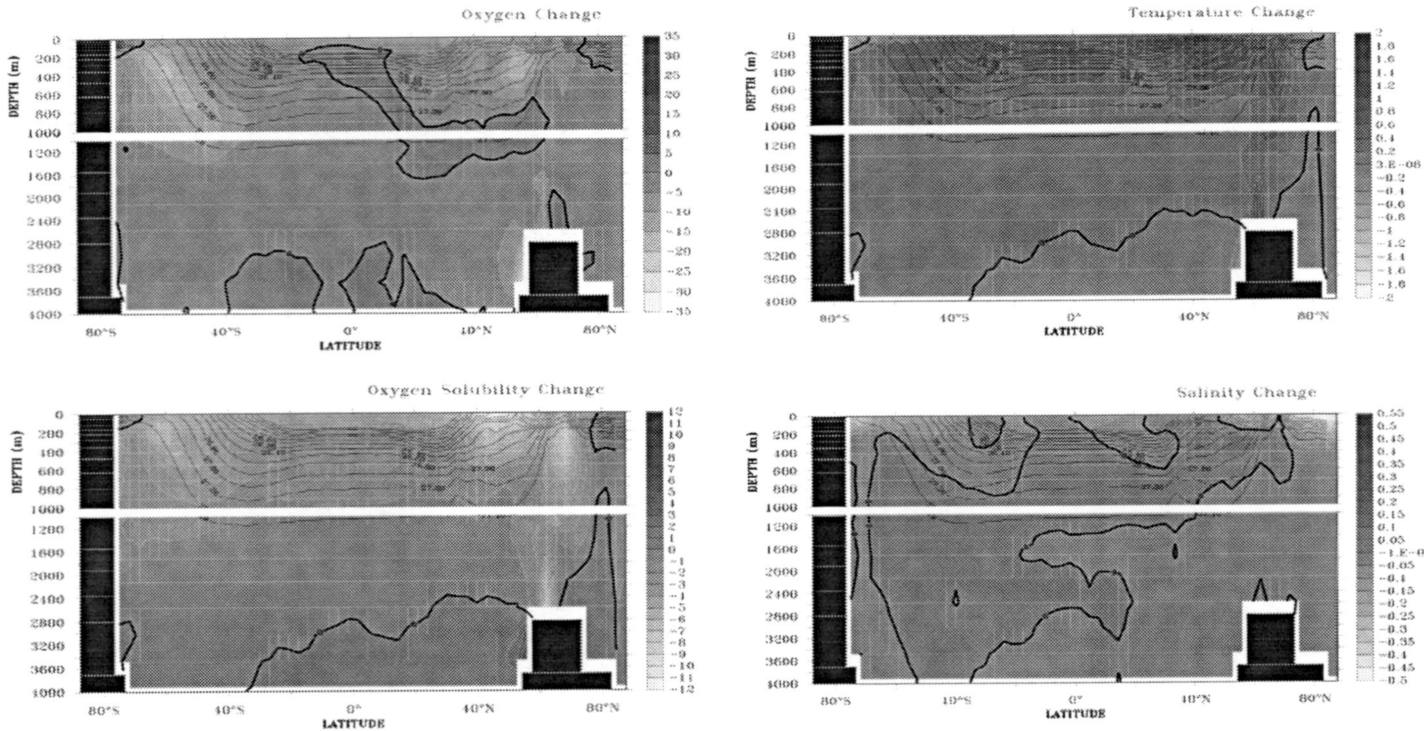

Figure 7. Zonal averaged change along depth surfaces between the transient and the control experiment during the period 2000-2010 for: a) oxygen (μmol kg^{-1}); b) solubility only change in dissolved oxygen (μmol/kg); c) temperature (°C); and d) salinity. Shown in these sections are the averaged potential density surfaces from the control experiment for the same period.

The oxygen anomaly created by this process then propagates northward into the sub-tropics with the subduction of both Antarctic Intermediate Water and AABW.

In addition to the prominent changes in the dissolved oxygen concentrations with global warming our model projects prominent changes to the temperature and salinity field in the ocean. In the Southern Ocean, the pattern of the zonal averaged change in temperature and salinity (Fig. 7c-d) differ from the oxygen changes. The projected temperature and salinity changes are greatest at the surface and decrease with depth while the dissolved oxygen decrease displays a prominent sub-surface maximum. The maximum temperature and salinity changes also occur further north than the latitude of the maximum oxygen decrease. The differences in the pattern of change in the temperature, salinity and dissolved oxygen fields imply they contain independent information on the response of the ocean to global warming. Additional analysis of the model simulations is needed to demonstrate how the information in these different tracers can be used to further our understanding of how the ocean is changing with global warming. Insight gain from such analysis would be useful to interpreting observed changes in temperature, salinity and dissolved oxygen in the ocean.

3. DISCUSSION AND FUTURE DIRECTIONS

3.1 MODEL LIMITATIONS

The projected dramatic changes in dissolved oxygen concentrations in the Southern Ocean with global warming demonstrate the sensitivity of this biogeochemical tracer to global warming. However, there are a number of limitations in the ocean model used for our global warming simulations, which may impact the response of dissolved oxygen to global warming.

One, simulating deep water formation like AABWis a challenge [*Doney and Hecht*, 2002] The low resolution OGCM used in this study cannot faithfully represent the local mesoscale processes of shelf water formation, downslope flow, and entrainment important to the formation of AABW. Furthermore, in our model deep water ventilation is restricted to the Weddell and Ross Seas in contrast to CFC observations which support additional formation sites in the Indian Ocean near the Amery Ice Shelf and Adélie Coast [*Rintoul and Bullister*, 1999]. Finally, in our model the deep water masses are lighter and warmer than the Southern Ocean observations and these errors suggest our model produces too little low salinity shelf water. All these limitations questioned our ability to simulated AABW formation and they drive the need to further investigate through both model and observations the sensitivity of AABW formation to processes like sea-ice formation, downslope flow and large-scale ocean circulation.

For the year 2000-2010, the changes in the ventilation of the upper ocean through changes in density and winter mixed layers is the process that dominating the projected oxygen declines in the Southern Ocean. For simulating the upper ocean dynamics, again our coarse resolution OGCM both horizontally and in depth is not ideal. Furthermore, the simple convective instability mechanism used in the model to simulated mixing in the upper ocean neglects the role wind mixing may have in enhancing winter mixing. In a strong wind region like the Southern Ocean, wind mixing can be an important process to deepening the winter mixed layer [*Wang and Matear*, 2001]. Such a limitation of our OGCM may explain the too shallow mixed layers in the sub-Antarctic zone and lead over-sensitive response of our model to changes in surface density cause by warming and freshening of the Southern Ocean with global warming.

Our model may over-estimate the magnitude of the response changes in ocean ventilation to global warming however; there is a general consensus in climate model simulations that the ocean will become more stratified with global warming. Under such a scenario one expects the upper ocean ventilation to decline and with it dissolved oxygen levels in the ocean interior. The project response of dissolved oxygen to changes in the AABW formation rate and upper ocean mixing under global warming clearly demonstrate that dissolved oxygen is sensitive to these changes and the emphasize the potential to use dissolved oxygen to monitor and detect such changes in the ocean.

3.2 OBSERVED CHANGES

In the Southern Ocean, observational evidence of multi-decadal changes in its water properties are mounting(e.g. *Gille*, 2002). Observations show a warming in the sub-Antarctic Mode Water (300-700 m) [*Bindoff and Church*, 1992], a freshening in the Antarctic Intermediate Water (AAIW) [*Wong et al.*, 1999] and a dissolved oxygen decline in the AAIW[*Matear et al.*, 2000]. The temperature and salinity changes are inferred to reflect the warming and freshening of the subducted sub-Antarctic Mode Water [*Bindoff and Church*, 1992] and Antarctic Intermediate Water [*Wong et al.*, 1999], respectively. The dissolved oxygen changes are interpreted as changes in the ventilation of the upper ocean [*Matear et al.*, 2000]. As demonstrated by *Rintoul and England* [2002], to change the water properties of the sub-Antarctic Mode Water through changes in the air-sea fluxes would required an unrealistic change in the heat and freshwater fluxes. A more plausible explanation of the observed changes is that a change in the ventilation of the upper ocean has changed the water properties of the mode and intermediate water. Reduce ventilation would tend to decrease the sub-surface dissolved oxygen levels while causing a freshening and warming of the surface water since the underlying water is warmer and saltier. Subsequent ekman transport of the anomalously water north would warm and freshen the water in the sub-Antarctic zone and warm and freshen the sub-Antarctic Mode Water. A more through analysis of the temperature, salinity and oxygen data in the Southern Ocean is needed to explore this possibility.

With the collection of the hydrographic data from the WOCE program we are now in great position to pursue the comparison of recent to historical data. Furthermore, dissolved oxygen is an ideal tracer for such a comparison because next to temperature and salinity, dissolved oxygen is the next most observed hydrographic parameter. With the implementation of the ARGO drifter program the oceanographic community is embarking on a program that will revolutionalize ocean observations [see chapter xx in this volume]. Through the ARGO program temperature and salinity profiles of the upper ocean will be autonomously collected and in regions like the sparsely sampled Southern Ocean they will dramatically increase the amount of observations. The addition of an oxygen sensor to some of the ARGO drifters would provide an important additional parameter for monitoring the ocean state. Exploiting these three tracers may enable one to separate changes in ocean ventilation from changes in the flux of heat and freshwater into the ocean. Such observations would be critical to monitoring the state of the ocean and providing observations to test climate model projections of global warming.

LIST OF REFERENCES

Aumont, O., J.C. Orr, D. Jamous, P. Monfray, O. Marti, and G. Madec, A degradation approach to accelerate simulations to steady state in a 3-D tracer transport model of the global ocean, *Climate Dynamics*, *14*, 101-116, 1998.

Bi, D.H., W.F. Budd, A.C. Hirst, and X.R. Wu, Collapse and reorganisation of the Southern Ocean overturning under global warming in a coupled model, *Geophysical Research Letters*, *28* (20), 3927-3930, 2001.

Bindoff, N.L., and J.A. Church, Warming of the water column in the southwest Pacific Ocean, *Nature*, *357*, 59-62, 1992.

Broecker, W., and T.-H. Peng, *Tracers in the Sea*, 690pp. pp., Columbia University, Palisades, New York, 1982.

Broecker, W.S., S.L. Peacock, S. Walker, R. Weiss, E. Fahrbach, M. Schroeder, M. U., C. Heinze, R. Key, T.-H. Peng, and S. Rubin, How much deep water is formed in the Southern Ocean?, *Journal of Geophysical Research*, *103* (C8), 15,833-15,843, 1998.

Conkright, M.E., R.A. Locarnini, H.E. Garcia, T.D. O'Brien, T.P. Boyer, C. Stephens, and J.I. Antonov, World OCEAN ATLAS 2001: Objective Analyses, Data Statistics and Figures, pp. 17, National Oceanographic Data Center,, Silver Spring, MD, 2002.

Cox, M.D., A primitive equation, three-dimensional model of the ocean., pp. 141, GFDL Ocean Group Technical Report, GFDL/Princeton University, Princeton N.J., 1984.

Doney, S.C., and M.W. Hecht, Antarctic bottom water formation and deep-water chlorofluorocarbon distributions in a global ocean climate model, *Journal of Physical Oceanography*, *32* (6), 1642-1666, 2002.

Dutay, J.-C., J.L. Bullister, S.C. Doney, J.C. Orr, R. Najjar, K. Caldiera, J.-M. Campin, H. Drange, M. Follows, Y. Gao, N. Gruber, M.W. Hecht, A. Ishida, F. Joos, K. Lindsay, G. Madec, E. Maier-Reimer, J.C. Marshall, R.J. Matear, P. Monfray, G.-K. Plattner, J. Sarmiento, R. Schiltzer, R. Slater, I.J. Totterdell, M.-F. Weirig, Y. Yamanaka, and A. Yool, Evaluation of ocean model ventilation with CFC-11: comparison of 13 global ocean models, *Ocean Modelling*, *4* (2), 89-120, 2002.

Eppley, R.W., Temperature and phytoplankton growth in the sea, *Fish. Bull. U.S.*, *70*, 1063-1085, 1972.

Farrell, J.W., T.F. Pedersen, S.E. Calvert, and B. Nielsen, Glacial-interglacial changes in nutrient utilization in the equatorial Pacific Ocean, *Nature*, *377*, 514-517, 1995.

Flato, G.M., and W.D. Hibler, On a simple sea-ice dynamics model for climate studies, *Ann. Glaciol.*, *14*, 72-77, 1990.

Gargett, A.E., Vertical eddy diffusivity in the ocean interior, *Journal of Marine Research*, *42*, 359-393, 1984.

Gent, P.R., J. Willebrand, T.J. McDougall, and J.C. McWilliams, Parameterizing Eddy-Induced Tracer Transports in Ocean Circulation Models, *Journal of Physical Oceanography*, *25*, 463-474, 1995.

Gille, S.T., Warming of the Southern Ocean since the 1950s, *Science*, *295* (5558), 1275-1277, 2002.

Goosse, H., and H. Renssen, A two-phase response of the Southern Ocean to an increase in greenhouse gas concentrations, *Geophysical Research Letters*, *28* (18), 3469-3472, 2001.

Gordon, H.B., and S.P. O'Farrell, Transient climate change in the CSIRO coupled model with dynamical sea ice., *Mon. Wea. Rev.*, *125*, 875-907, 1997.

Hirst, A.C., The Southern Ocean response to global warming in the CSIRO coupled ocean-atmosphere model, *Environmental Modeling and Software: Special issue on Modelling Global Climate Change*, *14*, 227-242, 1999.

Hirst, A.C., S.P. O'Farrell, and H.B. Gordon, Comparison of a coupled ocean-atmosphere model with and without oceanic eddy-induced advection. Part I: Ocean spinup and control integrations, *Journal of Climate*, *13*, 139-163, 2000.

Houghton, J.T., L.G.M. Filho, J. Bruce, H. Lee, B.A. Callander, E. Haites, N. Harris, and K. Maskell, Climate Change 1994, in *Radiative Forcing of Climate Change and An Evaluation of the IPCC IS92 Emission Scenarios*, pp. 339, Intergovermental Panel on Climate Change, Cambridge University Press, 1995.

Levitus, S., M.E. Conkright, J.L. Reid, and R.G. Najjar, Distribution of nitrate, phosphate and silicate in the world oceans, *Progress Oceanography*, *31*, 245-273, 1993.

Manabe, S., and R.J. Stouffer, Century-scale effects of increased atmospheric CO_2 on the oceanic-atmosphere system, *Nature*, *364*, 215, 1993.

Matear, R.J., and A.C. Hirst, Climate Change Feedback on the Future Oceanic CO_2 uptake, *Tellus*, *51B* (3), 722-733, 1999.

Matear, R.J., A.C. Hirst, and B.I. McNeil, Changes in dissolved oxygen in the Southern Ocean with climate change, *Geochemistry Geophysics Geosystems (http://gcubed.magnet.fsu.edu/main.html)*, *1* (November 21), 2000.

O'Farrell, S.P., Sensitivity study of a dynamical sea-ice model. The effect of the external stresses and land boundary conditions on ice thickenss distribution, *Journal of Geophysical Research, 103*, 15,751-15,782, 1998.

Orsi, A.H., G.C. Johnson, and J.L. Bullister, Circulation, mixing, and production of Antarctic Bottom Water, *Progress in Oceanography, 43* (1), 55-109, 1999.

Peng, T.-H., and W.S. Broecker, Seasonal variability of carbon dioxide, nutrients and oxygen in the northern North Atlantic surface water: observations and model, *Tellus, 39B*, 429-458, 1987.

Redfield, A., B. Ketchum, and F. Richards, The influence of organisms on the composition of sea water, in *The Sea*, edited by M. Hill, pp. 26-77, Interscience, New York, 1963.

Rintoul, S.R., and J.L. Bullister, A late winter hydrographic section from Tasmania to Antarctica, *Deep-Sea Research Part I-Oceanographic Research Papers, 46* (8), 1417-1454, 1999.

Rintoul, S.R., and M.H. England, Ekman transport dominates local air-sea fluxes in driving variability of subantarctic mode water, *Journal of Physical Oceanography, 32* (5), 1308-1321, 2002.

Roemmich, D., and C. Wunsch, Two transatlantic sections: Meridional circulation and the heat flux in the subtropical North Atlantic Ocean, *Deep Sea Research, 32A*, 619-664, 1985.

Sarmiento, J.L., T.M.C. Hughes, R.J. Stouffer, and S. Manabe, Simulated response of the ocean carbon cycle to anthropogenic climate warming, *Nature, 393*, 245-249, 1998.

Semtner, A.J., A model of thermodynamic growth of sea ice in numerical investigations of climate, *Journal of Physical Oceanography, 6*, 379-389, 1976.

Wang, W., and R.J. Matear, Modeling the upper ocean dynamics in the Subantarctic and Polar Frontal Zones in the Australian sector of the Southern Ocean, *Journal of Geophysical Research, 106*, 31,511-24, 2001.

Wong, A.P.S., N.L. Bindoff, and J.A. Church, Large-scale freshening of intermediate waters in the Pacific and Indian oceans, *Nature, 400* (6743), 440-443, 1999.

Yamanaka, Y., and E. Tajika, The role of the vertical fluxes of particulate organic matter and calcite in the oceanic carbon cycle: Studies using a ocean biogeochemical general circulation model, *Global Biogeochemical Cycles, 10*, 361-382, 1996.

GLACIAL-INTERGLACIAL CO_2 VARIATIONS

HEZI GILDOR
Department of Environmental Sciences,
The Weizmann Institute of Science, Rehovot, Israel
hezi.gildor@weizmann.ac.il

INTRODUCTION

Earth's climate undergoes variations on a wide range of time scales, from seasonal and interannual to glacial-interglacial and beyond. These variations are revealed in a variety of proxy records collected from ice-cores, ocean sediments and other sources (Bradley, 1999; Duplessy, 1999). The glacial-interglacial cycles are among the most striking phenomena in climate dynamics. While we are now at an interglacial, the last ice age was at its peak only about 20,000 years ago. Ice sheets covered much of North America and Europe, global average temperature was about 6° colder than today, and sea level was about 120 meters lower below today's. Various paleoclimate proxies (Petit et al., 1999; Imbrie et al., 1984) indicate that during the past 800,000 years or so (the late Pleistocene), the glacial-interglacial cycles were characterized by a pronounced 100,000 year (100 kyr) time scale, with additional weaker spectral peaks at 41 and 23 kyr (Figure 1) and an asymmetric saw-tooth structure (the slow build-up of the land glaciers and the relatively abrupt melting).

During at least the last four glacial-interglacial cycles, atmospheric CO_2 has undergone significant and fairly regular variations, with concentrations during glacial periods being about 80 ppm less than during warmer interglacial periods. During glaciation, atmospheric CO_2 gradually decreases, reaches a minimum of about 200 ppm during glacial maxima, then increases relatively quickly to a concentration of about 280 ppm at the end of glacial terminations (Figure 1).

1. WHY DO WE CARE?

The issue of glacial-interglacial variations in CO_2 revealed in polar ice cores has been an open question in glacial climate science for some 20 years now. Despite the effort invested to identify the mechanism responsible for these variations and the possible role it may play in global warming, a generally accepted explanation is still missing (Pedersen and Bertrand, 2000; Archer et al.,

M. Follows and T. Oguz (eds.), The Ocean Carbon Cycle and Climate, 317–352.

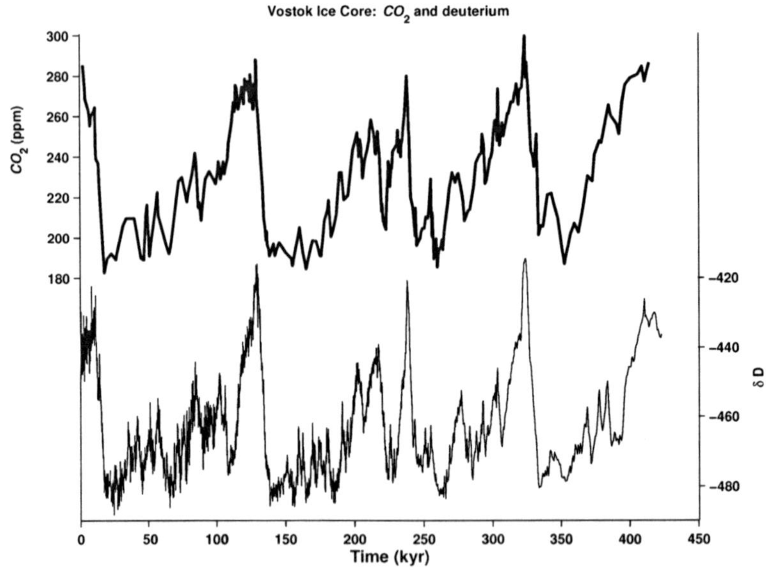

Figure 1. Time series (GT4 timescale for ice on the lower axis) of CO_2 and δD (Based on *Petit et al.*, 1999). Time is in kyr before present. δD (ratio of deuterium to hydrogen in ice) is a proxy for air temperature with colder temperature corresponding to a lower value. During peak glacials, temperature and CO_2 have a minimum.

2000b; Sigman and Boyle, 2000). Moreover, we still do not know what is the role of these glacial-interglacial CO_2 variations. Are they essential for the existence of the glacial-interglacial cycles via their greenhouse effect? Alternatively, are the CO_2 variations a feedback of the glacial cycles in the physical climate system that only serve to amplify the glacial variability?

The answers to the above questions are not yet clear (Archer et al., 2000b) although their importance is obvious. Some investigators used the relative phasing between CO_2 and other climate variable variations to address these questions; however, as shown below, this method is controversial (Alley et al., 2002). Numerous models suggest that even in the absence of CO_2 variations, glacial-interglacial cycles still exist (Gildor and Tziperman, 2001b; Berger et al., 1998) while in contrast, the CO_2 variations by themselves cannot drive these cycles (Loutre and Berger, 2000). Nevertheless, the variations in CO_2 contribute significantly to the amplitude of the glacial cycles (Figure 11) in both simple and hybrid models (Gildor and Tziperman, 2001a; Weaver et al., 1998). Further, the processes that contribute to the glacial-interglacial CO_2 variations are in operation today as well, therefore we need to understand them to better estimate the response of the climate system to anthropogenic increase of CO_2.

In this chapter I will review several climatic feedbacks that affect atmospheric CO_2 as well as few of the hypotheses to explain the glacial-interglacial CO_2 variations. I will not attempt to describe all the explanations that have been proposed or discuss their advantages and disadvantages. For that, the reader is referred to other reviews of the glacial-interglacial CO_2 cycles including those by Broecker and Peng (1998); Sigman and Boyle (2000); Pedersen and Bertrand (2000) and Archer et al. (2000b). Here I will concentrate mainly on the possibility that CO_2 variations are driven by the glacial cycles themselves rather than being an essential element for the existence of these cycles.

2. BRIEF REVIEW: THE CARBONATE SYSTEM IN SEAWATER

For completeness, I will outline a brief and simplified view of the carbonate system in sea water. More comprehensive discussions can be found in chemical oceanography texts and numerous review articles including Sarmiento (1993); Raven and Falkowski (1999); Pilson (1998); and Zeebe and Wolf-Gladrow (2001).

The direction of the local air-sea flux of CO_2 is determined by the difference in the partial pressure of CO_2 (pCO_2) in sea water and in the overlying atmosphere. pCO_2 in seawater depends on its solubility, α, and the concentration of dissolved CO_2, denoted $[CO_2]$, such that $pCO_2 = [CO_2]/\alpha$. The solubility is inversely related to temperature and salinity (Weiss, 1974).

When CO_2 is dissolved in water, some of the molecules hydrate to form carbonic acid, H_2CO_3, which further dissociate to bicarbonate and carbonate ions ($H_2CO_3^-$ and CO_3^{2-}) :

$$\begin{array}{ccc} [CO_2] + [H_2O] \Leftrightarrow [H_2CO_3] \Leftrightarrow & \left[HCO_3^-\right] & + [H^+] \\ & \Updownarrow & \\ & \left[CO_3^{2-}\right] & + [H^+] \end{array} \tag{1}$$

Here, $[CO_2]$ represents the concentration of unhydrated free CO_2 molecules in the solution. For practical reasons, it is common to treat $[CO_2]$ and $[H_2CO_3]$ together, represented by $[CO_2^\star]$ (Pilson, 1998).

These chemical reactions occur on timescales that are rapid relative to most relevant oceanographic processes. Hence, we often assume equilibrium between these species, and the partitioning of carbon between them is described by empirically determined dissociation constants, k_1 and k_2, which (like α) are functions of temperature and salinity.

$$k_1 = \frac{[HCO_3^-][H^+]}{[CO_2^\star]} \quad , \quad k_2 = \frac{[CO_3^{2-}][H^+]}{[HCO_3^-]} \tag{2}$$

The total CO_2 ($\sum CO_2$), or dissolved inorganic carbon, in seawater may be defined as

$$\Sigma CO_2 = [CO_2^\star] + \left[HCO_3^-\right] + \left[CO_3^{2-}\right] \tag{3}$$

Since $[CO_2^\star]$ is typically low in seawater ($\sim 0.5\%$), relative to $\left[HCO_3^-\right]$ and $\left[CO_3^{2-}\right]$ ($\sim 86.5\%$ and 13% respectively) (Zeebe and Wolf-Gladrow, 2001), we can make the following simplifying approximation (Holligan and Robertson, 1996; Broecker and Peng, 1982):

$$\Sigma CO_2 \approx \left[HCO_3^-\right] + \left[CO_3^{2-}\right] \tag{4}$$

Total alkalinity is determined by the concentration of strong ions (including Ca^{2+}) in sea water. It is a measure of the buffer capacity and represents the capacity of substances in sea water to react with hydrogen ions (during titration with strong acid) to the point when essentially all carbonate species are protonated. The global mean alkalinity is determined by the fluxes of salts into and out of the ocean. As a first approximation (Holligan and Robertson, 1996) we can write

$$A_T \approx A_C = 2\left[CO_3^{2-}\right] + \left[HCO_3^-\right] \tag{5}$$

A_C in (5) represents the carbonate alkalinity. The largest error in this approximation comes from neglecting the contribution of $[B(OH)_4^-]$ (boron ion) which contributes around 5% to the total alkalinity. The four equations (2), (4) and (5) have four unknowns ($[CO_2^\star]$, $[HCO_3^-]$, $[CO_3^{2-}]$, and $[H^+]$) which we may solve for in terms of k_1, k_2, A_T and $\sum CO_2$. Hence, the partial pressure of CO_2 in seawater may be approximated as a function of the local dissolved inorganic carbon and alkalinity:

$$pCO_2 \approx \frac{K_2}{\alpha K_1} \frac{(2\sum CO_2 - A_C)^2}{(A_C - \sum CO_2)} \tag{6}$$

We will use this relationship to estimate how variations in ocean temperature and salinity during glacial periods could have contributed to lower atmospheric CO_2. If a local air-sea gradient of pCO_2 exists, there is a net flux across the sea surface that equilibrates with the atmosphere on a timescale of about

9 months (depending on various physical factors, see section 4). Thus, on timescales of several years the partial pressure of CO_2 in the atmosphere and global surface oceans approaches an equilibrium in accord with (6) - i.e. the mean surface ocean and global atmospheric pCO_2 are approximately equal. However, surface ocean ΣCO_2 and A_T may deviate from equilibrium due to exchanges with the large, deep ocean reservoirs (and ultimately sedimentary and geologic interactions) and biological processes. Over some thousands of years, the deep ocean, surface ocean and atmospheric reservoirs of carbon are expected to adjust toward steady state.

There are three main ways in which the ocean's physical state and biochemical processes affect $\sum CO_2$, alkalinity, and therefore pCO_2 (Volk and Hoffert, 1985; Heinze et al., 1991; Sarmiento and Bender, 1994):

(i) The solubility pump (abiotic). The solubility of CO_2 in sea water is inversely related to temperature and salinity. Hence, at equilibrium with a given pCO_2, cooler waters are richer in ΣCO_2 than warmer waters. This in part explains why the cool, deep waters of the ocean are enriched in carbon relative to the warm surface. This effect has been termed the "solubility pump" of carbon (Volk and Hoffert, 1985). More on the solubility pump in section 5.

(ii) The organic (soft tissue) pump. Phytoplankton utilize carbon during photosynthesis to produce organic soft tissue material, which reduces $\sum CO_2$ in sea water but does not significantly affect alkalinity. A fraction of the organic material produced is ultimately exported to the deep ocean (after participating in the complex food web), where it is mostly remineralized by bacterial activity. In this process carbon is pumped from the surface to deep waters by the combined effect of photosynthesis, sinking and respiration. This increases the vertical gradient of $\sum CO_2$, reduces pCO_2 in the surface ocean and draws down CO_2 from the atmosphere.

(iii) The carbonate pump. Some planktonic species also produce calcium carbonate skeletal material (e.g. coccolithophorids) described schematically by the following reaction:

$$Ca^{2+} + 2HCO_3^- \Rightarrow CaCO_3 + CO_2 + H_2O$$

Although this process consumes additional $\sum CO_2$, it also affects the alkalinity (through the consumption of the Ca^{2+} ion) at a ratio of 1:2. Alkalinity is reduced and the capacity of seawater to hold carbon is decreased. Hence, the formation of this skeletal material acts to increase pCO_2 in surface water (by reducing the denominator in (6)) (Holligan and Robertson, 1996), which counteracts the effect of the soft tissue pump. It has been estimated that for every four carbon atoms removed from the surface water as soft organic tissue, one atom is removed as calcite (Broecker and Peng, 1982). This ratio of organic to inorganic carbon is known as the "Rain Ratio". Since much more soft tissue is produced than $CaCO_3$, the net effect of biological processes is to remove carbon from the surface, thus reducing pCO_2 in surface water and in the atmosphere.

3. GLACIATION AND TERRESTRIAL BIOSPHERE

Since about 98% of the combined atmosphere-ocean carbon inventory is in the ocean (Siegenthaler and Sarmiento, 1993; Sigman and Boyle, 2000), on centuries to millennial timescales, small changes in the large oceanic inventory of carbon can have a significant impact on atmospheric partial pressure. (On even longer timescales, interactions with very large geological reservoirs become significant.) But before searching the ocean for the cause of glacial-interglacial CO_2 variations, we should look at carbon storage on land.

The terrestrial biosphere (including soil) reservoir of carbon is about 2100 Gt C (Siegenthaler and Sarmiento, 1993; Sigman and Boyle, 2000), which is a few times more than the atmosphere reservoir which is approximately 600 Gt C (Sigman and Boyle, 2000). Production of organic matter on land will cause a reduction in the atmosphere-ocean content of CO_2, while oxidation of organic matter return CO_2 to the atmosphere-ocean reservoir. If anything, we expect the carbon reservoir on land to shrink during glacial times, as a result of ice sheets covering large areas and of dryness of the atmosphere. This can be verified by the carbon isotopic signature ($\delta^{13}C$) of the atmosphere-ocean reservoir. $\delta^{13}C$ of the terrestrial biosphere is about $-25‰$ while that of the atmosphere-ocean reservoir is about $0.5‰$. Based on $\delta^{13}C$ from calcite shells extracted from ocean sediments, $\delta^{13}C$ of the atmosphere-ocean reservoir was $0.35 - 0.40‰$ lower during glacial times (Broecker and Peng, 1998; Crowley, 1995). This indicates a transfer of about 500-650 Gt from the terrestrial to the atmosphere-ocean reservoir, so actually cause a change in atmospheric CO_2 of the opposite sign than required! After some equilibration with the ocean and interaction with $CaCO_2$ in the sediment (see section 8.2), we expect this transfer to increase atmospheric CO_2 by about 15 ppm. It must be noted that this is a conservative estimate. Other estimates based on pollen data suggest a transfer of 750-1050 Gt (Crowley, 1995).

4. AIR-SEA CO_2 FLUX

The ocean surface waters need time to equilibrate with atmospheric CO_2 (about 10 months for 50 m depth layer), and the efficiency of the carbon pumps depends on the degree of equilibration (Toggweiler et al., 2003a; Toggweiler et al., 2003b). It should be noted that the equilibration time for CO_2 is much longer compared with, for example, O_2 because CO_2 is only a small fraction of the $\sum CO_2$ in seawater (Zeebe and Wolf-Gladrow, 2001). If the air-sea gas exchange rate was infinite, surface and atmospheric pCO_2 would have been in equilibrium. Finite air-sea gas exchange prevents the ocean from reaching equilibrium with the atmosphere, and as we will see later, this has opposite effects on the solubility and biological pumps. At present, although pCO_2 of the ocean and the atmosphere are similar over a large fraction of the ocean surface, there are regions with significant deviations. There is a tropical band with positive $\triangle pCO_2$ (i.e., pCO_2 of the ocean is larger than pCO_2 of the atmosphere), another positive band in the Southern Ocean, and negative $\triangle pCO_2$ in the North Atlantic (e.g., Figure 4 of Takahashi et al. (1997)).

Some processes in the ocean displace the pCO_2 of the ocean away from equilibrium while others try to bring it close to equilibrium. Amongst the latter, we find the air-sea CO_2 flux and the degree of equilibration is in part determined by a combination of the air-sea gas exchange rate and the time that water spends in contact with the atmosphere. The former is affected by numerous factors including wind speed and the difference in pCO_2 of the atmosphere and the ocean, while the latter is affected by sea ice cover (which partially insulates the ocean from the atmosphere) and by ocean circulation.

4.1 WIND SPEED

The air-sea CO_2 flux (F_{CO_2}) is generally parameterized as

$$F_{CO_2} = K_g \triangle pCO_2$$

where K_g is the gas exchange coefficient (sometimes called the piston velocity or transfer velocity) and $\triangle pCO_2$ is the pCO_2 of seawater minus pCO_2 of air (air-sea difference in pCO_2) (Siegenthaler and Sarmiento, 1993; Najjar, 1991). K_g is mainly a function of wind speed but also depends on other factors such as ocean waves and temperature (see review by Asher and Wanninkhof, 1998). Large uncertainties of the exact dependency on wind speed prevent us to accurately estimate the air-sea flux (Siegenthaler and Sarmiento, 1993; Takahashi et al., 1997). Bates and Merlivat (2001) calculated an air-sea CO_2 flux using hourly wind data that was three times higher than a flux calculated using daily mean wind data. Further complicating the estimation of the air-sea flux are large local and temporal variation of $\triangle pCO_2$ (Siegenthaler and Sarmiento, 1993; Takahashi et al., 1997).

Finite gas exchange prevents the ocean to hold as much CO_2 as thermodynamical equilibrium allows (Toggweiler et al., 2003a). Considering climate change implications, there are numerous indications, like dust records extracted from ice cores (Keir, 1993a), that climate during glacial periods was windier. This might contribute to the draw down of atmospheric CO_2 by enhancing the solubility pump (Keir, 1993a).

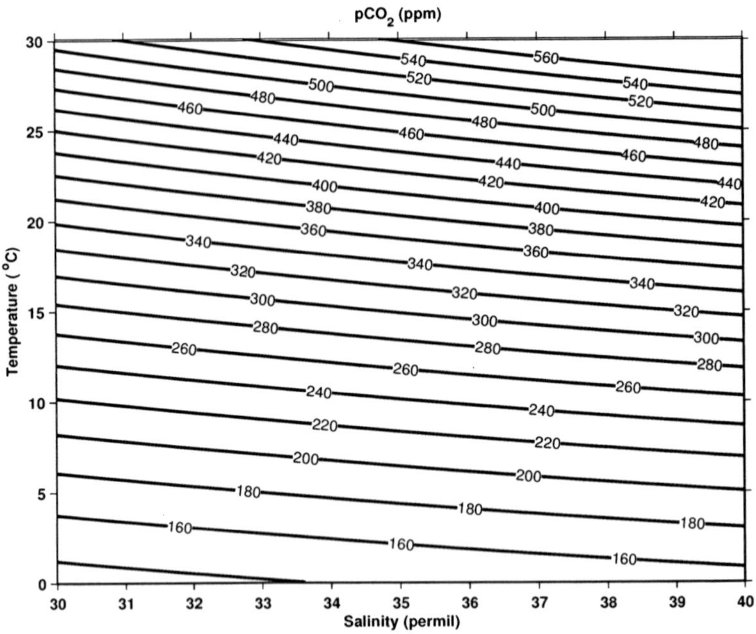

Figure 2. pCO_2 as a function of temperature and salinity for $\sum CO_2 = 2000\mu mol\, kg^{-1}$ and $A_C = 2200\mu eq\, kg^{-1}$.

4.2 SEA-ICE COVER

Sea-ice cover significantly reduces air-sea fluxes by isolating the ocean from the atmosphere. We will see in section 6.4 a mechanism which attributes the glacial-interglacial CO_2 variations to a larger sea ice cover in the Southern Ocean during glacials.

5. SOLUBILITY OF CO_2 IN THE OCEAN

The potential effect of solubility of CO_2 as function of temperature and salinity is seen from equation (6). Figure 2 shows pCO_2 as a function of temperature and salinity for $\sum CO_2 = 2000\mu mol\, kg^{-1}$ and $A_C = 2200\mu eq\, kg^{-1}$. Similarly, if we take two parcels of seawater in equilibrium with the same atmospheric pCO_2, a parcel at 2^oC should contain about 170 $\mu mole\, kg^{-1}$ more carbon than a parcel at 22^oC. If sinking water in the modern ocean were in full equilibrium with the atmosphere, it would account for almost half of the observed surface to deep $\sum CO_2$ difference (Toggweiler et al., 2003a). About 75% of the sensitivity of pCO_2 to temperature (Figure 2) is due to α while the rest is due to $K2/K1$. About 50% of the sensitivity of pCO_2 to salinity is due to $K2/K1$, 30% due to $\frac{[HCO_3^-]^2}{[CO_3^{2-}]}$ $\left(= \left[\frac{\left(2\sum CO_2 - A_T\right)^2}{\left(A_T - \sum CO_2\right)}\right]\right)$, and 20% due to α (Najjar, 1991). During glacials, the temperature of the ocean was lower (solubility of CO_2 increases) while on the other hand salinity was higher (solubility of CO_2 decreases).

Three simple box models used by groups from Harvard, Princeton, and Bern (know as the Harvardton Bears) in 1984 demonstrated that the high latitude oceans, despite their relative small area ($\sim$ 20% of the total), largely control atmospheric CO_2. The reason is that the deep ocean, where most of the ocean volume and carbon reside, is communicated with the surface ocean mainly in polar regions (specifically in regions of deep water formation, nowadays in the Southern Ocean and in the North Atlantic). In the low latitude oceans, stable stratification prevents efficient vertical mixing between surface and deep waters. In high latitudes, where isopycnal outcrop to the surface, water can move along these isopycnals without the need to overcome gravity.

The warm surface water, with large surface area but small volume compared to the deep ocean, can exchange carbon with the large, deep carbon reservoir either through ocean circulation (advection and vertical mixing of water) or through the atmosphere. The extent that high latitude ocean controls atmospheric CO_2 depends on the relative importance of these two carbon exchanges. Broecker and Peng (1998) illustrate this point in a simple way (see also Bacastow 1996; Broecker et al., 1999; Keir 1993b). Consider for the moment an abiotic ocean with a small volume, warm reservoir that occupies 80% of the ocean surface area, and a large volume, cold reservoir that occupies only 20% of the surface (Figure 3).

Lets assume that the deep, cold reservoir is so large that its $\sum CO_2$ does not change as a result of exchanges with the relatively small warm surface ocean and/or with the atmosphere. Therefore, its pCO_2 is kept constant at 280 ppm. When deep water upwells to the surface, the pCO_2 rises to 790 ppm as a results of the dependency of the solubility on temperature. At each moment, atmospheric pCO_2 is approximately an area-weighted average of the surface ocean pCO_2. If the rate of gas exchange with the atmosphere is low compared to the circulation rate, atmospheric CO_2 would be:

$$pCO_2^{atm} = 0.8 \times 790 + 0.2 \times 280 = 688ppm$$

In contrast, if gas exchange is very fast, the excess carbon will move through the atmosphere to the cold reservoir, and pCO_2 in the warm reservoir would be 280 ppm. If the rate of thermohaline circulation (THC) is 30 Sv and the rate of gas exchange is $6 \times 10^{-2} mol\, ppm^{-1} m^{-2} yr^{-1}$, we will get $pCO_2^{atm} = 324$ ppm while pCO_2 of the warm surface will be 335 ppm. For the circulation dominated case, the small reservoir of warm surface water (but with large area)

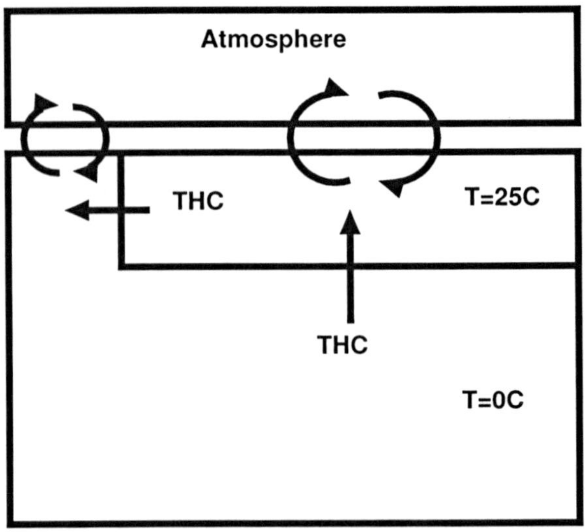

Figure 3. Simple abiotic model after Broecker and Peng, 1998. Warm reservoir at 25^oC occupying 80% of the surface area and cold reservoir (0^oC) occupying the rest 20%. The two reservoir communicate via thermohaline circulation (THC) and exchange CO_2 with the atmosphere.

significantly affect atmospheric CO_2. For the infinitely fast gas exchange case atmospheric CO_2 stabilizes at the pCO_2 of cold water.

Simple box models and general circulation models (GCMs) do not agree on the relative importance of the cold high latitudes ocean and the warm low and mid latitudes. We really do not know if the real ocean behaves more similar to box models or to GCMs, which is of course a crucial point which has been intensively debated in recent years (Broecker et al., 1999; Archer et al., 2000a; Bacastow, 1996; Toggweiler et al., 2003a; Toggweiler et al., 2003b). Toggweiler et al. (2003a) suggest that the ocean lies somewhere between the simple three-box model (discussed further below) and GCMs. The difference in the two types of models results from the extent of equilibration of the surface waters before sinking to the deep ocean. For example, the solubility pump in the three box model is more efficient than in the real ocean, because the area of the polar box is relatively large. On the other hand, The polar outcrops in the GCM is too small compared to the real ocean so the solubility pump is less efficient.

We can get a sense of the upper bound of the solubility effect. The temperature of the water around Antarctica is near the freezing point, so it cannot get much colder even during glacial periods. In the North Atlantic the water forming today is relatively warm (and saline), so it could have been few degrees colder than today. Assuming that half of the deep water originated from the North Atlantic, the deep ocean could have been about 2 degrees colder during the glacial maxima (at certain sites estimates based on temperatures of pore water suggest cooling of up to 4 degrees (Schrag et al., 1996)). A 2 degrees cooling will lower atmospheric CO_2 by about 20 ppm, and taking into account cooling of lower latitudes, an additional 10 ppm decrease could occur (Broecker and Peng, 1998).

Salinity variations have the opposite effect. During glacials, as more ice is stored on land, the ocean becomes more saline. For the last glacial maximum, with sea level about 120 m lower than present, salinity was 3% higher (Broecker and Peng, 1998). By itself, such an increase in salinity reduces the solubility and causes an increase in atmospheric pCO_2 by nearly 10 ppm (Broecker and Peng, 1998; Sigman and Boyle, 2000). Including the temperature effect we can explain a drawdown of about 20 ppm, comparable to the increase expected from the reduced storage of carbon on land. We still need to explain a drawdown of about 80 ppm by other mechanisms.

6. STRONGER ORGANIC PUMP

Biological production affects surface pCO_2 by changing $\sum CO_2$ and alkalinity. As mentioned earlier, since the organic pump is stronger than the carbonate pump, the net effect of biological production is to reduce surface pCO_2 and eventually atmospheric CO_2. As long as the rain ratio is constant, increased production results in reduced surface pCO_2. For simplicity, in the following we will assume that the rain ratio stays constant and will neglect the carbonate pump. We will return back to the carbonate pump in subsection 8.2.

The strength of the organic pump is modified by biological and physical processes. We will demonstrate it first using the classical three-box model of Sarmiento and Toggweiler (1984) and Siegenthaler and Wenk (1984) shown in Figure 4. We then will look at a more realistic four-box model (Toggweiler, 1999; Gildor and Tziperman, 2001a). First, it should be mentioned that an additional drawdown of about 20-30 ppm (Toggweiler, 1999) (to what can be achieved by a stronger organic pump) is generated by a processes called "$CaCO_2$ compensation" (Broecker and Peng, 1987), and this will be discussed in subsection 8.2.

6.1 NUTRIENT UTILIZATION AND INVENTORY

In addition to light and carbon, organisms need macro-nutrients (such as phosphate and nitrate) and micro-nutrients (such as iron) to produce organic matter. In large areas of the ocean, such as the subtropical gyres, biological productivity appears to be limited by the major nutrients, nitrate and phosphate (Broecker and Peng, 1982). In other regions (termed HNLC, High Nutrients Low Chlorophyll), notably the Southern Ocean, nitrate and phosphate

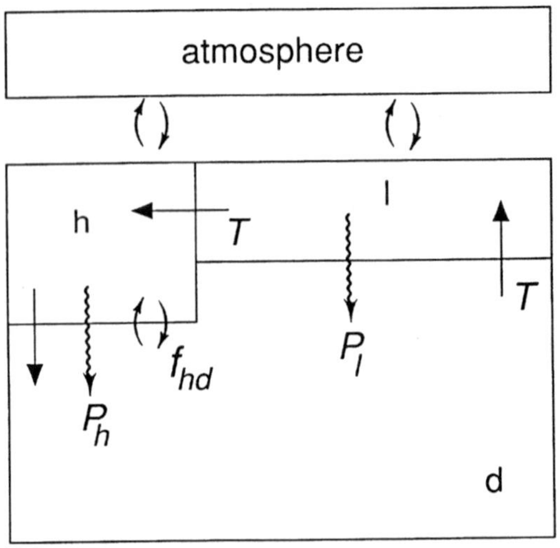

Figure 4. A box model for glacial-interglacial CO_2 variations. The boxes represent the upper mid-latitude ocean (l), the surface polar ocean (h) and the deep ocean (d), as well as the atmosphere. Taken from Toggweiler (1999).

are plentiful (Figure 5) and other factors, such as light or iron availability (Martin, 1990) limit phytoplankton growth (although other hypotheses exist (Evans and Parslow, 1985; Whitfield, 1993)). The degree of nutrient utilization in the ocean is a measure of the strength of the organic pump.

In the three-box model the ocean consists of two surface boxes and one deep box. Box l represents the low and mid latitude surface ocean, box h the high latitudes of both hemispheres, and box d the deep ocean. T represents the unidirectional conveyor circulation which starts by deep water formation in the North Atlantic while f_{hd} is a bidirectional mixing term representing the combined effect of all processes contributing to exchange of water between the surface and deep ocean like deep convection, diapycnal mixing, or isopycnal eddy stirring. Phosphate is the limiting nutrient in this model. Biota in the l box converts all available nutrients into sinking particles (P_l) so phosphate concentration in box l is zero. As biological production in the h box is limited by factors other than nutrients, the sinking flux from this box (P_h) is a free parameter and phosphate concentration in the high latitude box, PO_{4h} is a diagnostic. For a full description of the three-box model including its parameters the reader is referred to Toggweiler (1999).

One of the earliest theories for the glacial-interglacial CO_2 variations was increased utilization of the available nutrients in the Southern Ocean (Sarmiento and Toggweiler, 1984; Siegenthaler and Wenk, 1984). In the three-box model

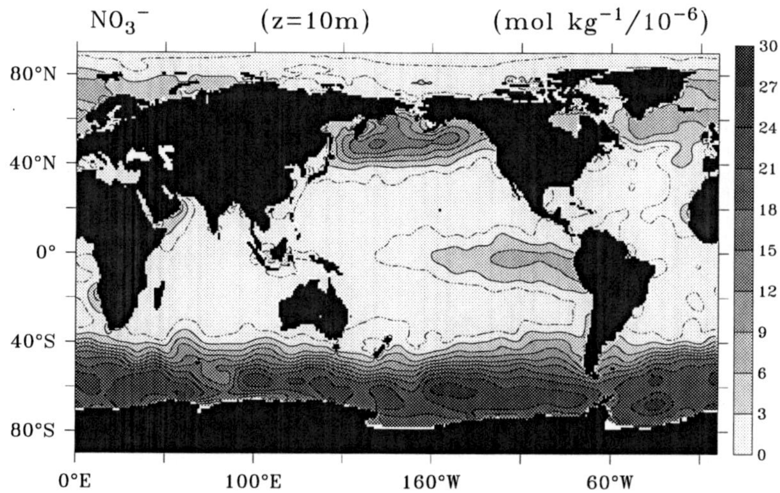

Figure 5. Annual climatology of the ocean distribution of nitrate at 10m depth. Contours are at three micromolar intervals from zero, with the addition of the one micromolar contour (climatology of Conkright et al. (1994)).

(and similar ones) of Figure 4, the biological pumps utilize all the nutrients in the low and mid latitudes box so strengthening the biological pumps must be done through processes at polar regions leading to nutrient depletions (Toggweiler et al., 2003b).

For the box model of Figure 4, one may write the total carbon ($\sum CO_2$) and phosphate (PO_4) balance for the deep box under steady-state conditions as a balance of advection (by the thermocline circulation term T) and mixing (by f_{hd}), plus terms that represent the sinking of particulate organic matter from the surface boxes to the deep box as part of the export production from the low and high latitude surface boxes, P_l and P_h:

$$0 = (f_{hd} + T) * (\sum CO_{2h} - \sum CO_{2d}) + Red_{c:p}(P_l + P_h) \quad (7)$$

$$0 = (f_{hd} + T) * (PO_{4h} - PO_{4d}) + (P_l + P_h) \quad (8)$$

in which $Red_{c:p}$ is the "Redfield Ratio" (see section 8.1) and P_l and P_h are in phosphate units. Extracting $P_l + P_h$ from (8) and substitute in (7), yields

$$\sum CO_{2d} - \sum CO_{2h} = Red_{c:p}(PO_{4d} - PO_{4h}) \quad (9)$$

Neglecting temperature and salinity differences (we are interested here in the role of the organic pump so we leave out solubility effects, (Toggweiler et al., 2003a)), for sufficiently fast gas exchange with the atmosphere, $\sum CO_{2h}$ and $\sum CO_{2l}$ are close to each other, so also

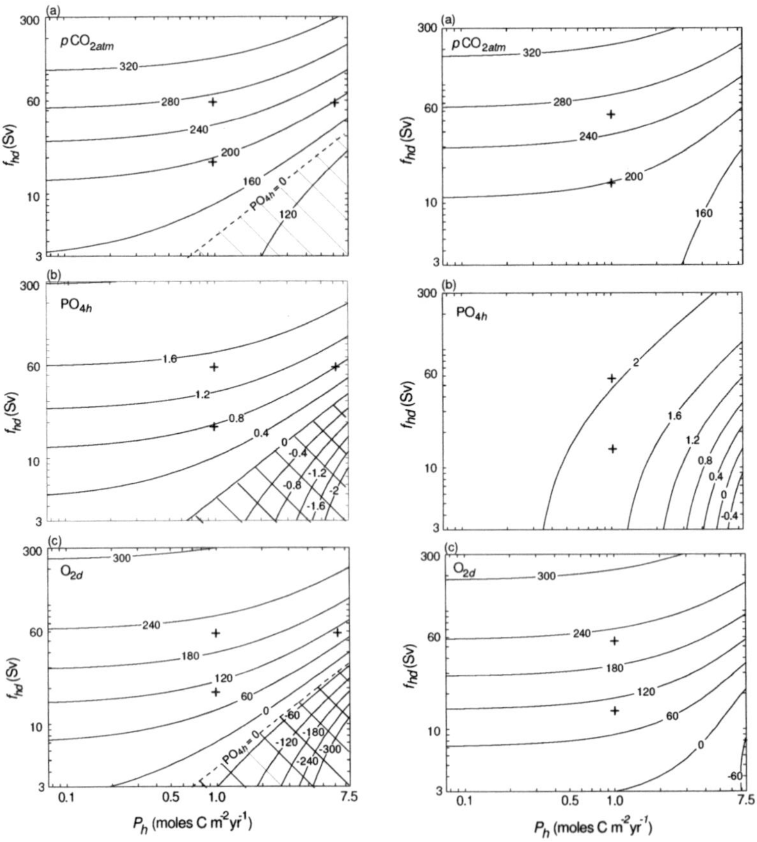

Figure 6. Results from the three (left) and four (right) box models. Taken from Toggweiler (1999).

$$\sum CO_{2d} - \sum CO_{2l} \sim Red_{c:p}(PO_{4d} - PO_{4h}) \qquad (10)$$

The left hand side of (10) is a measure for the strength of the organic pump. PO_{4d} is essentially constant which means that the strength of the organic pump under our assumptions is determined by PO_{4h}. f_{hd} brings to the surface water rich in both CO_2 and PO_4. Because biological production in the h box is not efficient, CO_2 can leak from box h to the atmosphere. Sarmiento and Toggweiler (1984) showed that by increasing P_l (reducing PO_{4h} and accordingly $\sum CO_2$) it is possible to reduce atmospheric CO_2 by reducing $\sum CO_2$ in box h thereby reducing CO_2 outgassing (Figure 6).

It was later suggested that increased iron supply in glacial periods, resulting from a windier and drier climate, may have enhanced the efficiency of the

Southern Ocean biological carbon pump where iron appears to limit nutrient utilizations (Martin, 1990).

Before continuing to explore another way to strengthen the organic pump in the three-box model, two additional comments must be made. First, changes in atmospheric iron transport may modulate the biological carbon pumps in another way. Iron supply to the ocean stimulates nitrogen fixation by cyanobacteria (Falkowski, 1997) leading to an increase in total inventory of nitrogen in the ocean, in particular in regions where nitrogen seems to be the limiting nutrient. Utilization of the added nitrogen, i.e. a strengthening of the biological pump, will cause drawdown of atmospheric CO_2. There are numerous problems with this hypothesis which involves a large shift in the global ocean nitrogen inventory. If the export production becomes very large, it should be accompanied by both significant oxygen depletion in the deep ocean, where remineralization occurs (Pedersen and Bertrand, 2000) and by more calcite preservation in the sediments (Archer et al., 2000b). There is no evidence for such changes. In addition, unless the N to P ratio changes (see subsection 8.1), an increase in nitrogen inventory without a parallel increase in phosphate will not enable enhanced production (Sigman and Boyle, 2000).

Second, there are other effects of climate change that can affect the oceanic inventory or utilization of nutrients, some of them might act to reduce biological production during glacial periods. For example, a decrease in sea-level will reduce the area of continental shelf which is generally more productive than the open ocean (Taylor, 1992) and is a region of denitrification. Increased sea-ice cover can reduce biological production by decreasing the available photosynthetically active radiation in polar regions where light, rather than nutrients, is likely to be the limiting factor on primary production (Taylor, 1992). (The last effect is taken into account in the model of section 7.)

6.2 OCEAN CIRCULATION AND VERTICAL MIXING IN THE SOUTHERN OCEAN

Reduced vertical mixing in the Southern Ocean during glacials is another one of the earliest mechanisms suggested to explain the low CO_2 levels during glacial periods. As also demonstrated by the three-box model, this is but another way to affect the strength of the biological organic pump, this time by physical change.

In equation (7) we assume $P_l = T * PO_{4d}$, which means that we assume all the surface nutrients in the low-latitude box are utilized by biological activity and converted to export production. Further we assume $P_l \gg P_h$ meaning again that the low-latitude biological activity is efficient, acts over a larger area relative to that of the high latitudes, and therefore utilizes all the available nutrients. Using these assumptions we have

$$\sum CO_{2d} - \sum CO_{2h} = Red_{c:p} \frac{T * PO_{4d}}{f_{hd} + T}. \tag{11}$$

Again neglecting temperature and salinity differences, for sufficiently fast gas exchange with the atmosphere, upper ocean $\sum CO_2$ in lower and higher latitudes are approximately equal, so that

$$\sum CO_{2d} - \sum CO_{2l} \sim Red_{c:p} \frac{T * PO_{4d}}{f_{hd} + T}, \tag{12}$$

which is the result we were after: the $\sum CO_2$ difference between the upper and deep ocean is controlled by vertical mixing. A reduction in vertical mixing in the high latitude Southern Ocean should result according to (12), in an increase in the concentration of the $\sum CO_2$ difference between the upper and the deep ocean. Taking the deep ocean concentration to be constant due to its large reservoir, this implies a reduction of the surface $\sum CO_2$ (and CO_2) concentration. Again, since f_{hd} brings to the surface CO_2-rich water, reduced vertical mixing will reduce leakage and lower atmospheric CO_2.

6.3 FOUR-BOX MODEL

The three-box model is not sufficiently realistic as it includes the high latitudes of both hemispheres in one "generic" polar box. This one box is the only connection between the deep ocean and the atmosphere. The connection is mediated through the concentration of nutrients in this single box. Therefore both ways to lower atmospheric CO_2 in the three box model, reduced vertical mixing or enhanced nutrients utilization, require significant reduction in polar nutrients as seen in the left panel of Figure 6. We note that different proxies ($\frac{^{13}C}{^{12}C}$, $\delta^{15}N$ and $\frac{^{231}Pa}{^{230}Th}$) tell different stories on the degree of nutrient utilization in the Southern Ocean (see (Broecker and Peng, 1998)). In contrast, lowering atmospheric CO_2 by reduced ventilation in the four box model requires very little reduction in Southern Ocean nutrients, as can be seen in the right panel of Figure 6.

The functioning of the two polar regions in the carbon cycle is significantly different. The added "northern" polar box in the four-box model of Toggweiler (1999) (Figure 7), representing the sinking region of the North Atlantic, makes this box model much more realistic. The water formed in the North Atlantic gets there from the low latitudes by the conveyor circulation. It therefore has a low PO_4 concentration as a result of depletion during its flow northward. During the time spent in low latitudes, biological production extracts significant amount of $\sum CO_2$ from the water. As they cool on their way northward, their

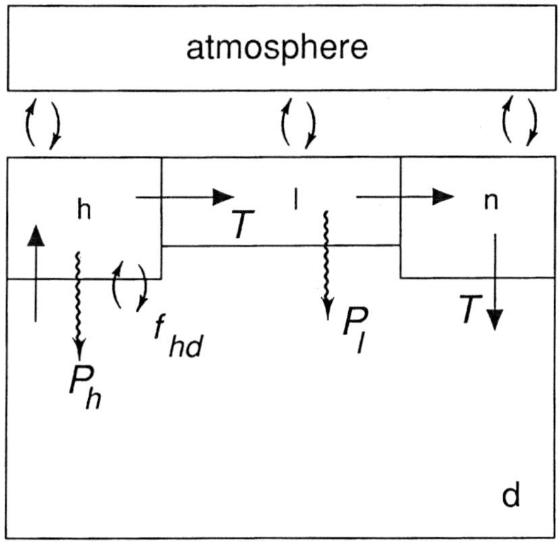

Figure 7. The four-box model of Toggweiler (1999). Box n represents the North Atlantic where deep water formed as part of the conveyor belt. These water upwelled in the Southern Ocean, represented by box h. See Toggweiler (1999) for this model parameters.

capacity to hold CO_2 increases and they absorb CO_2 from the atmosphere. In the four-box model, as well as in the real ocean, the North Atlantic is the region were the solubility effect causes the ocean to absorb CO_2 from the atmosphere. In this region, limited gas exchange prevents the solubility pump from being fully efficient (Keir, 1993a).

The Southern Ocean is where most of the water formed in the North Atlantic upwells to the surface. Part of the water upwelled continues as surface water flowing northward closing the conveyor belt, while part sinks back as new deep water. On the way southward, the water which originated in the North Atlantic accumulates remineralized CO_2, and has more CO_2 than its solubility enables. As it reaches the surface, CO_2 outgases to the atmosphere, increasing atmospheric CO_2. Reduced vertical mixing brings less CO_2-rich water to the surface, so less leaks out to the atmosphere. The rate of air-sea CO_2 exchange plays an important role here as well. By being finite, it reduces the CO_2 leakage thereby strengthening the organic pump. The reduced vertical mixing in the four box model has only a minor effect on the nutrient content in the h box which represents the Southern Ocean (upper right panel of Figure 6). Extended sea ice cover would have a similar effect by inhibiting the air-sea flux (see next section) just like biological production did (see previous section).

6.4 SEA-ICE COVER

As mentioned above, sea-ice cover reduces significantly air-sea fluxes by isolating the ocean from the atmosphere. Stephens and Keeling (2000) suggested that larger sea ice cover in the Southern Ocean during glacial times caused a reduction in air-sea CO_2 flux from the ocean to the atmosphere and contributed to the lower glacial CO_2 concentration. If during cold periods the CO_2-rich water spends less time in contact with the atmosphere, sinks and loses contact with the atmosphere before all the excess CO_2 can escape, we would expect a reduction in atmospheric CO_2. This mechanism does not requires any change in biological production but more efficiency in trapping the remineralized CO_2 in the deep ocean. Since O_2 equilibrates much faster than CO_2, it will be less affected, explaining why we do not find evidence of a lack of oxygen in the deep ocean.

In the model of Stephens and Keeling (2000) the ocean is divided into six boxes, and temperature, salinity, surface nutrients, sea ice cover and circulation patterns are prescribed. Note that all water is upwelled south of the polar front. Air-sea flux, biological production and carbonate sediment interaction are parameterized. Figure 8 demonstrates the sensitivity of the model to sea-ice cover around Antarctica and to the amount of water which flows northward from the Antarctic to the sub-Antarctic box. Using best-guess parameters, the model produces about 70 ppm drawdown of atmospheric CO_2. This mechanism, though, requires a very large sea cover, 99-100% of the area south of the polar front. Simulations using ocean GCM with active thermodynamic sea ice (in contrast to specifying the sea cover as in the Stephens and Keeling box model) indicates that even during the last glacial maximum (LGM), it is difficult to have such a large sea-ice cover. In the Southern Ocean, the wind pulls the ice apart, creating leads and cracks in the ice. The flux through these openings is very strong and partially compensates for the insulating effect of sea ice in the covered region (Maqueda and Rahmstorf, 2002).

7. SIMPLE BOX MODEL OF THE GLACIAL CYCLES INCLUDING GLACIAL- INTERGLACIAL CO_2 VARIATIONS

The last two mechanisms, like other models suggested in the past for the glacial-interglacial CO_2 variations, presented restricted glacial and interglacial steady state solutions (Stephens and Keeling, 2000; Toggweiler, 1999; Siegenthaler and Wenk, 1984) but could not explain the transition between these states. For example, the model of Toggweiler (1999) is not capable of explaining why the vertical mixing in the ocean should change. Similarly, Stephens and Keeling (2000) specified sea ice cover changes. The fitness of the mechanism for CO_2 variations in the global picture of glacial cycles is important.

Dealing with the transition requires a physical mechanism and model of the glacial-interglacial cycle that is sufficiently detailed. Such a physical mechanism and model for the 100 kyr glacial oscillations were presented in Gildor and Tziperman (2000; 2001b), which included simplified yet explicit model components for the oceanic meridional circulation, sea ice, land glaciers, and atmosphere. Using this model, Gildor and Tziperman (2001a) proposed a physical mechanism for both reduced vertical mixing and increased sea-ice cover. In this mechanism, the North Atlantic Deep Water cooled due to northern hemisphere glaciation, is transported southward by the thermohaline circulation, and subsequently cooling the deep water upwelling in the Southern Ocean. This, in turn, affects Southern Ocean stratification, reduces the rate of vertical mixing of the surface water with the deep water, and increases the sea ice cover.

7.1 MODEL DESCRIPTION

The global meridional box model, shown in Figure 9 and described in Gildor et al. (2002), is composed of ocean, atmosphere, sea-ice and land-ice submodels, including active ocean biogeochemistry and a variable atmospheric

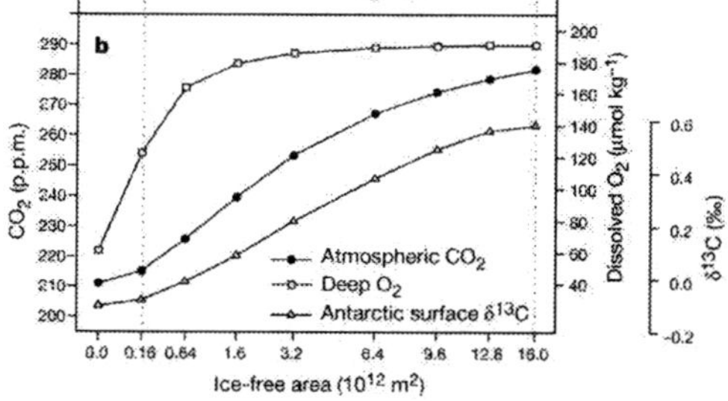

Figure 8. Steady-state model solutions for different ice coverages south of the Antarctic Polar Front (APF). We vary the exposed sea surface in the B and A boxes such that the same fraction of total surface is ice-free in each. The x-axis values represent the sum of exposed area in these two boxes, and are scaled by their square root to expand the left side of the plot. We use $1.6 \times 10^{13}\, m^{-2}$ as a modern estimate of the ice-free area south of the APF, and $1.6 \times 10^{11}\, m^2$ to illustrate a possible glacial value. **a**, Atmospheric CO_2 using modern-preindustrial parameters and different fractions (Fa) of Antarctic surface waters entering the subantarctic box. **b**, Atmospheric CO_2, deep O_2 , and Antarctic surface $\delta^{13}C$ using Fa = 0.3. (Figure 2 of Stephens and Keeling (2000), legend modified)

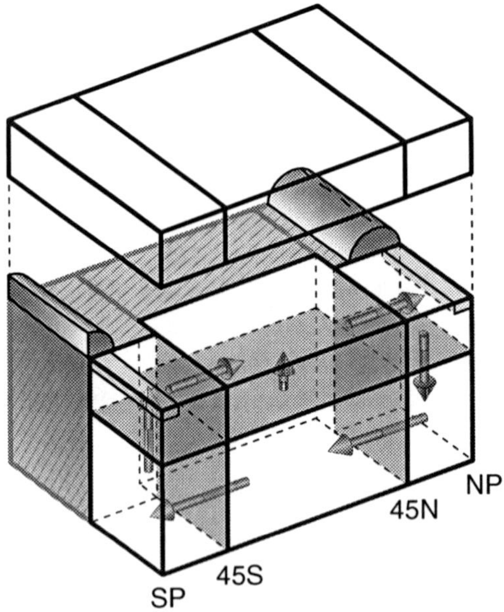

Figure 9. A schematic representation of the coupled box model used in this study, showing the atmospheric boxes (upper slab), the thermohaline circulation (arrows), the parabolic land ice sheets over land and a partial sea ice cover in the polar ocean boxes. All of these physical climate system components are allowed to vary based on the model equations. The ocean biogeochemistry and the atmospheric CO_2 fully interact with the physical climate components

CO_2 which may change due to exchanges with the ocean. The polar boxes represent the regions between 45° and the poles, while the mid-latitude box includes the regions between 45^oS and 45^oN. The meridional THC through the northern box is governed by the standard balance between horizontal friction and the meridional pressure gradient (Stommel, 1961). It is known that the THC dynamics in the Southern Ocean are different from those of the Northern Hemisphere. This upwelling is the result of the westerly winds, driving a northward Ekman drift and thus an upwelling in the Southern Ocean (Gnanadesikan, 1999; Toggweiler and Samuels, 1993). Because the exact dependence of upwelling on the intensity of the wind and on sea-ice cover is still unclear, the intensity of the upwelling (and thus of the THC) through the southern polar box is fixed to a specified value (of 16 Sv). The vertical mixing parameterization between the southern surface and deep boxes follows (Gargett, 1984), who proposed that the vertical mixing coefficient should be of the form $K_v = K_0(\frac{\partial \rho}{\partial z})^{-1}$.

Sea-ice forms in this model when the ocean water temperature decreases below a critical freezing temperature, and melts above that temperature. Sea ice presence and evolution affect the surface albedo, the salinity budget in the ocean, the air-sea heat flux, CO_2 flux and evaporation by partially insulating the ocean from the atmosphere.

The lower surface of each atmospheric box is a combination of ocean, land, land ice, and sea ice, each with its specified albedo. The average potential temperature of each atmospheric box is calculated based on the energy balance of the box. Atmospheric CO_2 is a prognostic variable which directly affects the long-wave emissivity.

The model for the northern land ice-sheets, whose slow evolution provides the 100 kyr time scale in this model, is zonally symmetric, assumes perfect plasticity (Weertman, 1976; Ghil, 1994), grows due to the precipitation in the polar boxes, and decreases due to ablation, ice-runoff, and calving processes.

The biochemistry model is similar to those used in 3D biochemical general circulation models (Maier-Reimer, 1993). The model includes total CO_2 ($\sum CO_2$), alkalinity (A_T), and PO_4 (taken to be the limiting nutrient) as prognostic variables that are used to calculate pCO_2 and therefore also the exchange of CO_2 with the atmosphere. The model is of the 'closed-system' type (Sigman and Boyle, 2000), so total alkalinity and PO_4 are constant during the integration although their distribution in the ocean can change. The biochemistry prognostic equations are similar to the advection-diffusion equations used for the temperature and salinity, except for an added source/sink term due to export production and remineralization. The Redfield ratio is assumed to remain constant and the rain ratio is treated as in Maier-Reimer (1993). The rate of export production depends on the latitude via the light factor, the amount of nutrients, and the ocean area not covered by sea ice. The sinking particles from each surface box are assumed to be completely remineralized in the deep box below. See Gildor et al. (2002) for a full description of this model.

7.2 A PHYSICAL MECHANISM FOR GLACIAL-INTERGLACIAL CHANGES IN SOUTHERN OCEAN VERTICAL MIXING AND SEA ICE COVER

Consider the box model results for one full glacial cycle (Figure 10a-f). The stratification in the Southern Ocean is composed of cold fresh upper water above salty, warmer water whose source is the North Atlantic Deep Water (NADW). During the stage of Northern Hemisphere ice sheets build-up (Figure 10a, at time 100 kyr to 30 kyr), the deep water formed in the North Atlantic becomes gradually colder (Figure 10b) as a result of general Northern Hemisphere cooling due to the increased albedo of the growing land glaciers. Similarly, the mixing of the NADW on its way south with cooler glacial-period surface water in the mid-latitudes results in the NADW arriving to the Southern Ocean colder and denser. This is in agreement with both proxy records (Schrag

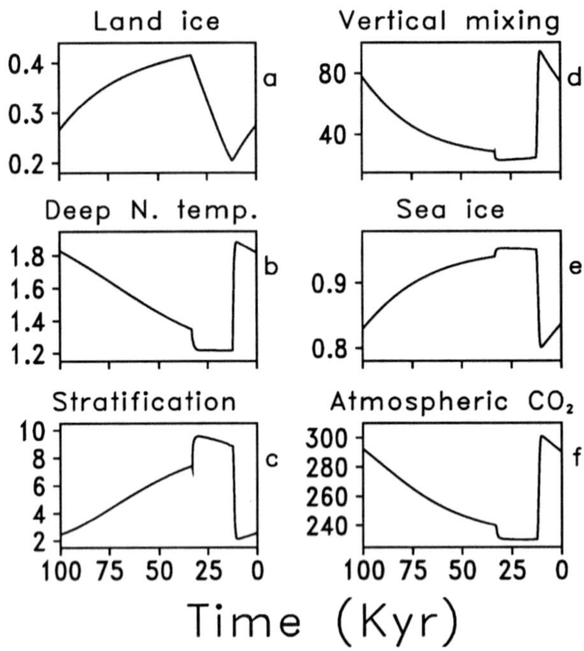

Figure 10. Model results for a complete glacial-interglacial cycle. (a) land ice in the Northern Hemisphere (fraction of northern land box area covered by glaciers), (b) temperature of the NADW flowing into the Southern Ocean (°C), (c) the stratification in the Southern Ocean (density difference between upper and deep water, $10^2 \times kg\,m^{-3}$), (d) vertical mixing in the Southern Ocean (Sv), (e) sea ice extent in the Southern Ocean (fraction of the Southern Ocean area), (f) atmospheric CO_2 in ppm.

et al., 1996) and general circulation model studies (Weaver et al., 1998) which indicate colder deep water in the Atlantic ocean during the last glacial maximum (LGM) and a colder outflow to the Southern Ocean. Because Antarctica is covered by ice even during interglacial periods, the temperature of the adjacent surface water in the southern box stays close to the freezing point. The Southern Ocean deep water cooling makes the model stratification in the Southern Ocean gradually more stable during glaciation (Figure 10c), which is consistent with proxy observations (Francois et al., 1997). Based on the above mentioned vertical mixing parameterization (Gargett, 1984), such a more stable stratification reduces the mixing between the CO_2 rich deep water and the surface water (Figure 10d). This is precisely the change of mixing deduced from proxy records (Sikes et al., 2000) and also the one needed to be arbitrarily specified by Toggweiler (1999) without the physical mechanism which we provide here. Note that a symmetrically opposite effect happens during deglaciation: the NADW warms, stratification in the Southern Ocean weakens, and the vertical mixing there strengthens.

Although Southern Ocean sea ice exists throughout the model glacial cycle, its meridional extension varies between glacial and inter-glacial states (Figure 10e), as well as with the seasonal cycle. During all phases of the glacial cycle, the NADW upwelling in the Southern Ocean is warmer than the surface water and acts to limit the sea ice extent there (Gordon, 1981). During the glaciation phase, the cooling of the NADW (Figure 10b) cools the water upwelling in the Southern Ocean (which is still warmer than the surface water). This cooling of the upwelling water, as well as the reduced vertical mixing of the surface water with the warm deep water (Figure 10d) both contribute to the growth of the Southern Ocean sea ice during glaciation (Figure 10e). The change in sea ice cover during the glacial-interglacial oscillation which is predicted by the proposed physical mechanisms and the model, is similar to that arbitrarily specified by Stephens and Keeling (2000), and explains part of the atmospheric CO_2 variations via the insulating effect of sea-ice cover on gas air-sea exchange.

The variations in the vertical mixing and in the sea-ice extent in the Southern Ocean induce together, via the ocean biochemistry, a difference of about 75 ppm in this atmospheric model CO_2 between glacial and interglacial periods (Figure 10f).

So far we have ignored the mechanism of the glacial cycle in the Northern Hemisphere. This Northern Hemisphere cycle is responsible for the changes in the temperature of the NADW reaching the Southern Ocean and causing the vertical mixing and sea ice extent changes in our proposed mechanism. The glacial cycle "sea-ice switch" mechanism of the physical climate system in this box model is described in detail elsewhere (Gildor and Tziperman, 2000; Gildor and Tziperman, 2001b). The physical mechanism for the changes in the Southern Ocean vertical mixing and sea-ice extent is independent of the details of the mechanism of glacial cycles in the Northern Hemisphere. It is implicitly assumed in the above discussion that the glaciation (and deglaciation) starts in the Northern Hemisphere (as it does in our physical model) and affects the Southern Hemisphere after a few hundreds years, once the NADW signal reaches the Southern Ocean. As mentioned in section 9, the phase relation between the two hemispheres is still debated.

7.3 THE ROLE OF CO_2 VARIATIONS

The glacial oscillations in this model exist due to a self-sustained internal variability of the physical climate system (Saltzman and Verbitsky, 1994; Kallen et al., 1979). The CO_2 changes are therefore not the driving force of the glacial-interglacial oscillation (Loutre and Berger, 2000) but rather are induced by the physical changes to the Southern Ocean stratification and vertical mixing. There is, however, an interesting two-way feedback between the physical

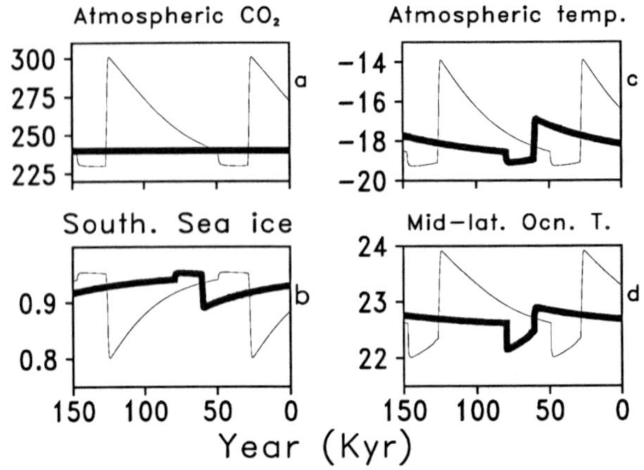

Figure 11. The amplification of the glacial-interglacial signal of the physical climate system by the atmospheric CO_2 variations. Shown are results from a physics only model using a fixed atmospheric CO_2 concentration (thick line) and results from a full coupled model with an active ocean biochemistry model (thin line). (a) atmospheric CO_2 (ppm), (b) Southern Ocean sea ice extent (fraction of the Southern Ocean area), (c) atmospheric temperature above the Southern Ocean (°C), (d) mid latitude upper ocean temperature.

components of the climate system and the ocean biochemistry. This feedback involves the amplification of the physical climate system glacial signal by the variations in the atmospheric CO_2. For example, the Southern Ocean mixing and sea ice changes induced by the glacial-interglacial cycle of the physical climate system cause variations in atmospheric CO_2 (Figure 11a) which, in turn, significantly amplify the glacial-interglacial cycle in the Southern Ocean sea-ice variability (Figure 11b). Through the sea ice albedo effect, this amplifies the variability of the southern atmospheric temperature relative to a fixed CO_2 scenario (Figure 11c). Mid-latitude atmospheric and ocean temperatures (Figure 11d) are also directly affected by the direct radiative forcing of the variable atmospheric CO_2.

8. OTHER HYPOTHESES FOR GLACIAL-INTERGLACIAL CO_2 VARIATIONS

The previous sections (besides 6.1) describe mechanisms for glacial- interglacial CO_2 variation, which can be described as "physical". Variations in the physical components of the climate system drive the CO_2 variations which only act as amplifier. In this section I will describe a few hypotheses which are more "biological" or "chemical". The "$CaCO_3$ compensation" mentioned in section 8.2 is a way to enhance the efficiency of the mechanism discussed above that strengthens the organic pump by trapping more remineralized carbon in the deep ocean.

8.1 VARIATIONS IN REDFIELD RATIO

A potentially important ecological effect which can affect atmospheric CO_2 involves changes to the stoichiometry of biological processes. Redfield noted that the ratio of dissolved inorganic nutrients in seawater closely reflects the ratio in bulk oceanic organic matter, with the ratio of carbon to nitrate to phosphate, $C : N : P = 106 : 16 : 1$ (Redfield et al., 1963) (Other studies of the composition of seawater (Anderson and Sarmiento, 1994; Takahashi et al., 1985) suggest slightly different average values.) This ratio, known as the "Redfield Ratio", reflects the relative need for each element in living matter. The efficiency of the biological pump of carbon from the surface to the deep ocean is highly sensitive to the "Redfield ratio" (Heinze et al., 1991). This ratio represents bulk oceanic organic matter and is not necessarily true on a species by species basis so a significant shift in the ecological structure stimulated by climatic shifts might possibly affects the biological pump of carbon in the ocean. Could ecological change really lead to a alteration of this ratio on significant scales?

There is already evidence that climate change can significantly affect the structure of ecosystems on decadal time scales (Francis et al., 1998; Roemmich and McGown, 1995) with implications for nutrient drawdown stoichiometry. Recent time-series observations of the biogeochemical environment near Hawaii have shown a local trend in favor of nitrogen fixing organisms, with higher than Redfield $N : P$ ratios, over the past decade (Karl, 1999). (Nitrogen fixation has supported as much as half of the local biological export production in this period.) Observations in the Southern Ocean show that local stratification dictates the dominant phytoplankton species. The difference in the relative drawdown of carbon and nutrients between these phytoplankton species can be more than 50%, with higher $C : P$ ratios associated with species living in less stratified water (Arrigo et al., 1999). So variations in the Redfield ratio seem plausible. Nevertheless, as with the theory of increased nitrogen inventory or utilization, increased production by the amount needed to explain glacial CO_2 should cause anoxia in the deep ocean, and evidence for that have not been found. In addition, the response of ocean ecosystems to different climates is, as yet, unknown. Our current understanding of the oceanic ecosystem and its response to changes in external forcing is not adequate to confidently predict either its response to a changing climate or the consequences for the ocean carbon cycle (Sarmiento et al., 1998; Falkowski et al., 2000).

8.2 ALKALINITY SCENARIOS

Before mentioning a few hypotheses involving variations in alkalinity, we require a short guide to the processes which affect the alkalinity distribution in the ocean, the concentration of $CaCO_3$ in the sediments, and few definitions.

The solubility of $CaCO_3$ increases with decreasing temperature and increasing pressure. Within the deep ocean, where temperature variations are relatively small, the effect of pressure on $CaCO_3$ is dominant (Broecker and Peng, 1982). Based on equations (4) and (5) we can write $[CO_3^{2-}] \approx A_T - \sum CO_2$. In other words, the saturation is determined by the distribution of A_T and $\sum CO_2$.

In thermodynamical equilibrium, the rate of dissolution is equal to the rate of precipitation, i.e.

$$CaCO_3(s) \rightleftharpoons Ca^{2+}(aq) + CO_3^{2-}(aq)$$

and the water is said to be saturated with respect to $CaCO_3$. We can measure the concentrations of Ca^{2+} and CO_3^{2-} for this equilibrium and since the concentration of Ca^{2+} in the ocean is nearly constant, the saturation with regard to calcite will be determined by the concentration of CO_3^{2-} compared with the concentration of CO_3^{2-} at saturation.

A few definitions are necessary: **Saturation depth** is the depth where the concentration of CO_3^{2-} is equal to CO_3^{2-} concentration at calcite saturation. The **compensation depth** is the depth below no $CaCO_3$ is found in the sediments. If thermodynamical equilibrium was the whole story, $CaCO_3$ below the saturation depth should spontaneously dissolve, and if the dissolution were fast enough, the **compensation depth** would coincide with the saturation depth. Since the calcite shells sink fast relative to the dissolution rates, they do not spend enough time in the undersaturation zone before they get to the bottom. The rate of dissolution depends on the difference between the in-situ CO_3^{2-} concentration and the CO_3^{2-} concentration in saturation. Obviously, this difference is small near the saturation depth and increases at greater depths in the ocean. The **lysocline** is the depth where dissolution starts to be detectable by sharp decrease in the $CaCO_3$ concentration. It can be separated from the saturation depth by a few hundred meters (Zeebe and Wolf-Gladrow, 2001) and it is located at different depths in different ocean basins as a result of variations in $[CO_3^{2-}]$ (which in turn, is a function of biological processes and ocean circulation).

In the real ocean the situation is even more complicated. For some reason, although the surface ocean is supersaturated with respect to $CaCO_3$, inorganic $CaCO_3$ is hardly produced (Pilson, 1998). Moreover, part of the $CaCO_3$ produced by organisms dissolves in the upper, saturated ocean (Milliman, 1993). In addition, dissolution can also occur inside the sediment, when decomposition of organic material adds CO_2 to the pore water, according to

$$CaCO_3 + CO_2 + H_2O \rightleftharpoons 2HCO_3^- + Ca^{2+}.$$

Alkalinity is largely determined by production and dissolution of $CaCO_3$ inside the ocean on one hand, and continental weathering followed by river runoff on the other hand. In steady state, supply and removal of alkalinity will balance each other. Although this steady state can be disrupted in many ways, such as by a perturbation in the the supply of alkalinity from land which may result from climate change, there are processes in the ocean which exert negative feedback on such an imbalance, working to restore a steady state. As the production rate of $CaCO_3$ in the ocean is larger than the rate of supply, the ocean compensate for this by dissolution of part of the produced $CaCO_3$. This is called "$CaCO_3$ **compensation**" (Broecker and Peng, 1998; Broecker and Peng, 1987). This processes can affect the drawdown of CO_2 by mechanisms such as enhanced biological organic production. Assume that the ocean absorbs CO_2 from the atmosphere. $\sum CO_2$ in the ocean will increase and $[CO_3^{2-}]$ will decrease since $[CO_3^{2-}] \approx A_T - \sum CO_2$ and air-sea CO_2 flux does not affect the alkalinity. With lower $[CO_3^{2-}]$, the saturation depth and the lysocline will be shallower and more $CaCO_3$ will dissolve. Since Ca^{2+} input exceeds Ca^{2+} removal ocean alkalinity will increase until the $[CO_3^{2-}]$ rises to close to its original value and balance is restored. Higher alkalinity means reduced pCO_2, so in this scenario $CaCO_3$ compensation contributes additional drawdown of atmospheric CO_2 to mechanisms based on a stronger organic pump. Note that this is a transient effect, without a significant shift in the steady-state depth of the lysocline (Sigman and Boyle, 2000).

Assume now that some perturbation causes an increase in ocean alkalinity. It will also cause an increase in $[CO_3^{2-}]$. This will deepen the saturation depth, and since the saturation depth and the lysocline are coupled, the lysocline will deepen as well, so more calcite will be buried restoring the balance between alkalinity input and burial. Model results suggest that the lysocline should deepen by about 1 km for every 25 ppm reduction in atmospheric CO_2 (Sigman and Boyle, 2000). Sediment records tell us that variations in the lysocline depth weren't large enough to support theories which invoke significant variations in alkalinity as a main source for glacial-interglacial CO_2 variations (Sigman and Boyle, 2000). Proxy records for variations in the pH of bottom water are not conclusive (Anderson and Archer, 2002; Sanyal et al., 1995).

8.2.1 Glaciation, sea level and coral reef exposure. The coral reef hypothesis (Opdyke and Walker, 1992) suggests that the glacial-interglacial CO_2 variations are the result of variations in reef carbonate deposition. At present, about 50% of $CaCO_3$ precipitation occurs in coral reefs and shallow waters (Milliman, 1993). When sea level declines as a results of ice accumulation on land, coral reefs are exposed, undergo dissolution and increase the alkalinity of the ocean, hence reducing pCO_2. During deglaciation, sea level rises, covering larger shelf area and enabling coral growth. The carbonate deposition reduces the alkalinity in surface waters and causing an increase in pCO_2. However, the lysocline record should show evidence of these alkalinity variations. As mentioned above, there is evidence that variations in lysocline depth were less than 1 km, excluding alkalinity change as the main driver for CO_2 variations.

8.2.2 Variations in the "Rain Ratio". Some mechanisms implicated past changes in the rain ratio as a leading influence on glacial-interglacial CO_2 variations (Berger and Keir, 1984; Archer and Maier-Reimer, 1994; Holligan and Robertson, 1996). Phytoplankton that produce calcite shells (mainly coccolithophorids) compete for nutrients against those which produce silicate shells (diatoms). Silica is a limiting factor for diatoms and it is usually depleted before NO_3^- and PO_4^{3-} (Archer et al., 2000b). As in the case of iron, more intense winds during glacial periods might transport more silica to the surface ocean, stimulating diatom production at the expense of coccolithophorids (Treguer and Pondaven, 2000; Harrison, 2000). In addition, it was demonstrated that availability of iron affect the Si:N uptake ration by phytoplankton (Franck et al., 2000; Takeda, 1998). Matsumoto et al. (2002) suggest that increased iron supply to the Southern Ocean during glacials resulted in reduced Si:N export ratio there. The excess silicic acid could then leak to lower latitudes and support diatom growth there at the expense of coccolithophorids thus reducing the rain ratio. If during glacial periods the rain ratio was smaller, i.e. if less $CaCO_3$ were produced, surface alkalinity would be higher and pCO_2 lower. Again, if less $CaCO_3$ is produced, less is buried and to restore steady state in the ocean between alkalinity input and output, the lysocline has to deepen.

A way to avoid the lysocline depth constraint was suggested by Archer and Maier-Reimer (1994). It is based on the possibility that dissolution can also occur inside the sediment as a result of remineralization of soft organic matter. This dissolution can occur even above the saturation depth. Remineralization of organic matter produces CO_2, and leads to the following reaction:

$$CO_2 + CaCO_3 + H_2O \Leftrightarrow 2HCO_3^- + Ca^{2+}$$

The lysocline, the depth where significant dissolution occurs, will be **above** the saturation depth. The degree of possible decoupling between the lysocline and the saturation depth is a matter of debate and different among models; the model of Sigman et al. (1998) demonstrates strong coupling while the model of Archer et al. (2000) is more flexible. The needed variations in the rain ratio are large and we have no physiological or other explanation to justify such a change, although variation in silicate transport to the ocean might provide a partial explanation. In addition, these variations in the rain ratio should be spatially uniform, otherwise we should have seen "bumps" in the glacial lysocline, and no such bumps have been found (Broecker and Peng, 1998).

8.2.3 Carbon storage in the polar North Atlantic. In a recent paper, Taylor et al. (2002) suggest that during glacial periods, local variations in the rain ratio cause accumulation of carbon in the deep polar North Atlantic and Arctic oceans. Part of the "excess" organic carbon arrived from land eroded by the advancing glaciers, hence no additional productivity is required. As a result of the southward shift of the sites of deep water formation, this regions become more isolated from the rest of the ocean and so a global signature in the lysocline is not expected.

9. HOW DO WE EVALUATE COMPETING THEORIES?

When there are few theories to explain past variations of climatic variables, we have to find a way to verify or falsify them. We use numerous proxies to estimate environmental variables such as temperature, circulation and nutrient utilization in the past. We then use these estimates as constraints to distinguish between different "depictions" of past climates and between different mechanisms proposed to answer questions such as what caused glacial-interglacial CO_2 variations. As the only way to estimate past conditions, these proxy records are most valuable. Nevertheless, they should be used with care. First, one should keep in mind that these records are extracted from a limited number of locations (like ocean sediment cores and ice cores) and they do not necessarily represent global or even regional conditions. Second, each proxy record is affected by many factors and processes and therefore each record has more than one source of variability. Not surprisingly, occasionally different proxies tell different - and sometimes contradictory - stories, as was noted above in some cases.

As an example, consider the absence of clear phasing between CO_2 and ice volume variations. Although some interpretations of proxy records infer that CO_2 lags southern hemisphere temperature but leads northern hemisphere warming during deglaciation (Monnin et al., 2001; Fischer et al., 1999; Mudelsee, 2001), other interpretations (Alley et al., 2002) as well as some models (Gildor and Tziperman, 2001a) suggest the opposite. The difficulty in constructing clear phasing is partly a result of millennial scale variability superimposed on the glacial cycle signal (Alley et al., 2002) and partly because each proxy for ice volume (like $\delta^{18}O$ from planktonic and benthic foraminifera) is also affected by other factors such as temperature, salinity and pH (Mix et al., 2001). The rate of sea level rise (which represents the ice volume) is an highly debated issue and an active area of research (Mitrovica, 2003) with some studies suggesting that the marine $\delta^{18}O$ (common proxy record for ice volume) lags sea level by few thousands years (Clark and Mix, 2000).

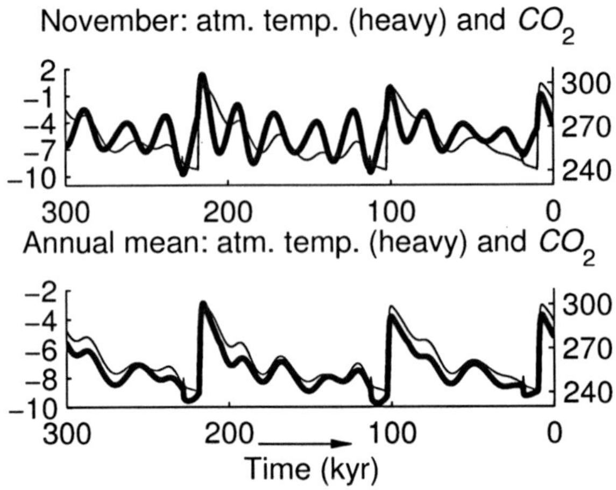

Figure 12. Seasonal effects on phase relations: atmospheric CO_2 (light solid, right axes, in ppm) and temperature (heavy, left axes, in ^oC) in the southernmost of the three atmospheric boxes. Upper panel: consecutive Novembers and lower panel: annual means.

Gildor and Ghil (2002) demonstrated using a box model (described in section 7) of the climate system that the phase relation between atmospheric CO_2 and southern hemisphere temperature can be opposite during different seasons and, moreover, that the phase relation can change during different stages of the glacial cycle (Figure 12). Isotopic signatures of the temperature in ice-cores record the condition in the time of precipitation or condensation, thus it may favor one phase relation during certain stages over another depending on the seasonality of precipitation. We cannot expect to get a reliable lead-lag relationship in the near future (Alley et al., 2002). Still, one finds in the literature numerous cases where theories related to glacial-interglacial cycles are tested based on such lead-lag relationships (Broecker and Henderson, 1998). As an example, the coral reef hypothesis has been discarded partly because CO_2 rise is expected to lag melting of the ice sheets.

In addition, it is also common to neglect the uncertainties which accompanied the measurements and interpretation of proxy records and the uncertainties in our models (LeGrand and Alverson, 2001). When these are taken into account, it actually seems that the constraints put by the available proxies are not strong enough to enable us to choose amongst competing theories (LeGrand and Alverson, 2001). So other considerations, such as the simplicity and plausibility of the mechanism, and the way it fits in the big picture are also necessary (Alley et al., 2002). For further reading on these issues, the reader is refer to Alley et al. (2002); Mix et al. (2001); Crowley (1999) and Legrand and Alverson (2001). When one encounters in the literature discussions based on the phasing relation between proxy records (Broecker and Henderson, 1998) and/or discarding of mechanisms based on the violation of the (current!) interpretation of a single proxy record, keep in mind the above warnings and base your judgment on how the mechanism fits into the big picture, and how plausible it is.

10. SUMMARY

1 We have seen numerous explanations for the glacial-interglacial CO_2 variations. Simple change in ocean circulation leading to reduced water formation in the Southern Ocean explain the glacial-interglacial CO_2 variations in simple models of Toggweiler (1999) and Gildor and Tziperman (2001). The CO_2 variations are induced by the cycle in the physical system but amplify them.

2 We see that different types of model have different sensitivities to the suggested mechanisms. In simple box models, atmospheric pCO_2 is especially sensitive to high latitude ocean scenarios (see section 5); in general circulation models, low latitudes are important as well (Bacastow, 1996; Archer et al., 2000a). So far, only simple models were able to reproduce the full amplitude of the glacial-interglacial variations, but this does not necessarily mean that they are better than GCMs or more correct. We understand some of the differences between box models and GCMs (Toggweiler et al., 2003a; Toggweiler et al., 2003b) and have to improve our GCMs to the point where they can reproduce the lowering of atmospheric CO_2 during glacials. One step forward in trying to reconcile the differences between box models and GCMs is the work of Follows et al. (2002). They show that variability in the volume and characteristics of the sub-tropical ventilated thermocline affect the distribution of carbon between the ocean and the atmosphere. For example, colder water in this reservoir which is partially insulated from the atmosphere, can store more carbon during cold periods. This carbon reservoir and the processes which affect its volume and characteristics are not resolved in simple models. Another cause for difference is the treatment of diapycnal mixing in GCMs which seems to be too high (Oschlies, 2001).

3 Could it be that these variations are not the result of an exclusive process but due to a combination of the processes reviewed? The conceptual model of Sigman and Boyle (2000) combines few of the ideas mentioned above. Models such as the one used by Archer et al. (2000b) try to answer this question qualitatively, but they are still not good enough to give us a conclusive answer.

4 Current global biogeochemical models typically parameterize the biological processes using crude empirical relations. We still lack fundamental understanding of the factors controlling the ecological system (like the rain and/or Redfield ratios) in the ocean. We need to move towards more explicit, prognostic, representations of ocean ecology.

5 We need to find new proxies to reconstruct more reliably past climates and to help us distinguish between competing hypotheses.

ACKNOWLEDGMENTS

I would like to thank Mick Follows and Temel Oguz for the opportunity to present this material at the NATO Advanced Study Institute on "The ocean Carbon Cycle and Climate" and thank NATO for their financial support. I also thank J. R. Toggweiler and an anonymous reviewer for their most constructive comments.

REFERENCES

Alley, R. B., Brook, E. J., and S. Anandakrishnan, A northern lead in the orbital band: north-south phasing of ice-age events. *Quat. Sci. Rev.*, 21:431–441, 2002.

Anderson, D. M. and D. Archer, Glacial-interglacial stability of ocean pH inferred from foraminifer dissolution rates. *Nature*, 416:70–73, 2002.

Anderson, L. and J. Sarmiento, Redfield ratios of remineralization determined by nutrient data analysis. *Global Biogeochem. Cycles*, 8:65–80, 1994.

Archer, D., Eshel, G., Winguth, A., Broecker, W., Pierrehumbert, R., Tobis, M., and R. Jacob, Atmospheric pCO_2 sensitivity to the biological pump in the ocean. *Global Biogeochem. Cycles*, 14:1219–1230, 2000a.

Archer, D. and E. Maier-Reimer, Effect of deep-sea sedimentary calcite preservation on atmospheric CO_2 concentration. *Nature*, 367:260–263, 1994.

Archer, D., Winguth, A., Lea, D., and N. Mahowald, What caused the glacial/interglacial atmospheric pCO_2 cycles? *Rev. Geophys.*, 38:159–189, 2000b.

Arrigo, K. R., Robinson, D. H., Worthen, D. L., Dunbar, R. B., DiTullio, G. R., VanWoert, M., and M. P. Lizotte, Phytoplankton community structure and the drawdown of nutrients and CO_2 in the Southern Ocean. *Science*, 283:365–367, 1999.

Asher, W. and R. Wanninkhof, Transient tracers and air-sea gas transfer. *J. Geophys. Res.*, 103:15939–15958, 1998.

Bacastow, R., The effect of temperature change of the warm surface waters of the ocean on atmospheric CO_2. *Global Biogeochem. Cycles*, 10:319–333, 1996.

Bates, N. and L. Merlivat, The influence of short-term wind variability on air-sea CO_2 exchange. *Geophys. Res. Lett.*, 28:3281–3284, 2001.

Berger, A., Loutre, M. F., and H. Gallee, Sensitivity of the LLN climate model to the astronomical and CO_2 forcings over the last 200 ky. *Clim. Dyn.*, 14:615–629, 1998.

Berger, W. and R. Keir, Glacial-Holocene changes in atmospheric CO_2 and the deep-sea records. In Hansen, J. and Takahashi, T., editors, *Climate processes and climate sensitivity*, volume 29 of *Geophysical Monograph Series*, 337–351. American Geophysical Union, 1984.

Bradley, R. S., *Paleoclimatology, reconstructing climates of the Quaternary*, volume 64 of *International Geophysics Series*. Academic Press, second edition, 1999.

Broecker, W. and G. Henderson, The sequence of events surrounding termination II and their implication for the cause of glacial-interglacial CO_2 changes. *Paleoceanography*, 13:352–364, 1998.

Broecker, W., Lynch-Stieglitz, J., Archer, D., Hofmann, M., Maier-Reimer, E., Marchal, O., Stocker, T., and N. Gruber, How strong is the Harvadton-Bear constraint? *Global Biogeochem. Cycles*, 13:817–820, 1999.

Broecker, W. S. and T. H. Peng, *Tracers in the sea*. Lamont-Doherty Earth Observatory of Columbia University, 1982.

Broecker, W. S. and T. H. Peng, The role of $CaCO_3$ compensation in the glacial to interglacial CO_2 change. *Global Biogeochem. Cycles*, 1:15–29, 1987.

Broecker, W. S. and T. H. Peng, *Greenhouse puzzles*. Lamont-Doherty Earth Observatory of Columbia University, second edition, 1998.

Clark, P. U. and A. C. Mix, Global change: Ice sheets by volume. *Nature*, 406:689–690, 2000.

Conkright, M., Levitus, S., and T. Boyer, World ocean atlas 1994 volume 1: Nutrients. NOAA Atlas NESDIS 1, U.S. Department of Commerce, Washington, D.C., 1994.

Crowley, T. J., Ice age terrestrial carbon change revisited. *Global Biogeochem. Cycles*, 9:377–389, 1995.

Crowley, T. J., Paleomyths I have known, in Holland, W.R, S. Joussaume, and F. David, Editors, *Modeling the earth's climate and its variability*, 377–422, NATO ASI Series. Elsevier, 1999.

Duplessy, J., Past Climate Changes, in Holland, W.R, S. Joussaume, and F. David, Editors, *Modeling the earth's climate and its variability*, 315–376, NATO ASI Series. Elsevier, 1999.

Evans, G. T. and Parslow, J. S. A model of annual plankton cycles. *Biological Oceanography*, 3:327–347, 1985.

Falkowski, P., Evolution of the nitrogen cycle and its influence on the biological sequestration of CO_2 in the ocean. *Nature*, 387:272–275, 1997.

Falkowski, P., Scholes, R., Boyle, E., Canadell, J., Canfield, D., Elser, J., Gruber, N., Hibbard, K., Hogberg, P., Linder, S., Mackenzie, F., B. Moore III, B. M., Pedersen, T., Rosenthal, Y., Seitzinger, S., Smetacek, V., and W. Steffen, The global carbon cycle: A test of our knowledge of earth as a system. *Science*, 290:291–296, 2000.

Fischer, H., Wahlen, M., Smith, J., Mastroianni, D., and B. Deck, Ice core records of atmospheric CO_2 around the last three glacial terminations. *Science*, 283:712–714, 1999.

Follows, M., Ito, T., and J. Marotzke, The wind-driven, subtropical gyres and the solubility pump of CO_2. *Global Biogeochem. Cycles*, 16 (4): Art. No. 1113, 2002.

Francis, R., Hare, S., Hollowed, A., and W. Wooster, Effects of interdecadal climate variability on the oceanic ecosystems of the NE Pacific. *Fish. Oceanogr.*, 7:1–21, 1998.

Franck, V. M., Brzezinski, M. A., Coale, K. H., and D. M. Nelson, Iron and silicic acid concentrations regulate si uptake north and south of the polar frontal zone in the pacific sector of the Southern Ocean. *Deep-Sea Res.*, 47:3315–3338, 2000.

Francois, R., Altabet, M. A., Yu, E., Sigman, D. M., Bacon, M. P., Frank, M., Bohrmann, G., Bareille, G., and L. D. Labeyrie, Contribution of Southern Ocean surface-water stratification to low atmospheric CO_2 concentrations during the last glacial perios. *Nature*, 389:929–935, 1997.

Gargett, A., Vertical eddy diffusivity in the ocean interior. *J. Mar. Res.*, 42:359–393, 1984.

Ghil, M., Cryothermodynamics: the chaotic dynamics of paleoclimate. *Physica D*, 77:130–159, 1994.

Gildor, H. and M. Ghil, Phase relations between climate proxy records: Potential effect of seasonal precipitation changes. *Geophys. Res. Lett.*, Art. No. 1024, 2002.

Gildor, H. and E. Tziperman, Sea ice as the glacial cycles climate switch: role of seasonal and orbital forcing. *Paleoceanography*, 15:605–615, 2000.

Gildor, H. and E. Tziperman, Physical mechanisms behind biogeochemical glacial-interglacial CO_2 variations. *Geophys. Res. Lett.*, 28:2421–2424, 2001a.

Gildor, H. and E. Tziperman, A sea-ice climate-switch mechanism for the 100 kyr glacial cycles. *J. Geophys. Res.*, 106:9117–9133, 2001b.

Gildor, H., Tziperman, E., and J. R. Toggweiler, The sea-ice switch mechanism and glacial-interglacial CO_2 variations. *Global Biogeochem. Cycles*, Art. No. 1032, 2002.

Gnanadesikan, A., A simple predictive model for the structure of the oceanic pycnocline. *Science*, 283:2077–2079, 1999.

Gordon, A., Seasonality of Southern Ocean sea ice. *J. Geophys. Res.*, 86:4193–4197, 1981.

Harrison, K. G., Role of increased marine silica input on paleo-pCO_2 levels. *Paleoceanography*, 15:292–298, 2000.

Heinze, C., Maier-Reimer, E., and K. Winn, Glacial pCO_2 reduction by the world ocean: experiments with the Hamburg carbon model. *Paleoceanography*, 4:395–430, 1991.

Holligan, P. M. and J. E. Robertson, Significance of ocean carbonate budgets for the global carbon cycle. *Global Change Biology*, 2:85–95, 1996.

Imbrie, J., Hays, J., Martinson, D., McIntyre, A., Mix, A., Morley, J., Pisias, N., Prell, W., and N. Shackleton, The orbital theoty of Pleistocene climate: Support from a revised chronology of the marine $\delta^{18}O$ record, in *Milankovitch and Climate, Part I*, 269–305. D. Reidel, 1984.

Kallen, E., Crafoord, C., and M. Ghil, Free oscillations in a climate model with ice-sheet dynamics. *J. Atmos. Sci.*, 36:2292–2303, 1979.

Karl, D., A sea of change: Biogeochemical variability in the North Pacific subtropical gyre. *Ecosystems*, 2:181–214, 1999.

Keir, R. S., Are atmospheric CO_2 content and Pleistocene climate connected by wind-speed over a polar Mediterranean-Sea. *Global And Planetary Change*, 8:59–68, 1993a.

Keir, R. S., Cold surface ocean ventilation and its effect on atmospheric CO_2. *J. Geophys. Res.*, 98:849–856, 1993b.

LeGrand, P. and K. Alverson, Variations in atmospheric CO_2 during glacial cycles from an inverse ocean modeling perspective. *Paleoceanography*, 16:604–616, 2001.

Loutre, M. and A. Berger, No glacial-interglacial cycle in the ice volume simulated under a constant astronomical forcing and a variable CO_2. *Geophys. Res. Lett.*, 27:783–786, 2000.

Maier-Reimer, E., Geochemical cycles in an ocean general circulation model. preindustrial tracer distribution. *Global Biogeochem. Cycles*, 7:645–677, 1993.

Martin, J. H., Glacial-interglacial CO_2 change: the iron hypothesis. *Paleoceanography*, 5:1–13, 1990.

Matsumoto, K., Sarmiento, J. L., and M. A. Brzezinski, Silicic acid leakage from the Southern Ocean: A possible explanation for glacial atmospheric pCO_2. *Global Biogeochem. Cycles*, 16, Art. No.–1031, 2002.

Milliman, J. D., Production and accumulation of calcium carbonate in the ocean: Budget of a nonsteady state. *Global Biogeochem. Cycles*, 7:927–957, 1993.

Mitrovica, J. X., Recent controversies in predicting post-glacial sea-level change. *Quat. Sci. Rev.*, 22:127–133, 2003.

Mix, A. C., Bard, E., and R. Schneider, Environmental processes of the ice age: land, oceans, glaciers (EPILOG). *Quat. Sci. Rev.*, 20:627–657, 2001.

Monnin, E., Indermuhle, A., Dallenbach, A., Fluckiger, J., Stauffer, B., Stocker, T. F., Raynaud, D., and J. Barnola, Atmospheric CO_2 concentartions over the last glacial termination. *Science*, 291:112–114, 2001.

Maqueda, M. A. M. and S. Rahmstorf, Did antarctic sea-ice expansion cause glacial CO_2 decline? *Geophys. Res. Lett.*, 29, Art. No. 1011, 2002.

Mudelsee, M., The phase relations among atmospheric CO_2 content, temperature and global ice volume over the past 420 ka. *Quat. Sci. Rev.*, 20:583–589, 2001.

Najjar, R. G., Marine biogeochemistry. In Trenberth, K. E., editor, *Climate System Modeling*, 241–280. Cambridge Univ. Press, 1991.

Opdyke, B. N. and J. C. G. Walker, Return of the coral reef hypothesis: Basin to shelf partitioning of $CaCO_3$ and its effect on atmospheric CO_2. *Geology*, 20:733–736, 1992.

Oschlies, A., Model-derived estimates of new production: New results point towards lower values. *Deep-Sea Res.*, 48:2173–2197, 2001.

Pedersen, T. and P. Bertrand, Influence of oceanic rheostats and amplifiers on atmopsheric CO_2 content during the Late Quaternary. *Quat. Sci. Rev.*, 19:273–283, 2000.

Petit, J. R., Jouzel, J., Raynaud, D., Barkov, N. I., Barnola, J. M., Basile, I., Bender, M., Chappellaz, J., Davis, M., Delaygue, G., Delmotte, M., Kotlyakov, V. M., Legrand, M., Lipenkov, V. Y., Lorius, C., Pepin, L., Ritz, C., Saltzman, E., and M. Stievenard, Climate and atmospheric history of the past 420,000 years from the Vostok ice core, Antarctica. *Nature*, 399:429–436, 1999.

Pilson, M. E. Q., *An introduction to the chemistry of the sea*. Prentice Hall, 1998.

Raven, J. and P. Falkowski, Oceanic sinks for atmospheric CO_2. *Plant Cell And Environment*, 22:741–755, 1999.

Redfield, A., Ketchum, B., and F. Richards, The influence of organisms on the composition of sea water. In Hill, M., editor, *The Sea, Vol 2*, 26–77. Interscience, New York, 1963.

Roemmich, D. and J. McGown, Climatic warming and the decline of zooplankton in the California current. *Science*, 267:1324–1326, 1995.

Saltzman, B. and M. Verbitsky, CO_2 and glacial cycles. *Nature*, 367:419, 1994.

Sanyal, A. N., Hemming, G., Hanson, G. H., and W. S. Broecker, Evidence for a higher pH in the glacial ocean from boron isotopes in foraminifera. *Nature*, 373:234–236, 1995.

Sarmiento, J., Ocean carbon-cycle. *Chemical and Engineering News*, 71:30–43, 1993.

Sarmiento, J. L. and M. Bender, Carbon biogeochemistry and climate change. *Photosynthesis Research*, 39:209–234, 1994.

Sarmiento, J. L., Hughes, T. M. C., Stouffer, R. J., and S. Manabe, Simulated response of the ocean carbon cycle to anthropogenic climate warming. *Nature*, 393:245–252, 1998.

Sarmiento, J. L. and J. R. Toggweiler, A new model for the role of the oceans in determining atmospheric pCO_2. *Nature*, 308:621–624, 1984.

Schrag, D., Hampt, G., and D. Murray, Pore fluid constraints on the temperature and oxygen isotopic composition of the glacial ocean. *Science*, 272:1930–1932, 1996.

Siegenthaler, U. and J. Sarmiento, Atmospheric carbon dioxide and the ocean. *Nature*, 365:119–125, 1993.

Siegenthaler, U. and T. Wenk, Rapid atmospheric CO_2 variations and ocean circulation. *Nature*, 308:624–625, 1984.

Sigman, D. M. and E. A. Boyle, Glacial/interglacial variations in atmospheric carbon dioxide. *Nature*, 407:859–869, 2000.

Sigman, D. M., McCorkle, D. C., and W. R. Martin, The calcite lysocline as a constraint on glacial/interglacial low-latitude production changes. *Global Biogeochem. Cycles*, 12:409–427, 1998.

Sikes, E. L., Samson, C. R., Guilderson, T. P., and W. R. Howard, Old radiocarbon ages in the southwest Pacific Ocean during the last glacial period and deglaciation. *Nature*, 405:555–559, 2000.

Stephens, B. B. and R. Keeling, The influence of Antarctic sea ice on glacial-interglacial CO_2 variations. *Nature*, 404:171–174, 2000.

Stommel, H., Thermohaline convection with two stable regimes of flow. *Tellus*, 13:224–230, 1961.

Takahashi, T., Broecker, W., and S. Langer, Redfield ratio based on chemical data from isopycnal surfaces. *J. Geophys. Res.*, 90:6907–6924, 1985.

Takahashi, T., Feely, R. A., Weiss, R. F., Wanninkhof, R. H., Chipman, D. W., Sutherland, S. C., and T. T. Takahashi, Global air-sea flux of CO_2: An estimate based on measurements of sea-air pCO_2 difference. *Proc. Natl. Acad. Sci. U.S.A.*, 94:8292–8299, 1997.

Takeda, S., Influence of iron availability on nutrient consumption ratio of diatoms in oceanic waters. *Nature*, 393:774–777, 1998.

Taylor, J., Tranter, M., and G. Munhoven, Carbon cycling and burial in the glacially influenced polar North Atlantic. *Paleoceanography*, 17, Art. No. 1001, 2002.

Taylor, N., The role of the ocean in the global carbon cycle. Part 2: Climate feedbacks. *Weather*, 47:237–241, 1992.

Toggweiler, J. R., Variation of atmospheric CO_2 by ventilation of the ocean's deepest water. *Paleoceanography*, 14:572–588, 1999.

Toggweiler, J. R., Gnanadesikan, A., Carson, S., Murnane, R., and J. L. Sarmiento, Representation of the carbon cycle in box models and GCMs: 1. Solubility pump. *Global Biogeochem. Cycles*, 17, Art. No.–1026, 2003a.

Toggweiler, J. R., Murnane, R., Carson, S., Gnanadesikan, A., and J. L. Sarmiento, Representation of the carbon cycle in box models and GCMs: 2. Organic pump. *Global Biogeochem. Cycles*, 17, Art. No.–1027, 2003b.

Toggweiler, J. R. and B. Samuels, Is the magnitude of the deep outflow from the Atlantic ocean actually governed by the southern hemisphere winds? In Heimann, M., editor, *The Global Carbon Cycle*, 303–331. NATO ASI Series, Springer-Verlag, Berlin, 1993.

Treguer, P. and P. Pondaven, Silica control of carbon dioxide. *Nature*, 406:358–359, 2000.

Volk, T. and M. Hoffert, Ocean carbon pumps: analysis of relative strengths and efficiencies in ocean-driven atmospheric CO_2 changes. In Sundquist, E. T. and Broecker, W. S., editors, *The carbon cycle and atmospheric CO_2: natural variations Archean to present*, volume 32 of *Geophysical monograph*, pages 99–110. American Geophysical Union, 1985.

Weaver, A. J., Eby, M., Fanning, A., and E. C. Wiebe, Simulated influence of carbon dioxide, orbital forcing and ice sheets on the climate of the Last Glacial Maximum. *Nature*, 394:847–853, 1998.

Weertman, J., Milankovitch solar radiation variations and ice age ice sheet sizes. *Nature*, 261:17–20, 1976.

Weiss, R. F., Carbon dioxide in water and sea water: the solubility of non-ideal gas. *Mar. Chem.*, 2:203–215, 1974.

Whitfield, M., Ocean biology, trace metals and climate, in *Modeling oceanic climate interaction*, 337–371. NATO ASI Series, Springer-Verlag, 1993.

Zeebe, R. and D. Wolf-Gladrow, *CO_2 in seawater: equilibrium, kinetics, isotopoes*, volume 65 of *Elsevier Oceanography Series*. Elsevier, 2001.

DATA ASSIMILATION INTO MARINE CARBON MODELS

RICHARD J. MATEAR
CSIRO Marine Research and the Antarctic Cooperative Research Center, Hobart, Tasmania, Australia.
richard.matear@marine.csiro.au

INTRODUCTION

Data assimilation is a method for combining and synthesizing data to enhance the interpretation of the data[*Winguth et al.*, 1999], to build a consistent picture of a system [*Schlitzer*, 2002], to estimate model parameters [*Spitz et al.*, 1998] and to enable the extrapolation of observations in space and time [*Eknes and Evensen*, 2002]. The field of data assimilation is well known in both meteorology and physical oceanography where it routinely provides the tool to make short-term forecasts of weather and upper ocean currents [*Ghil and Malanotte-Rizzoli*, 1991, *Thacker*, 1988; *Tziperman and Thacker*, 1989a]. At present, a comprehensive international program (Global Ocean Data Assimilation Experiment) is being developed and implemented to build a real time ocean prediction system [http://www.usgodae.fnmoc.navy.mil/]. The ocean prediction system is well advanced for predicting ocean ocean circulation with a comprehensive plan to collect physical data from in situ instruments (ARGO profiling floats) and satellites, and to assimilate the data into models.

The application of data assimilation to tackle marine carbon questions is a recent development and the field of research is only in its infancy. Marine carbon models attempt to simulate complex physical, chemical and biological processes. This requires parameterizing processes often with poorly determined model parameters. The assimilation of data into carbon

M. Follows and T. Oguz (eds.), The Ocean Carbon Cycle and Climate, 353–395.

models provides a framework to calibrate these models and enable one to build and refine prognostic models of carbon and nutrient cycling in the ocean. The successful calibration of carbon models requires a realistic model and sufficient data to constrain this model. Data assimilation studies can address both these issues by determining the simplest model that is consistent with the available observations and providing insight into what observations are critical to building and constraining more realistic models. Data assimilation should become a valuable tool to further our understanding of carbon cycling in the ocean and to provide real time estimates of the air-sea CO_2 fluxes.

This chapter seeks to demonstrate the ability of data assimilation to advance carbon cycle research and to hopefully stimulate further applications of data assimilation in marine carbon cycle research. In this chapter, I will first review the basic concept of data assimilation. Next, I will present three examples of data assimilation that use data to build, calibrate and quantify the dynamic behaviour of the marine carbon and nutrient cycles. I have chosen to present 3 examples to demonstrate the diverse range of carbon cycle research questions that can be tackled within the framework of data assimilation. None of the chosen examples is meant to be a complete summary of the individual studies but rather they serve as a platform to demonstrate the set-up, application, benefits and limitations of data assimilation. In the first example, atmospheric and oceanographic data is used to constrain air-sea fluxes of CO_2 [*Roy et al.*, 2003]. In the second example an adjoint data assimilation method is applied to the phosphate tracer equation to investigate nutrient cycling in the eastern Equatorial Pacific ocean [*Matear and Holloway*, 1995]. In the third example, data assimilation is used to determine the model parameters of an ecosystem model [*Matear*, 1995]. Finally, I will conclude the paper with summary of the advantages and limitations of data assimilation.

For a brief review of inverse theory and its application to assimilating data into the linear and non-linear problems presented in this chapter refer to appendix A. For further reading, a recent book on data assimilation summaries the mathematics of data assimilation and presents several examples of using data assimilation to investigate biogeochemical cycling [*Kasibhatla*, 2000]. Some additional examples of the data assimilation into marine carbon models are *Athias et al.*, 2000; *Gloor et al.*, 2003; *Prunet et al.*, 1996].

In the future, the monitoring of the marine carbon cycle will lead to diversifying the range of observations collected, increasing the density of in situ observations, and increasing the measurements made from satellites. In this enhanced data environment there will be a pressing need to synthesize these datasets in a consistent manner. In this synthesis, data assimilation should play an important role by providing a powerful method for combining and interpreting data. Coupled with the increased data volume, there will be the continued increase in the complexity of marine carbon models to make them more realistic. Again, as this occurs we will need to continually access the data needs required to quantify these new processes and improve their representations in marine carbon cycle models. Again data assimilation techniques have the potential to make valuable contributions to this effort.

1. GENERAL CONCEPT OF DATA ASSIMILATION

The essential elements of data assimilation are a model that describes the behaviour of the system (forward model), observations that provide information on the systems dynamics and a technique for incorporating the observations into the model (inverse method). The term inversion is often used to describe the process of using data constraints to infer model parameters. The general modelling approach is to prescribe the model parameters and perform a model simulation to produce model output that is then compared to observable quantities. This process is often referred to as forward modelling where the forward model describes the link between the model parameters and the observable fields. In the application of data assimilation the process is reversed, and one utilizes the observations to "invert" the forward model to obtain the underlying model parameters. A "cost function" is used to incorporate the observable fields as misfits with respect to the estimated fields of forward model simulation. Then an optimization algorithm is used minimize the cost functions and estimate parameters of the forward model. The performance of data assimilation is limited by both the suitability of the forward model in representing the system and the data used to constrain the problem.

In the following sections, three different data assimilation examples are presented. Each example has its unique forward model, cost function and algorithm for solving the inverse problem.

2. EXAMPLE ONE: ESTIMATING AIR-SEA CO_2 FLUXES

CO_2 is increasing in the atmosphere but less than half of the annual anthropogenic CO_2 emissions remain in the atmosphere because they are taken up by the oceans and the terrestrial biosphere. To predict the future climate impact of rising atmospheric CO_2 levels we need to understand the how the ocean and the terrestrial biosphere control atmospheric CO_2 levels.

For the ocean, this includes determining the magnitude and spatial distribution of the oceanic CO_2 fluxes.

2.1 ASSIMILATION APPROACH

Assimilation of atmospheric CO_2 measurements into an atmospheric transport model provides one method to estimate surface CO_2 fluxes from the atmosphere into the ocean and terrestrial biosphere. The method relies on dividing the surface of the earth into regions and on an atmospheric transport model that relates the surface CO_2 fluxes in these regions to the atmospheric CO_2 concentrations. The method estimates the surface CO_2 fluxes that are consistent with the spatial and temporal patterns of atmospheric CO_2 concentrations under prescribed atmospheric transport of CO_2. The method has been widely used to provide regional surface CO_2 fluxe estimates [*Bousquet et al.*, 2000; *Cias et al.*, 1995; *Rayner et al.*, 1999; *Tans et al.*, 1990].

Early studies noted that the simulated inter-hemispheric CO_2 gradient from atmospheric transport models and prescribed fossil fuel sources was substantially larger than observed (Fig. 1) [*Tans et al.*, 1990].

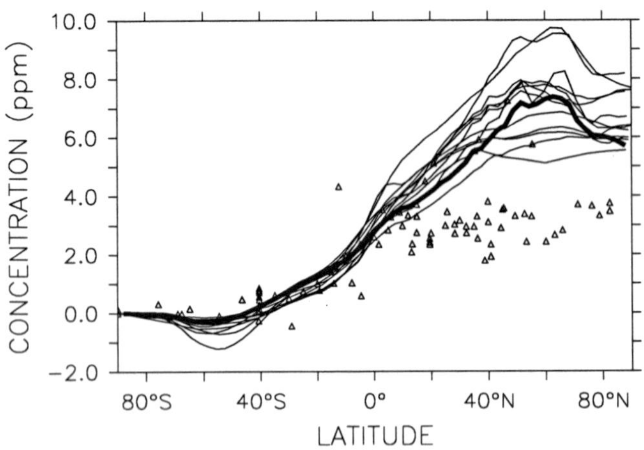

Figure 1. The observed (Δ) annual mean atmospheric CO2 concentrations (ppm) and the simulated (lines) interhemispheric gradients from a range of transport models [Gurney et al., 2002]. The simulated gradients were produced from the best estimates of fossil fuel, terrestrial [Randerson et al., 1999] and oceanic [T99 in Takahashi et al., 2002] source fields. The bold line represents the atmospheric transport model, CRC-MATCH, used in this example. Note: All data are normalised to the South Pole.

The observed inter-hemispheric gradient is predominantly forced by the asymmetry in fossil fuel combustion (approximately 95% occurs in the

Northern Hemisphere). Inverse models were able to reproduce observed CO_2 gradients by estimating a much larger northern hemispheric CO_2 sink (usually over land) than the southern hemispheric sink [*Bousquet et al.*, 2000; *Rayner and Law*, 1999; *Roy et al.*, 2003]. However, ocean CO_2 flux estimates based on either the observed CO_2 partial pressure differences between the surface ocean and the atmosphere (ΔpCO_2) [*Metzl et al.*, 1999; *Takahashi et al.*, 2002; chapter XX in this book] or global ocean carbon models simulations [*Matear and Hirst*, 1999; chapter YY in this book] have produced much larger Southern Ocean CO_2 uptakes than these atmospheric inversions.

2.2 A REVISED APPROACH

The advent of a more comprehensive ocean ΔpCO_2 compilation [*Takahashi et al.*, 2002] enables one to combine the independent atmospheric concentration data and oceanic CO_2 flux estimates in an inversion that estimates surface fluxes that are consistent with both datasets. The southern extratropics (20°S – 90°S) is a particularly interesting region for such a study since the almost complete ocean coverage reduces complications that might arise from terrestrial-atmosphere CO_2 fluxes.

Presented here is a summary of a study [*Roy et al.*, 2003], which used the inversion technique to combine atmospheric data and ocean flux estimates to compare and assess their consistency in the extratropical Southern Ocean. Model simulations demonstrate how critical but uncertain the CO_2 uptake is in the Southern Ocean [*Matear*, 2001]. In this relatively data poor region, it is essential that all oceanic and atmospheric data are used to estimate the CO_2 fluxes.

A time-dependent inversion of the surface CO_2 fluxes using the atmospheric CO_2 data from the 1980-97 period was set-up by dividing the earth's surface into 26 regions (Fig. 2). For each region one will determine the monthly surface fluxes CO_2 out of the atmosphere over the 1978-97 period (20 years). The total number of unknowns in our problem is 20x12x26=6240. One incorporates the geographical distribution of fluxes within the ocean regions by using the T99 compilation of *Takahashi et al.*, [2002] (Fig. 2). This is achieved by fixing the initial source pattern as in T99 and solving for the offsets from this distribution within each ocean region. The atmospheric data used includes the monthly-mean CO_2 concentrations from 65 stations marked in Figure 2. The data is a preliminary version of the dataset used by *Bousquet et al.*, [2000] and here one uses data from the period 1980-1997 (18 years). The potential number of data points to constrain our inverse problem is 18x12x65=14040 but since most of the 65 stations do not have observations for the entire the period the total number of data constraints is only about 7000. The O_2/N_2 record from Cape Grim, Tasmania [*Langenfelds et al.*, 1999]is also used to constrain the global CO_2 uptake by the terrestrial biosphere and the oceans [*Battle et al.*, 2000].

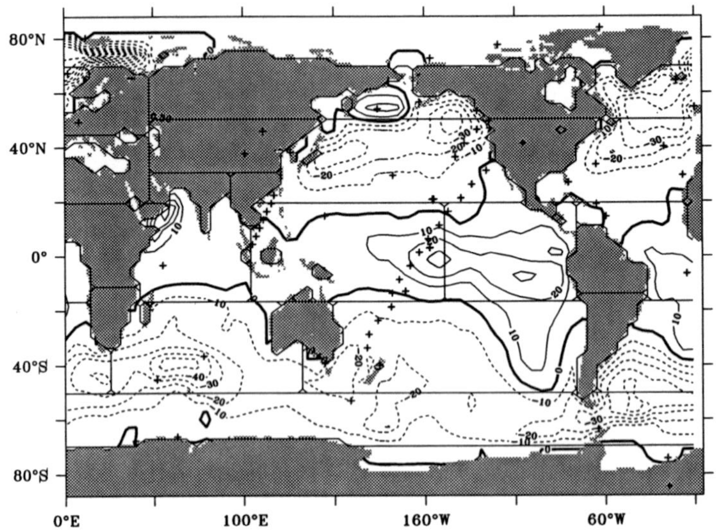

Figure 2. Annual mean ocean flux estimate (g m^{-2} yr^{-1}) T99 in Takahashi et al. (2002), which is used as the initial ocean sources in the inversion. The dashed contour lines denote negative values, which indicate a flux into the ocean (atmospheric sink). The inversion solves for the 26 regions (boxes) and uses atmospheric CO_2 concentrations from the 65 marked atmospheric stations (+).

To build the inverse problem we use a Green function approach [*Morse and Feshbach*, 1953]. For each of the surface sources we individually prescribe a 0.1Gt/month C pulse of duration one-month as input into an atmospheric transport model. Starting at the onset of the pulse, the resulting atmospheric concentrations at the measurement sites are stored for 3 years. For each station, the resulting 3-year time series relates how the surface flux from a region influences the future atmospheric CO_2 at the measurement site ($c_j(t)$). For the ith source region and the jth station this can be written as

$$r_{j,i}(t)\,x_i = c_j(t)\ ,\ r_{j,i}(t)=0 \text{ except for period } t=t_1 \text{ to } t_1+3 \qquad (1)$$

where t is time in years and $r_{j,i}$ is the response function of station j to source i generated with the 3 year simulation of the atmospheric transport model. Shown in Figure 3 is the response of the Cape Grim station in Tasmania to the unit source from the Southern Ocean and North Atlantic regions. Two things to note from this figure are: 1) sources from any region influence the future concentration at a site; 2) the atmosphere rapidly mixes away the impact of the source so that after a short period (≈3 years) the pulse sources is uniformly mixed in the atmosphere.

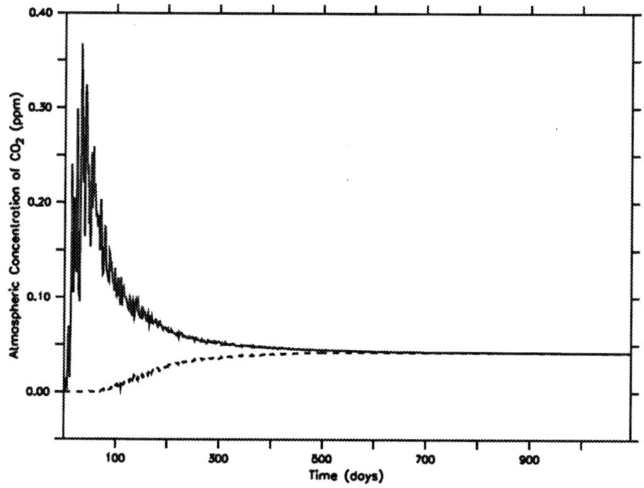

Figure 3. The response of the Cape Grim Station (40°S and 140°E) to a one month source of 0.1 Gt C to the atmosphere from the Southern Ocean region (solid line) and the North Atlantic region (dashed line).

The important step in the developing the inverse problem is that with the exception of the earth's surface there are no other sources or sinks of atmospheric CO_2. Therefore, the atmospheric CO_2 concentration at any one site is the sum of the contribution for each individual sources (it is a called a linear problem). Mathematically this means at station j the time series of the atmospheric concentration is

$$c_j(t) = \sum_{i=1}^{Nsources} r_{j,i}(t) x_i \text{ where } x \text{ is the unknown source to be estimated} \quad (2)$$

For all sites we can write the mapping of the sources to a concentration at a site as

$$\boldsymbol{c(t)} = \boldsymbol{R(t)x} \quad (3)$$

where

$$\mathbf{c}(t) = \{c_1(t), c_2(t), \ldots, c_{65}(t)\}^T$$

$$\mathbf{x} = \{x_1, x_2, \ldots, x_{6240}\}^T$$

$$\mathbf{R} = \begin{bmatrix} r_{1,1}(t) & \ldots & r_{1,6240}(t) \\ . & . & . \\ r_{65,1}(t) & \ldots & r_{65,6240}(t) \end{bmatrix}$$

By discretizing the problem using a constant time sampling of one month **c**(t) and **R**(t) can be expanded into vector **b** and matrix **A** respectively and the problem can be written as a matrix equation

$$\mathbf{b} = \mathbf{A}\mathbf{x} \tag{4}$$

To solve the problem, we minimizing the following cost function

$$\phi = (\mathbf{b} - \mathbf{A}\mathbf{x})^T \mathbf{Q} (\mathbf{b} - \mathbf{A}\mathbf{x}) + (\mathbf{x} - \mathbf{x}^b)^T \mathbf{W}^{-1} (\mathbf{x} - \mathbf{x}^b) \tag{5}$$

The first term includes the data constraints of trying to reproduce the observed atmospheric data and the second term includes constraints on finding a solution close to an initial guess ($\mathbf{x}^b$). The later term helps improve the behaviour of the inversion and ensure the result solution for **x** is reasonable (see Appendix A for details). The T99 ocean flux estimates in Takahashi et al. [2002] are used as an initial ocean CO_2 uptake estimate (Fig. 2). The weighting matrices **W** and **Q** incorporate the uncertainties in the both the initial flux estimates and observational data, respectively, into the inversion [*Roy et al.*, 2003].

Here, **Q** is a diagonal matrix with diagonal elements of the matrix set to inverse of the estimated variance of that particular observation. Hence, the more accurate a particular observation is the greater the relative importance that observation is given in the inversion. Estimated variance of observation reflects both instrumental error in the observation and high frequency variability, and potential biases in the observation. The monthly mean variance in the observations was estimated from the observed time-series of atmospheric CO_2. The value differed between sites with square root of the variance ranging between 0.3 and 5.2 ppm. By using a diagonal matrix one assumes there is no spatial or temporal correlation in the variance within or among stations. From an instrumental standpoint this is true since one would not expect instrumental errors at one site to correlate with other observing sites. However, nearby stations may have correlated observational biases. It is difficult to determine these correlated observational biases and they are ignored.

W is also diagonal matrix with the diagonal elements of the matrix set to the inverse of the prescribed variance in the source region fluxes. The prescribed variance of the sources regions fluxes attempts to include estimated uncertainties in the CO_2 fluxes from a region. For the results presented here the square root of the variance for land regions was 1.2 Gt C/yr and for ocean regions was 0.5 to 1.0 Gt C/yr depending on the size of the region. The higher uncertainty in the land regions reflects the greater uncertainty in the land sources than the ocean sources. The **W** matrix serves two purposes: 1) to incorporate the relative uncertainty of the different sources regions; 2) to provide an additional constraint to the inverse problem to ensure the solution is realistic. There is a fair amount of subjectivity in selecting these bounds, which does have significant consequences on the final results. However, the **W** does provide an ideal way to probe the sensitivity of the inverse problem to various scenarios. In the problem presented here, **W** was modified to force all ocean sources to be consistent with T99. This allows one to investigate how the land sources adjust prescribed to T99 ocean fluxes and examine how the simulated CO_2 concentrations at the observation sites compare to the observations.

With the incorporation of the data and parameter weighting matrices, **Q** and **W**, respectively, the solution that minimizes our cost function is

$$\mathbf{x} = \mathbf{x}^{b} + [\mathbf{A}^{T}\mathbf{Q}\mathbf{A} + \mathbf{W}^{-1}]^{-1}\,\mathbf{A}^{T}\mathbf{Q}(\mathbf{b} - \mathbf{A}\mathbf{x}^{b}) \quad (6)$$

The equation is derived by finding the solution to **x** that makes $\partial\phi/\partial\mathbf{x} = 0$. The rules given in Appendix B can use used to derived equation 6.

2.3 RESULTS AND DISCUSSION

Using the inverse model *Roy et al.*, 2003] calculated monthly surface flux from the oceans and terrestrial biosphere. For the 1990 to 1997 period, the inversion matches the inter-hemispheric CO_2 gradient by increasing the total annual mean northern hemisphere uptake (land and ocean) from 0.9 to 2.4 GtC yr^{-1}, primarily by increasing the terrestrial biosphere uptake. The inversion also reduces the total southern hemispheric uptake from 1.8 Gt C yr^{1} to 1.1 GtC yr^{-1}. The magnitude of the extratropical southern hemisphere (20°S-90°S)ocean and land uptake predicted by the inversion during the 1990's (1991-97) equalled the T99 ocean flux estimate of 1.8 GtC yr^{-1}, however, the spatial pattern is inconsistent with T99 fluxes. The inversion redistributes the uptake between the mid and high latitudes (Fig. 4) by reducing the high-latitude Southern Ocean (50°S - 90°S) uptake by 0.4 Gt C/yr and increasing mid-latitude uptake by the same amount.

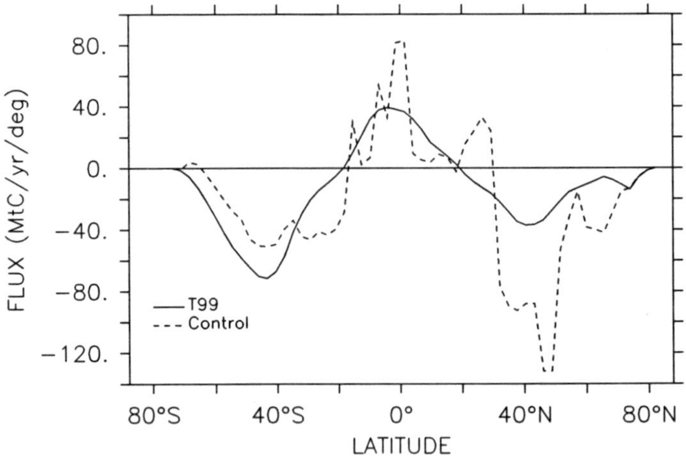

Figure 4. The zonal annual mean (1991-1997) CO_2 uptake (MtC/yr/deg) for the T99 ocean flux estimate in Takahashi et al. (2002) used as the initial ocean flux estimate in the inversion, and the predicted combined land and ocean (total) uptake of the inversion Note: A negative flux denotes uptake of atmospheric CO_2 by the either the land or ocean.

The atmospheric CO_2 data appears inconsistent with the southern hemisphere ocean fluxes from T99. Resolving this discrepancy is an important issue. There are several possibilities that may reconcile the differences between the atmospheric and ocean observations.

1. The under-sampling of the Southern Ocean during winter may explain the overestimate the southern ocean CO_2 uptake of the T99 ocean fluxes. This possibility is supported by new observations that show the Southern Ocean is a CO_2 source to the atmosphere during the winter [N. Metzl and B. Tilbrook personal communication].
2. Additional southern hemisphere atmospheric stations with improved measurement frequency are needed to improve the data coverage and reduce the uncertainty of the atmospheric inversion.
3. The inclusion of only one Southern Ocean region in the atmospheric inversion needs to be modified by dividing the Southern Ocean region into several oceanographically sensible regions.
4. The atmospheric inversion should be combine with the ocean inversion approach of *Gruber et al.*, [2001] to better exploit ocean observations that can be used to constrain the oceanic uptake of CO_2.

The list of potential improvements highlight two key limitations of the data assimilation, the quality of the data and suitability of the model used to assimilate the data. The first two items in the list suggest that improved data coverage may resolve the present inconsistency between atmospheric and ocean observations. While the third item suggests that a modification of the atmospheric inversion model used may resolve the situation. Finally, the fourth item suggests that completely different model should be used to assimilate ocean observations. A model that uses estimates of the accumulation of anthropogenic CO_2 in the ocean rather than data derived estimates of the air-sea CO_2 fluxes to constrain the ocean fluxes. All four of these items need to be pursued to improve our estimates of the Southern Ocean CO_2 uptake.

3. EXAMPLE TWO: ADJOINT DATA ASSIMILATION OF PHOSPHATE DATA TO INVESTIGATE NUTRIENT CYCLING IN THE EASTERN EQUATORIAL PACIFIC

An important element to modelling the oceanic carbon and nutrient cycles is the export of organic matter from the euphotic zone and the subsequent remineralization of this material with depth. The downward transport of organic carbon out of the euphotic zone by export production lowers the concentration of dissolved inorganic carbon (DIC) and nutrients in the euphotic zone and enables the ocean to reduce the atmospheric CO_2 concentration by sequestering carbon into the deep ocean. Understanding the mechanism of this transport is important to predicting the effect of a change in climate and the resulting change in ocean circulation upon the atmospheric CO_2 level.

In the ocean, organic matter can be divided into three classes: large particulate organic matter (POM), suspended POM and dissolved organic matter (DOM). The separation among the three classes of organic matter is somewhat arbitrary but important because they characterize distinctively different processes for the transport of organic matter in the ocean.

The large POM is generally classified as organic matter with diameter greater than 50 μm and it is primarily sampled by sediment traps. POM has a sinking rate of the order of 100 m d^{-1} [*Fowler and Knauer*, 1986]. At this sinking rate, POM is nearly unaffected by ocean currents and it sinks out of the euphotic zone to be degraded directly beneath where it was produced. This process provides a direct connection between productivity in the euphotic layer and the remineralization of organic matter below.

Suspended POM is measured by filtering the sea water [Bishop *et al.*, 1987] and its typical size is between 1 - 50 μm in diameter. DOM is operationally defined as all organic matter that will pass through a fine mesh filter of size 0.1 to 1 μm. In contrast to POM transport, the suspended POM and DOM are transported by the ocean currents enabling highly productive ocean regions to export organic matter horizontally to less productive regions.

Numerical models of oceanic nutrient cycling with only a POM component to transport organic matter can reproduce the observed surface phosphate concentrations but tend to substantially overestimate the phosphate concentration in the thermocline the eastern equatorial Pacific [*Bacastow and Maier-Reimer*, 1991; *Najjar et al.*, 1992]. *Najjar et al.*, [1992] coined the term "nutrient trapping" to describe this feature in their simulations. By adding a DOM pool to their model *Najjar et al.*, [1992] removed nutrient trapping and reproduced the observed phosphate concentrations in the thermocline of the eastern Equatorial Pacific.

By using the mathematical formalism based on the adjoint method [*Le Dimet and Talagrand*, 1986; *Morse and Feshbach*, 1953; *Thacker*, 1988; *Tziperman and Thacker*, 1989b], *Matear and Holloway*, [1995] demonstrated how data assimilation can be used to investigate nutrient cycling in the equatorial Pacific. Their study developed a model of the steady-state phosphate cycle in the North Pacific using historical phosphate measurements and an initial guess of the ocean circulation based on the Large-Scale Geostrophic (LSG) ocean general circulation model (OCGM) [*Bacastow and Maier-Reimer*, 1991]. Forward simulations with this model produced the nutrient trapping feature in the eastern Equatorial Pacific. The phosphate measurements were assimilated into a model using a conservation equation. By formulating the problem as an inverse problem, one can explore ways to improve the model-data agreement of phosphate without including dissolved organic phosphorus (DOP).

3.1 METHOD

The three components to data assimilation – 1) forward model; 2) cost function; and 3) inverse solver are presented in this section.

Forward Model

The evolution of the phosphate field was modeled with the conservation equation for phosphate

$$\frac{\partial P}{\partial t} + \nabla_3 \cdot (\mathbf{u} P) = \frac{\partial}{\partial z} \cdot (K_z \frac{\partial P}{\partial z}) + Q \tag{7}$$

where P was the phosphate concentration (μmol kg^{-1}), Q represents the biological sources minus sinks of phosphate, that is the conversion of P into particulate organic phosphorus (POP) and subsequent remineralization of POP back into P, K_v is coefficients of vertical mixing (m/s^2), and $\mathbf{u}$ is the three-dimensional velocity field (m/s). For our study, the circulation information comes from a 2-year average of the LSG simulation of

Bacastow and Maier-Reimer, 1991]. The circulation information includes the 3-dimensional velocity field, **u**, and monthly fields of convective mixing.

In the equation 7, K_z represents vertical mixing, which is parameterized by the occurrence of convective mixing in the 2-year simulation. Convective mixing occurs when one layer of the model is denser than the underlying layer. When this occurs the two layers are homogenized. In our K_v parameterization, the simple homogenization of two layers was applied from the top of the ocean downward whenever a model layer was unstable with respect to the underlying layer for more than half the year (six or more convective events occur in a year with a time step of one month). With the exception of convection, no other explicit mixing terms were used. However, the upwind advection scheme used in the finite differencing of equation 7 is numerically diffusive with the average numerical horizontal and vertical diffusivity coefficients for the North Pacific generally exceeding 2 x 10^6 cm^2/s and 0.5 cm^2/s, respectively.

Prescribing the horizontal velocities, the vertical velocities were determined by employing the conservation of volume.

$$\nabla_3 \mathbf{u} = 0 \tag{8}$$

The vertical velocities, w, were calculated column wise starting at the ocean bottom with no vertical flow through the ocean bottom. The vertical velocities through the air-sea interface represent the net freshwater fluxes due to evaporation minus precipitation.

To simulate the net nutrient uptake in the euphotic zone (z_e), export production (EP) in mol kg^{-1} yr^{-1} was defined as:

$$EP = LC \frac{R \cdot P}{0.4 \mu mol\ kg^{-1} + P} \tag{9}$$

$$Q = -EP \qquad \text{for } z < z_e . \tag{10}$$

where LC was the non-dimensional light intensity factor (as given by Bacastow and Maier-Reimer, 1990), R was an unknown scaling factor for export production in μmol kg-1 yr-1, and z_e was the depth of the euphotic layer (set to 100 m). To parameterize the remineralization of the sinking particles below the euphotic layer we use:

$$Q = \frac{\partial}{\partial z}[EP \cdot (\frac{z}{z_e})^{-C}] \cdot z_e \quad \text{for } z > z_e \tag{11}$$

where C defined the unknown remineralization length scale

The velocity field (**u**), the initial three-dimensional phosphate distribution (P^0), and the biogeochemical parameters R and C form a set of independent parameters. With these parameters the evolution of phosphate in the ocean can be simulated. These parameters were given initial values and they were modified in a way to minimize the following cost function.

Cost Function

The cost function, J, is used to characterize the desired features of model solution. The following is a brief description of the five terms used in the cost function that are summarized in Table 1.

Table 1. Summary of Terms used in the Cost Function

Term	Definition	Mathematical Form
1	penalize deviations from steady state	$\frac{1}{N}\sum_{i=1}^{N}(P_i^0 - P_i^2)^2 \frac{dv_i}{V}$
2	deviations from observed phosphate data	$\frac{1}{N_b}\sum_{i=1}^{N_b}(P_i^0 - P_i^*)^2 \frac{1}{\sigma_i^2}$
3	deviations from the model freshwater fluxes	$\frac{1}{N_w}\sum_{i=1}^{N_w}(w_i - w_i^*)^2$
4	deviation from the general circulation model's horizontal velocity field	$\frac{1}{N_u}\sum_{i=1}^{N_u}(u_i - u_i^*)^2 + \frac{1}{N_v}\sum_{i=1}^{N_v}(v_i - v_i^*)^2$
5	deviation from geostrophy	$\frac{1}{N_u}\sum_{i=1}^{N_u}(u_{zi} - u_{zi}^*)^2 + \frac{1}{N_v}\sum_{i=1}^{N_v}(v_{zi} - v_{zi}^*)^2$

N is the number of grid points where phosphate is conserved P^0 is the initial model phosphate concentration,

P^2 is the phosphate concentration after the forward integration of the model

dv_i is the volume represented by the ith grid point

V is the total volume of all the ocean grid points where phosphate is conserved

N_b is the number of phosphate measurements used in the inversion (it includes the phosphate measurements for all the grid points (N) plus the phosphate concentrations along the two open boundaries, South Pacific and Berring Strait)

P^* is the observed phosphate concentrations

σ is the error covariance term used to weight the agreement between the model and observations. The error covariances used in this penalty term were diagonal and only depth-dependent. They were derived from the observations at Station P in the Alaska Gyre (50°N and 145°W).

N_w is the number of surface vertical velocities used in the inversion

w is the vertical velocity of the inverse solution

w^* is the original vertical velocities from the LSG ocean circulation model

N_u and N_v are the number of u and v components used in the inversion

u and v are the horizontal velocities of the inverse solution

u^* and v^* are the horizontal velocities from the LSG ocean circulation model

u_z and v_z are the vertical shear of the horizontal velocities of the inverse solution

u^*_z and v^*_z are vertical shears of the horizontal velocities of LSG ocean circulation model

1. The simulated phosphate concentrations were forced to be close to steady-state.
2. The simulated phosphate concentrations were required to be close to the observations.
3. For inversions where the horizontal velocity field was modified, the fresh water fluxes across the ocean surface, parameterized in the model as vertical velocities at the ocean surface, were maintained to be nearly consistent with values used in the original LSG model.
4. The data assimilating model searched for a solution near the original velocity field specified by the OGCM.
5. To maintain dynamically consistency with the original OGCM, modifications to the original velocity field were constrained to be geostrophic.

The third, fourth and fifth terms in the cost function were included to force the inverse solution velocity fields to resemble the original fields. Inverse Solver

Inverse Solution

Adjoint model: To efficiently minimize a cost function with a large number of unknowns we use the adjoint of the conservation of phosphate and volume equations (equations 7 and 8) to compute the gradient of the cost function with respect to the independent model parameters [*Matear and Holloway*, 1995]. The conservation equations for phosphate and volume applied to every grid point in our model can be written in the form

$$F_i = 0, \quad \text{for i=1, to 2*N}$$

where N was the number of grid points used in the model. The Lagrangian equation L is given by

$$L = J + \sum_{i=1}^{N} \lambda_i F_i \tag{12}$$

where λ_i were the unknown Lagrange multipliers and J was the cost function defined in in the previous section. L=L(**x**, λ) was a function of the model parameters $\boldsymbol{x}$ (w,P^2 ,P^0,R, C, u, v), and the Lagrangian multipliers $\boldsymbol{\lambda}$. The minimum of J is a stationary point of L, and all partial derivatives of L with respect to model parameters, $\boldsymbol{x}_i$, and Lagrange multipliers, λ_i, vanish:

$$\frac{\partial L}{\partial \lambda_i} = F_i = 0 \tag{13}$$

$$\frac{\partial L}{\partial x_i} = \frac{\partial J_i}{\partial x_i} + \sum_{j=1}^{N} \lambda_j \frac{\partial F_j}{\partial x_i} = 0 \tag{14}$$

Differentiation with respect to the Lagrange multipliers recovered the model equations, and differentiation with respect to the state variables, **x**, lead to a set of equations defining the adjoint model and the gradient of the cost function. The Lagrange multipliers λ were calculated from (20) using the dependent variables, w and P^2 (see *Matear and Holloway*, 1995 for details). Introducing the calculated λ into (20) for the independent variables (P^0, R, C, u, v) then yields the gradient of J with respect to the model parameters. Appendix A discusses how this information can be used to minimize the cost function.

Solving the Inverse Problem

The model calculations proceeded according to the following steps:

1) Initialize the model by specifying the values of the independent variables (P^0, R, C, u, v).

2) Determine the vertical velocities using the conservation of mass equation.

3) Calculate the value of the cost function J.

4) Calculate the Lagrange multipliers λ by solving the adjoint equations and determine the gradient of J with respect to the model parameters (*Matear and Holloway*, 1995 for details).

5) Use the gradient of J in a descent algorithm to obtain a new improved set of model parameters. The gradient of J was passed to the descent algorithm (truncated Newton conjugate-gradient algorithm, TNPACK, Schlick and Fogelson, (1992)) to obtain a new, improved set of model parameters with a smaller value of the cost function J.

6) Repeat steps 2 through 5 until a minimum of J is found.

3.2 RESULTS AND DISCUSSION

In the first scenario, the velocity field was fixed (cost function only used terms 1 and 2) and optimal solutions were found for export production scaling factor (R), the length scale of remineralization of organic matter (C) and the steady-state phosphate field. The optimal solutions from this scenario did not produce realistic values for the export production particularly in the equatorial region (Fig. 5) and the length scale of remineralization of organic matter was too deep (C=0.62). The results are consistent with *Najjar et al.*, 1992] and *Bacastow and Maier-Reimer*, [1991] conclusion that a model with only POM transport of organic matter is unable to reproduce the observations without a significant pool of DOM. This scenario shows how the choice of the model influences the results of the data assimilation. By fixing the ocean circulation one creates a model that cannot produce both realistic phosphate and export production fields.

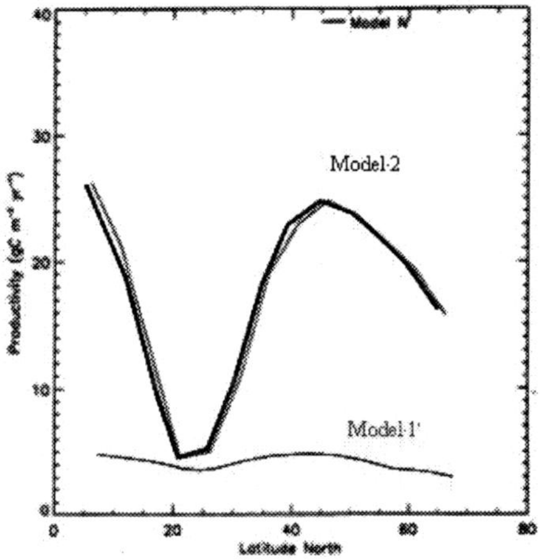

Figure 5. Model estimate of the zonal averaged export production in the North Pacific with prescribed circulation (model 1) and with a model where the circulation was determined from data assimilation (model 2).

Within the data assimilation framework one can probe the sensitivity of the results to the applied model by using different models. In the second scenario, the velocity field was allowed to change (all terms in the cost function used) and optimal solution was found for the phosphate and the velocity fields. The optimal solution from this scenario did produce realistic values for export production (Fig.5) and the phosphate field without significantly modifying the circulation fields used in the first scenario

(Fig.6). The solution shows that a model with only POM transport of organic phosphate can reproduce the observations of export production and phosphate concentration in the eastern equatorial Pacific through a slight modification of the circulation. Models with improved ocean dynamics in the equatorial Pacific region do not generate nutrient trapping and produce circulation features in the equatorial Pacific [*Aumont et al.*, 1999] consistent with the results from this inversion study.

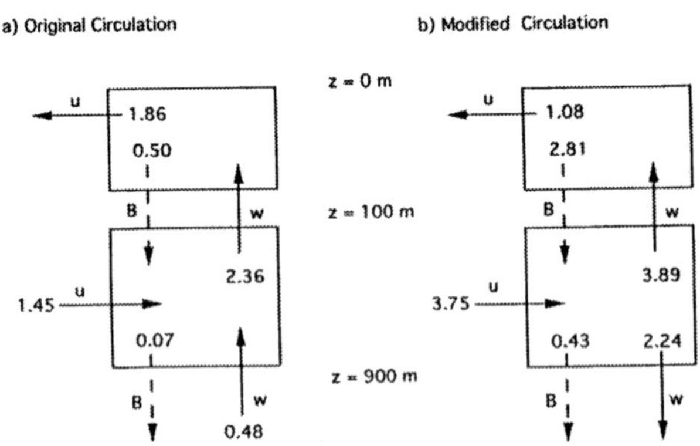

Figure 6. Phosphate budget in the eastern equatorial Pacific for the solution with fixed circulation and for the solution with variable circulation.

From the results of inversions with a variable circulation field, two points must be stressed. First, different velocity fields can produce the same tracer distribution. To reduce the non-uniqueness, additional independent tracers would be needed. Second, the phosphate field was very sensitive to the prescribed circulation field; that is, small changes in the circulation produced large changes in the phosphate field. The sensitivity of the tracer fields to the velocity fields suggests that the tracer fields possess significant information on the fine structure of the flow. However, to take advantage of this information one must provide additional information on the large-scale features of the flow. Mathematically, several different tracer fields should be sufficient to determine the large-scale information [*Fiadeiro and Veronis*, 1984]; unfortunately most observable tracer fields are not independent and thus on their on they are unable to constrain the large scale flow field. Constraints such as geostrophy provide a natural way to introduce information on the large scale flow field and rely on the tracer fields to determine the finer structure of the flow. *Schlitzer*, (2002) has expanded on the approach presented here to include additional tracers (Temperature, Salinity, Oxygen, Phosphate, Silicate, Dissolved Inorganic Carbon and Alkalinity). Assimilating data into the conservation equations for these tracers he obtains a steady state circulation field; export production field; and parameters for the remineralization organic carbon, inorganic carbon, and opal that are consistent with the observed tracer fields. He shows that the export production predicted by the inverse model in the Southern Ocean is much greater than estimates derived from ocean color satellite data. Schlitzer attributes this discrepancy to the underestimate of export production from the ocean color data. However, additional effort is needed to resolve this issue.

4. EXAMPLE THREE: PARAMETER ESTIMATION OF AN ECOSYSTEM MODEL

Marine biota play an important role in nutrient and carbon cycling in the ocean. Ecosystem models have been developed to describe the dynamics of the lowest trophic levels of the ecosystem and simulate carbon and nutrient cycling in the ocean. Generally, ecosystem models prescribe a set of biotic (phytoplankton, zooplankton) and abiotic (eg. Nitrate) compartments and a set of equations that describe the flow of material between compartments. To parameterize these flows, the ecosystem model requires a large number of model parameters to be specified. The assimilation of data into ecosystem models can be used to calibrate these parameters, to determine the simplest model that is consistent with the available observations and to provide insight into what observations are critical to building and constraining more realistic models.

4.1 METHOD

Ecosystem Model

The ecosystem model presented here simulates the nitrate (N), phytoplankton (P) and zooplankton (Z) concentrations in the mixed layer. The model is very simple but it describes how phytoplankton and zooplankton influence the nutrient cycle in the upper ocean. For this application, horizontal advection and diffusion were ignored and the pelagic ecosystem was assumed to consist of a homogeneous mixed layer overlying a deeper abiotic layer. The phytoplankton and zooplankton are confined and homogeneously distributed in the upper layer. The mixed layer dynamics were not explicitly modeled, rather observed mixed layer depths were used to describe the temporal evolution of the mixed layer depth (H)

$$\frac{dH}{dt} = \zeta(t) \quad (15)$$

Following Evans and Parslow (1985), only the zooplankton was assumed to be able to maintain its position in the mixed layer and therefore was concentrated as the mixed layer shallowed. A shallowing of the mixed layer ($\zeta < 0$) introduced no new water into the mixed layer and the concentrations of nitrate and phytoplankton components remained constant, while the concentration of zooplankton increased as all the zooplankton remain in the mixed layer. The deepening of the mixed layer ($\zeta > 0$) generated mixing with the water underlying the mixed layer to dilute the concentrations of nitrate, phytoplankton and zooplankton in the mixed layer. To capture this asymmetry in the model, one defined $\zeta^{+}(t) = \max(\zeta(t), 0)$.

For the model simulation, all three components were described in units of μmol/kg Nitrogen. The equations describing the temporal evolution of nitrate, phytoplankton and zooplankton concentrations were

$$\begin{aligned}
\frac{dN}{dt} &= -[\sigma(t,H)Q_1 - \mu_1]P + \frac{m+\zeta^{+}(t)}{H}(N_0 - N) + \\
&(1-\gamma_2)\gamma_5 \frac{g(P-P_0)Z}{K_3 + P - P_0} + (1-\gamma_4)\mu_5 Z \\
\frac{dP}{dt} &= [\sigma(t,H)Q_1 - \mu_1]P - \frac{g(P-P_0)Z}{K_3 + P - P_0} - \frac{m+\zeta^{+}(t)}{H}P \\
\frac{dZ}{dt} &= \frac{\gamma_2 g(P-P_0)Z}{K_3 + P - P_0} - \mu_5 Z - \frac{\zeta(t)}{H}Z
\end{aligned} \quad (16)$$

where N_o was the nitrate concentration one metre below the depth of the mixed layer. The ecosystem model was based on the Evans and Parslow (1985) model with the modification that a fraction of the grazed phytoplankton and of the zooplankton mortality were returned directly to the nitrate component. For the phytoplankton growth rate, the non-dimensionalized nutrient limitation term was

$$Q_1 = \frac{N}{K_1 + N} \quad (17)$$

The light-limited growth rate, in units of d^{-1}, averaged over the depth of the mixed layer was obtained by integrating over one day /night cycle

$$\sigma(t,H) = 2\frac{1}{H}\int_0^{\tau}\int_0^{H} F(I)dzdt \tag{18}$$

where *F(I)* was a function describing the phytoplankton photosynthesis-irradiance relationship (the *P-I* curve), 2τ was the day-length (calculated from Brock, 1981) and *I* was the photosynthetically active radiation (*PAR*) as a function of depth below the surface of the water. The function F(I) was defined by the Smith function (Smith 1936)

$$F(I) = \frac{Vp(T)\alpha I}{(Vp(T) + \alpha^2 I^2)^{1/2}} \tag{19}$$

where V_P was the growth rate as $I \rightarrow \infty$ and α was the initial slope of the *P-I* curve. The maximum temperature-dependent growth was defined following Eppley (1972)

$$V_p(T) = 0.6(1.066)^T \tag{20}$$

and light intensity at any depth was given by

$$I(z,t) = PAR\ I(0,t)\ exp(-k_w z - \int_0^z -k_c P dz) \tag{21}$$

where *I(0,t)* was the incident radiation observed immediately below the surface of the water, PAR was the fraction of that radiation that can be used by the phytoplankton for photosynthesis, *z* was depth in meters; *P* was the concentration of phytoplankton in the mixed layer, and k_w and k_c were light attenuation constants due to water and phytoplankton respectively (Evans and Parslow, 1985). The variation of *I(0,t)* with time was assumed to be triangular which allowed one to analytically integrate equation (18) (Evans and Parslow, 1985).

As described above, the ecosystem model required 14 model parameters to be specified for the model simulation, which are summarized in Table 2.

Model Integration

The full set of differential equations describing the ecosystem models was solved using a fourth-order Runge-Kutta algorithm. Testing of the algorithm showed that a one day time step was sufficient to produce an accurate simulation of the evolution of the three components. The simulations were run for three years and a day, with the third year of data being used by the optimization routine. Typically by the third year, the model achieved a steady state annual cycle, which was not sensitive to the initial conditions. At steady-state, the concentrations of the modeled components on the first day of the third year were identical to the concentrations on the first day of the fourth year.

Table 2. Model parameters required by the 3-compartment ecosystem model along with the initial guess of the parameter value and uncertainty.

	Symbol	A priori value	Units
Phytoplankton (P) parameters			
Specific mortality rate	μ_1	0.024 ± 0.05	d^{-1}
Half Saturation constant for nutrient uptake	K_1, K_2	1.0 ± 1.0	μmol/kg
Light attenuation by phytoplankton	K_c	0.06	m^{-1} (μmol/kg)$^{-1}$
Light attenuation by water	K_w	0.04	m^{-1}
Initial Slope of P-I curve	α	0.025 ± 0.01	$(W\ m^{-2}\ d)^{-1}$
Photosynthetically active radiation	PAR	0.5	
Zooplankton (Z) and Mesozooplankton (M) parameters			
Assimilation efficiency of Z	γ_2	0.5 ± 0.1	
Detrial fraction of Z mortality	γ_4	0.33 ± 0.1	
Fraction of Z grazing excreted as nitrogen metabolites	γ_5	0.6 ± 0.05	
Specific mortality rate of Z	μ_5	0.07 ± 0.02	d^{-1}
Maximum growth rate of Z	g	1.0 ± 0.2	d^{-1}
Half Saturation for ingestion of Z	K_3	0.5 ± 0.3	μmol/kg
Grazing threshold of Z	P_o	0.05 ± 0.05	μmol/kg
Physical Mixing parameters			
Diffusion Rate	m	3.0 ± 0.5	m d^{-1}
Total Number of model parameters			14

The ecosystem model was applied at Station P, a North Pacific time-series site (50°N and 145°W). A large amount of data on biological, chemical and physical oceanography is available from Station P. For this study, the data sets are separated into data used to force the model and data used to constrain the model parameters.

Forcing Data

The forcing terms of the ecosystem model were the annual cycles of solar radiation, mixed layer depth, mixed layer temperature and nitrate concentration below the mixed layer.

Multi-year monthly observations of total solar radiation at Station P reported by Dobson and Smith (1985) were averaged and interpolated to give the daily averaged values of incident solar radiation (Figure 7a). The daily mean solar radiation measurements were converted to peak values for noon by assuming the daily variation in light intensity was triangular. The photosynthetically active radiation (PAR) just below the surface at noon was assumed to be a constant fraction of the incident solar radiation. In reality, the ratio of PAR to total radiation is affected by sun zenith angle, water vapor content and aerosol optical thickness (Baker and Fouin, 1987). However, such complications were not considered necessary for the scope of the present study and a constant value was used.

Seasonal variations in the mixed layer depth (Fig. 7b) were based on the average mixed layer depth observed at Station P between 1970 and 1980. The daily changes in the mixed layer depth, calculated from finite centre differencing of Figure 7b, determined the amount of entrainment or detrainment occurring at the base of the mixed layer (Figure 7c). The high frequency variability observed in Figure 7c reflects the importance of episodic events on the temporal evolution of the mixed layer (Large et al., 1986), which was still evident even after averaging eleven years of data. The averaged mixed layer temperatures for this period of time were determined by integrating the observed temperature measurements over the depth of the mixed layer (Figure 7d).

Nitrate concentrations collected from Station P for the period of 1970 to 1981 provide a good sampling of the nitrate concentration in the upper 200 m of the ocean. From this data, the seasonal cycle of nitrate concentration in the upper ocean was determined.

Constraining the Parameter Optimization

The nitrate concentration in the mixed layer at station P was relatively high with the seasonal cycle varying between 9 μmol/kg and 16 μmol/kg. An estimate of the errors associated with the data was determined from the standard deviation of the observed surface nitrate concentrations (1970 to 1981) from the seasonal mean (Figure 8a).

A series of studies on the seasonal variation of the planktonic food web [*Anderson et al.*, 1977; *McAllister*, 1961; *Stephens*, 1964; *Stephens*, 1966; *Stephens*, 1968; *Stephens*, 1970] show Station P to lack a spring bloom of phytoplankton. At Station P the typical concentration of phytoplankton for all seasons varies between 0.3 and 0.4 mg Chl a m^{-3}. In contrast, in the subarctic North Atlantic, even the most remote oceanic sites have phytoplankton blooms every year of greater than 1.0 mg Chl a m^{-3} (Parson and Lalli, 1988) when the major nutrient levels decrease to zero. Although there is no spring bloom at Station P, there is a spring peak in primary production followed by a summer peak in zooplankton biomass (Parslow, 1981). In general at Station P, the phytoplankton are uniformly distributed in the mixed layer and there is no subsurface maximum (Anderson et al 1977). For the model, the phytoplankton concentration was set to 0.15 ± 0.05μmol/kg N which is determined by using a mg C / mg *Chl a* value of 50 and N/C value of 16/106. The error reflected the uncertainty in the *Chl a* concentrations and the uncertainty in converting the phytoplankton Chl a concentration to a carbon and then nitrogen concentration.

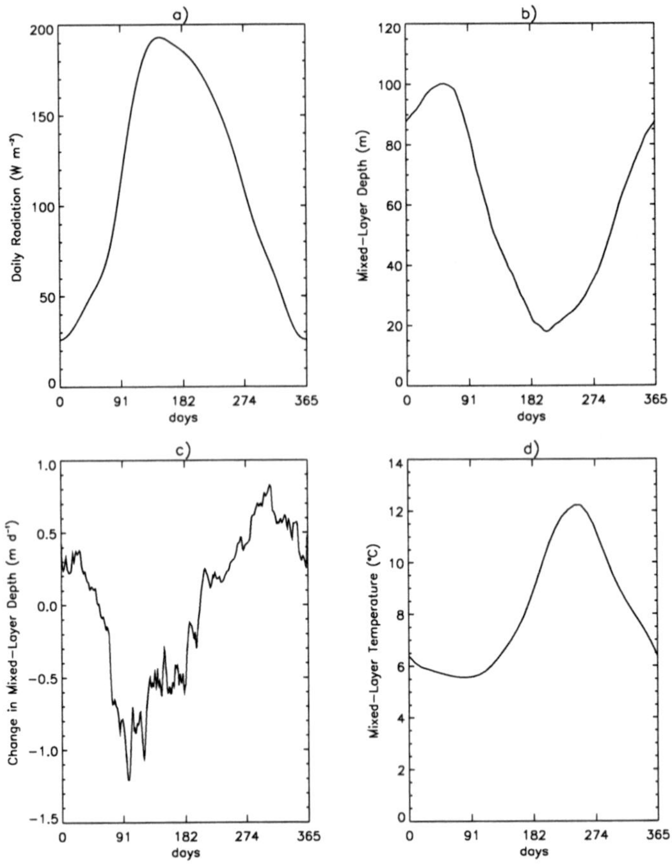

Figure 7. Observations from Station P for a) daily incident solar radiation, b) depth of the mixed layer (m), c) the rate of the change of the mixed layer depth (m d^{-1}), and d) temperature of the mixed layer (°C) [Matear, 1995].

The net phytoplankton productivity (NPP) was defined as the difference between the photosynthetic rate of phytoplankton minus the respiration rate of phytoplankton. Estimates of NPP are available at Station P over a period of several years (McAllister, 1969; and Parslow, 1981). To constrain the model, the estimated NPP of Parslow (1981) was used, with a correction, to account for the production occurring below the mixed layer in the summer (Fig. 8c). The differences between McAllister and Parslow estimates of NPP were used to assign the errors in the net phytoplankton productivity (NPP) measurements (Fig. 8c).

Cost Function

The cost function was used to describe the desired features of the ecosystem model. For our optimization problem, the general cost function (J) was given by the following equation:

$$J = \frac{1}{2}\sum_{i=1}^{n}(N_i - N_i^*)^2 \frac{1}{\delta N_i^2} + \frac{1}{2}\sum_{i=1}^{n}(P_i - P_i^*)^2 \frac{1}{\delta P_i^2} +$$
$$\frac{1}{2}\sum_{i=1}^{n}(NPP_i - NPP_i^*)^2 \frac{1}{\delta NPP_i^2} + \qquad (22)$$
$$\frac{1}{2}\sum_{j=1}^{nc}(C_1^j - C_{366}^j \frac{1}{\delta T^4} + \frac{1}{2}\sum_{i=1}^{m}(x_i - x_i^*)^2 \frac{1}{\delta x_i^2}$$

where N^*, P^*, and NPP^* referred to the observations with prescribed uncertainties δN, δ*P*, and δ*NPP*; n referred to the number of data observations in a year (n=365); nc referred to the number of components in the model; m referred to the number of model parameters; C_j referred to the different components in the model (i.e. N, P, Z, ...); δT=0.01 μmol/kg; and x referred to the model parameters. The cost function was composed of three penalty terms - data misfit penalty (terms 1-3), steady-state penalty (term 4) and the parameter penalty (term 5). The data misfit penalty required that the model output approximately reproduced the annual cycle of nitrate (N), phytoplankton (P), and net primary productivity (NPP). The steady-state penalty forced the model to produce a steady-state output, with only a seasonal cycle, by requiring the model components at the start of the fourth year to be approximately equal to the values at the start of the third year. The a priori parameter penalty on the values of the model parameters (x^*) incorporated information on the expected value of the model parameters. By setting δx_I in this penalty term to zero, one forced the value of x_i to be equal to the a priori value.

The parameter penalty term (term 5) was included to try and force the model parameters to be within acceptable limits. The a priori values, along with their prescribed uncertainty, are given in Table 2. These values were derived from previous applications of the ecosystem models [*Evans and Parslow*, 1985; *Fasham et al.*, 1990; *Frost*, 1987; *Frost*, 1993]. In this study, the parameters *PAR*, K_w and K_c were fixed at their a priori values.

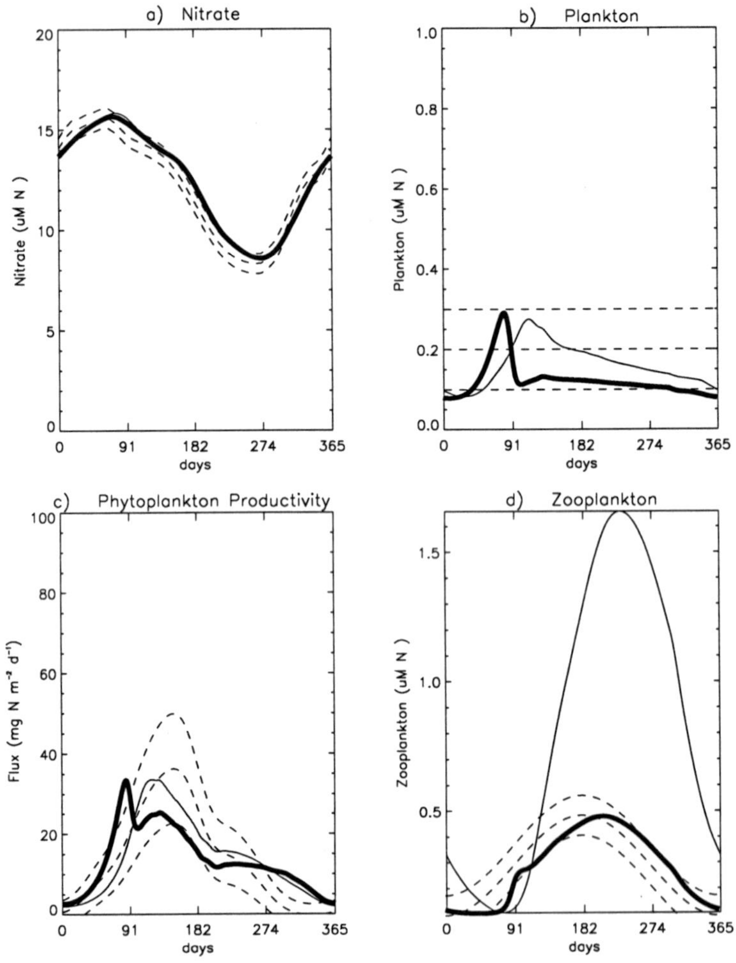

Figure 8. Model simulation from the optimal solution obtained from assimilating data into an ecosystem model without a zooplankton constraint (thin black line) and with a zooplankton constraint (thick black line). In the figures, the dashed lines denote the observed values and their estimated $\pm$ 1 standard deviation error. For nitrate simulation there is no difference between the models. In these plots day 0 denotes January 1 with the spring bloom occurring in March ($\approx$ day 90).

Optimization Algorithm

A stochastic method, simulated annealing, was used to determine the optimal model parameters. The simulated annealing technique is well suited to handle optimization problems with strong non-linearity and a small number of unknown parameters [*Krüger*, 1993]. The advantage of simulated annealing is that it is independent of the structure and analytical properties of the cost function and only requires the evaluation of the cost function. Also, the method is independent of the initial guess because the algorithm enforces a randomization of the initial guess. The major disadvantage is that the stochastic nature of the algorithm requires a large amount of computer time to reach an acceptable final state. Furthermore, one cannot guarantee convergence to a global minimum of the cost function.

The factors controlling the numerical demands of simulated annealing are the complexity of the ecosystem model, which needs to be integrated many thousands of times, and the number of parameters to optimize. Simulated annealing algorithm is well suited for our ecosystem optimizations because it is numerically efficient to integrate the ecosystem models and because the number of parameters to be optimized is small.

4.2 RESULTS AND DISCUSSION

The optimization successfully reproduces the observed nitrate, phytoplankton and NPP observations (Fig. 8, thin black line), but it produces very high concentrations of zooplankton in the mixed layer. The correlation matrix calculated from the error-covariance matrix (see Appendix A) gives the correlation coefficients between all pairs of optimal model parameters (Table 3). This table reveals that the parameters separate into 4 highly correlated parameters ($|r| > 0.8$), and 7 independent parameters. The formulated optimization problem is only able to uniquely determine 8 independent parameters. The set of highly correlated parameters, μ_1, K_1, θ and α are systematically related and therefore the solution determined for these parameters is not unique.

There is very limited data on the microzooplankton concentrations at Station P (Boyd et al., 1994), but the optimization predicted considerably greater concentrations than expected. To investigate the effect of a lower zooplankton concentration on the model, we add the following term

$$\frac{1}{2}\sum_{i=1}^{n}(Z_i - Z_i^*)^2 \frac{1}{\delta Z_i^2} \tag{23}$$

to the cost function (equation 22), where Z^* referred to the artificial zooplankton observations with prescribed uncertainties δZ. The artificial constraint on the zooplankton concentration is shown in Fig. 8d. With this added constraint, the model still produces a good fit with the data (Fig 8, thick black line). The model generates a small increase in phytoplankton concentrations from 0.1 to 0.3 μmol/kg in the spring and a slightly reduced NPP in spring as compared to the model without the zooplankton constraint (Fig. 8c).

Table 3. The correlation coefficients between the optimal parameters of model I.
Model Parameteres

	K_1	μ_1	θ	K_3	g	γ_2	μ_5	α	m	γ_5	γ_4
K_1	**1.00**	**-0.96**	**0.95**	-0.38	0.37	0.28	-0.52	**-0.97**	0.60	0.45	-0.13
μ_1	**-0.96**	**1.00**	**-0.83**	0.12	-0.38	-0.39	0.71	**0.88**	-0.54	-0.41	0.10
θ	**0.95**	**-0.83**	**1.00**	-0.63	0.32	0.14	-0.25	**-0.99**	0.58	0.44	-0.15
K_3	-0.38	0.12	-0.63	**1.00**	0.11	0.31	-0.55	0.53	-0.42	-0.26	0.04
g	0.37	-0.38	0.32	0.11	**1.00**	0.12	-0.21	-0.32	0.27	0.18	-0.03
γ_2	0.28	-0.39	0.14	0.31	0.12	**1.00**	-0.53	-0.21	-0.01	0.20	0.19
μ_5	-0.52	0.71	-0.25	-0.55	-0.21	-0.53	**1.00**	0.37	-0.11	-0.13	0.15
α	**-0.97**	**0.88**	**-0.99**	0.53	-0.32	-0.21	0.37	**1.00**	-0.49	-0.41	0.18
m	0.60	-0.54	0.58	-0.42	0.27	-0.01	-0.11	-0.49	**1.00**	0.49	0.11
γ_5	0.45	-0.41	0.44	-0.26	0.18	0.20	-0.13	-0.41	0.49	**1.00**	0.13
γ_4	-0.13	0.10	-0.15	0.04	-0.03	0.19	0.15	0.18	0.11	0.13	**1.00**

This example shows how data assimilation can be used to assess how well the data constrains the model parameters. It shows the inadequacy of the data on two fronts. One, by only relying on the nitrate, phytoplankton and NPP data, the optimization does not produce acceptable zooplankton concentrations. Two, the data could not uniquely constrain all the model parameters of the ecosystem model.

Parameter optimization of an ecosystem model provides a framework to explore both model complexity and the information content of different observation strategies. It is a direct technique for determining model parameters in a way that produces results that are consistent with the observed data. The optimization problem can be formulated in a very general manner and is easily modified to include additional data constraints as well as constraints on the model parameters provided by direct measurements. The automation of the process for determining the model parameters enables one to test a variety of ecosystem model configurations and how suitable the data is at constraining the model parameters. It also provides a framework to explore how additional datasets would impact the ability to constrain the model parameters. In doing this one could identify the key datasets necessary for calibrating an ecosystem model.

5. SUMMARY

This chapter has presented three different data assimilation examples. The examples serve to illustrate: 1) how to set-up and solve a data assimilation problem; 2) the potential of using data assimilation to tackle scientific questions relevant to marine carbon cycling; 3) the benefits and limitations of using data assimilation in these problems. In these examples, a variety of datasets are assimilated into both linear and non-linear models.

Unlike the more conventional "forward" modelling, data assimilation intimately links the model to the data. It is this link between the model and data, which is data assimilation's power and its limitation. The power of data assimilation is that it can exploit diverse datasets, and that the generated results are dependent on the datasets assimilated. The limitation of data assimilation is that the results are dependent on the quality of the data used, the consistency of the formulated model with the data, and the ability of inverting the data to obtain the underlying parameters of the formulated model.

To expand on the power and limitation of data assimilation lets look at the three examples presented in the chapter.

The data assimilation technique relies exclusively on the quality of the data to tell us something about the system we are investigating. Hence the results are only as good as the data we have available. The definition of "good" data not only reflects the accuracy of the measurements, but more importantly the ability of a particular dataset to constrain the processes one is investigating. Example 3 showed that the zooplankton concentration in the mixed layer was poorly constrained by the observations assimilated into ecosystem model. Hence, additional constraints were necessary to produce realistic zooplankton concentrations. This example also showed that the data available to assimilate into the ecosystem model was insufficient to constrain all the parameters of the ecosystem model. To uniquely determine all the parameters of the ecosystem model additional data is required. What this data needs to be is an open question but one that could easily be pursued using data assimilation by exploring the impact of artificial datasets on the solution. Example 1 suggested that one potential cause of the inconsistency between the atmospheric and the ocean flux data is lack winter flux data from the Southern Ocean. By acquiring more winter data and assimilating into the inverse model one could verify whether this could resolve the inconsistency.

Beside the quality of the data, the success of any data assimilation relates to the model used to assimilate the data. The model must represent the data that will be assimilated. Example 2 showed how the results changed as the model changed. In this example, when a prescribed ocean circulation field was used the assimilation of phosphate data could not produce realistic estimates of export production. It was only through modifying the inverse model to allow the circulation to change that both the simulated phosphate and export production fields became consistent with the observations. In example 1, a suggested cause of the inconsistency between the ocean and atmospheric data is the use of only one source region for the Southern Ocean. The division of the Southern Ocean into several regions would be an obvious study to pursue to test the sensitivity of the results to the resolution used for the Southern Ocean.

Furthermore, inherent in any data assimilation application is the difficulty in inverting the data to obtain information on the underlying model parameters. This inverting of the data tends to amplify the errors in the data with the potential to produce non-unique results. All the examples presented suffer from this problem and they attempt to reduce its impact by adding constraints to the cost function to make the problem more invertible. As shown in Appendix A (equation A17-A20) this reduces the impact that errors in the data have on the error in the optimal solution. In example 1, the initial estimates of the surface sources of CO_2 are included in the cost function to keep the solution realistic. In example 3, ecosystem model parameter constraints are included in the cost function to incorporate our present knowledge of the value of these parameters. This added term to the cost function makes the final solution more realistic. As better estimates of the model parameters become available this information can be incorporated into the data assimilation to improve the optimal solution.

Although the examples of data assimilation suffered from the limitations discussed above one is still capable of producing useful results from these studies. In example 1, it was shown how data assimilation combined different datasets (atmosphere concentration data and ocean flux measurements) to test their consistency and to improve the oceanic estimate of the CO_2 uptake. Example 2 demonstrated how data assimilation can investigate the sensitivity of the simulated phosphate field to biogeochemical parameterizations and to the ocean circulation. In example 3, the ability of time-series observations to constrain an ecosystem model was probed and the study showed the available data was insufficient to uniquely constrain the model parameters of a simple ecosystem model.

In the future, one expects an expansion of the range of observations collected to monitor the marine carbon cycle and an increase in the amount of data collected. With increased amounts of data, new methods will be required to interpret them. Data assimilation should play an important role in this task by providing a powerful method for combining and exploiting data. In conjunction with increased data, there will be the need to increase the complexity of marine carbon models to make them more realistic and consistent with the new observations. We will continually need to assess the data required to understand these new processes and improve their representation in models. Data assimilation studies have the potential to make valuable contributions to this effort by identifying the data needed to calibrate and build more realistic models, and by determining the model complexity that is appropriate given the available data.

By providing a bridge between modelling and field observations, data assimilation has the potential to influence both of these fields, and advance ocean carbon cycle research.

APPENDIX A: SOLVING THE INVERSE PROBLEM

To provide some background theory on data assimilation, this appendix briefly reviews the formulation and solution of both linear and non-linear problems. The process is known as an inversion since it reverses the natural direction of causality. In the following presentation I will use matrix notation to represent the problem. With this notation a **bold capital letter** will represent a matrix and **bold lower case letter** a vector. A "T" after a matrix or a vector denotes the transpose and a "-1" after a matrix denotes the inverse of a matrix. Two useful references on inverse theory are *Lawson and Hanson*, 1974 and *Menke*, 1984.

A.1 LINEAR PROBLEM

General Approach: The simplest application of data assimilation involves solving a linear inverse problem (see example one). In matrix form one can write the forward problem as

$$\mathbf{Ax} = \mathbf{b}$$

where **A** represents the forward mapping of the model parameters, vector **x**, on to the data, vector **b**. The make-up of **A**, **x** and **b** are

$$\begin{bmatrix} \leftarrow a_1 \rightarrow \\ \leftarrow a_2 \rightarrow \\ \vdots \\ \leftarrow a_N \rightarrow \end{bmatrix} \begin{bmatrix} x_1 \\ x_2 \\ \vdots \\ x_M \end{bmatrix} = \begin{bmatrix} b_1 \\ b_2 \\ \vdots \\ b_N \end{bmatrix} \tag{A1}$$

where N is the number of data points in **b** and M is the number of unknowns in **x**. For the following discussion we will assume the N equations ($a_1$to a_N) are independent (i.e. no linear combination of the equations can reproduce one of the equations. For this situation three different cases can occur.

1. M=N, one unique solution
2. $M<N$, overdetermined case and no one solution satisfies all the data, need to minimize the fit to the data (i.e. least squares fit)

3. M> N, underdetermined case and an infinite number of solutions exist. To obtain one solution additional constraints must be applied

Now consider the inverse solution that uses the inverse of matrix **A**, $\mathbf{A}^{-1}$ to map the data vector **b** back to the unknown vector **x**.

$$\mathbf{x} = \mathbf{A}^{-1}\mathbf{b} \tag{A2}$$

This is the generalized solution but a unique inverse of matrix **A** will only exist when N=M.

For the case when there are more data points than unknowns (N> M) one finds the solution for **x** than minimize the cost function ϕ in a least squares sense. In matrix notation this is

$$\begin{aligned}\phi &= \|\mathbf{Ax}-\mathbf{b}\| = (\mathbf{Ax}-\mathbf{b})^{T}(\mathbf{Ax}-\mathbf{b}) \\ &= \mathbf{x}^{T}\mathbf{A}^{T}\mathbf{Ax} - 2\mathbf{x}^{T}\mathbf{A}^{T}\mathbf{b} + \mathbf{x}^{T}\mathbf{x} + \mathbf{b}^{T}\mathbf{b}\ .\end{aligned} \tag{A3}$$

The solution for **x** is obtained by minimizing ϕ with respect to **x** (the rules for vector differentiation can be found in Appendix B) which requires

$$\frac{\partial\phi}{\partial\vec{\mathbf{x}}} = 2\mathbf{A}^{T}\mathbf{Ax} - 2\mathbf{x}^{T}\mathbf{A}^{T}\mathbf{b} = 0\ . \tag{A4}$$

From this equation, **x** is determine as

$$\begin{aligned}\mathbf{A}^{T}\mathbf{Ax} &= \mathbf{A}^{T}\mathbf{b} \\ \mathbf{x} &= (\mathbf{A}^{T}\mathbf{A})^{-1}\mathbf{A}^{T}\mathbf{b}\ .\end{aligned} \tag{A5}$$

For the case where M>N, there are more unknowns than equations and an infinite number of solutions exist. A convenient solution is to select the "smallest model" which is the one where **x** has the smallest magnitude. For this problem one minimizes the following cost function

$$\begin{aligned}\phi &= \|\mathbf{x}\| + \lambda(\mathbf{Ax}-\mathbf{b}) \\ &= \mathbf{x}^{T}\mathbf{x} + \lambda(\mathbf{Ax}-\mathbf{b})\ .\end{aligned} \tag{A6}$$

Lagrangian multiplier, λ are included in the cost function to force the solution to satisfy the equations that **Ax=b.** The solution to the problem is again obtained by differentiating the cost function with respect to both **x** and λ.

$$\frac{\partial \phi}{\partial \mathbf{x}} = 2\mathbf{x} - 2\lambda^T\mathbf{A} = 0$$
$$= \mathbf{x} - \mathbf{A}^T\lambda = 0$$
$$\mathbf{x} = -\mathbf{A}^T\lambda \tag{A7}$$

$$\frac{\partial \phi}{\partial \lambda} = \mathbf{Ax} - \mathbf{b} = 0$$
$$\mathbf{Ax} = \mathbf{b} \tag{A8}$$

One can combine the two equations to generate a new matrix equation that generalizes the system for solving an equation with equality constraints (for our case this is **Ax=b**).

$$\begin{bmatrix} \mathbf{I} & \mathbf{A}^{\mathrm{T}} \\ \mathbf{A} & \mathbf{0} \end{bmatrix} \begin{bmatrix} \mathbf{x} \\ \lambda \end{bmatrix} = \begin{bmatrix} 0 \\ \mathbf{b} \end{bmatrix} \tag{A9}$$

where **I** is the Identity matrix (diagonal elements are one and off-diagonal elements are zero), By substituting **x** from equations A7 into equation A8 one can gets the following equation

$$-\mathbf{AA}^T\lambda = \mathbf{b}$$
$$\lambda = -(\mathbf{AA}^T)^{-1}\mathbf{b} \quad . \tag{A10}$$

Substituting λ into equation A7 one obtains the solution

$$\mathbf{x} = \mathbf{A}^T(\mathbf{AA}^T)^{-1}\mathbf{b} \tag{A11}$$

Singular Value Decomposition: Singular Value Decomposition (SVD) of matrix **A** provides convenient way of solving the three cases presented in the previous section. There are several advantages of using SVD to solve the inverse problem. One, it handles both the overdetermined and underdetermined problem. Two, it provides the rank of matrix **A** along with insight into the ability to invert matrix **A** and solve the inverse problem. Three, it provide an easy way to assess the impact of data errors on the inverse solution and improve the quality of the inverse results.

In the previous discussion it was assumed **A** was made up of N independent equations. With SVD this assumption is no longer necessary since it will provide a solution for all three cases presented in the previous section. For forward problem, if there are k independent equations among the N simultaneous equations then matrix **A** has rank k and SVD of **A** is

$$\begin{array}{ccccccc} \mathbf{A} & = & \mathbf{U} & & \Lambda & & \mathbf{V}^{\mathrm{T}} \\ [\mathrm{N} \times \mathrm{M}] & = & [\mathrm{N} \times \mathrm{K}] & & [\mathrm{K} \times \mathrm{K}] & & [\mathrm{K} \times \mathrm{M}] \end{array} \quad \text{(A12)}$$

where

$$\mathbf{U} = \begin{bmatrix} | & | & | & & | \\ u_1 & u_2 & u_3 & \ldots & u_k \\ | & | & | & & | \end{bmatrix}$$

$$\Lambda = \begin{bmatrix} \lambda_1 & 0 & 0 & 0 \\ 0 & \lambda_2 & 0 & 0 \\ 0 & 0 & \ldots & 0 \\ 0 & 0 & 0 & \lambda_k \end{bmatrix}$$

$$\mathbf{V} = \begin{bmatrix} | & | & | & & | \\ v_1 & v_2 & v_3 & \ldots & v_k \\ | & | & | & & | \end{bmatrix} . \quad \text{(A13)}$$

The diagonal matrix **Λ** contains k singular values arranged in descending order occupying the diagonal elements of this matrix. The matrices **U** and **V** contain k orthogonal eigenvectors that span the data and model space, respectively. Four important properties of the eigenvector matrices are

$$\mathbf{U}^T\mathbf{U} = \mathbf{I}$$

$$\mathbf{V}^T\mathbf{V} = \mathbf{I}$$

$$\text{for k} < \text{M (case 2, underdetermined) } \mathbf{U}\mathbf{U}^{\mathrm{T}} = \mathbf{I}$$

$$\text{for k} > \text{M (case 3, overdetermined) } \mathbf{V}\mathbf{V}^{\mathrm{T}} = \mathbf{I}$$

$$\text{for k} = \text{M (case 1, one unique solution) } \mathbf{U}\mathbf{U}^{\mathrm{T}} = \mathbf{I}, \mathbf{V}\mathbf{V}^{\mathrm{T}} = \mathbf{I} .$$

From the SVD of matrix **A**, the generalized inverse of **A** is

$$\mathbf{A}^{-1} = \mathbf{V}\Lambda^{-1}\mathbf{U}^{\mathrm{T}} \quad \text{(A14)}$$

and the solution to the inverse problem is

$$\mathbf{x} = \mathbf{V}\Lambda^{-1}\mathbf{U}^{\mathrm{T}}\mathbf{b} \tag{A15}$$

The equation above will give the least squares solution for the overdetermined case and the smallest model solution for the underdetermined case. (Can be demonstrated by substituting equation A15 into the solution for the least squares and smallest model solution presented in the previous section)

To demonstrate how errors in **b** mapped to errors in **x** we can use the SVD of matrix **A.** Consider the forward problem where the $\mathbf{x}^{true}$ is prescribed and corresponding values for **b** are found, $\mathbf{b}^{true}$. The following equation represents this forward mapping

$$\mathbf{U}\Lambda\mathbf{V}^{\mathrm{T}}\mathbf{x}^{true} = \mathbf{b}^{true} \quad . \tag{A16}$$

If we define the energy of **b** as $\mathbf{b}^{\mathrm{T}}\mathbf{b}$ then

$$\begin{aligned} \mathbf{b}^{\mathrm{T}}\mathbf{b} &= \mathbf{x}^{\mathrm{T}}\mathbf{V}\Lambda\mathbf{U}^{\mathrm{T}}\mathbf{U}\Lambda\mathbf{V}^{\mathrm{T}}\mathbf{x} = \mathbf{x}^{\mathrm{T}}\mathbf{V}\Lambda^{2}\mathbf{V}^{\mathrm{T}}\mathbf{x} \\ &= \hat{\mathbf{x}}^{\mathrm{T}}\Lambda^{2}\hat{\mathbf{x}} \end{aligned}$$

The equation above shows that large singular values have the most influence on the energy of **b**. The energy of **b** explained by a given singular value is $\lambda_i^2 / \Sigma\, \lambda_k^2$. Therefore, the smaller the singular value the less influence it has on **b**. If there are data errors, $\delta\mathbf{b}$, in vector **b**, the inversion will map these to the solution of **x** as

$$\begin{aligned} \mathbf{x} &= \mathbf{V}\Lambda^{-1}\mathbf{U}^{\mathrm{T}}(\mathbf{b}^{true} + \delta\mathbf{b}) \\ &= \mathbf{x}^{true} + \mathbf{V}\Lambda^{-1}\mathbf{U}^{\mathrm{T}}\delta\mathbf{b} \quad . \end{aligned} \tag{A17}$$

The above equation shows that in contrast to the forward problem (A16), in the inverse problem it is small singular values, which cause the greatest amplification of errors in the data.

$$\begin{aligned} \mathbf{x}^{\mathrm{T}}\mathbf{x} &= \mathbf{x}^{true\,\mathrm{T}}\mathbf{x} + \delta\mathbf{b}^{\mathrm{T}}\mathbf{U}\Lambda^{-1}\mathbf{V}^{\mathrm{T}}\mathbf{V}\Lambda^{-1}\mathbf{U}^{\mathrm{T}}\delta\mathbf{b} \\ &= \mathbf{x}^{true\,\mathrm{I}}\mathbf{x} + \delta\mathbf{b}^{\mathrm{T}}U\Lambda^{-2}U^{\mathrm{T}}\delta\mathbf{b}^{\mathrm{T}} \\ &= \mathbf{x}^{true\,\mathrm{T}}\mathbf{x} + \delta\hat{\mathbf{b}}^{\mathrm{T}}\Lambda^{-2}\delta\hat{\mathbf{b}}^{\mathrm{T}} \end{aligned} \tag{A18}$$

Therefore, as the singular values approach zero they cause a large amplification of the errors in **b** in solution **x** but have little influence on how solution **x** fits the data, **b**. To reduce the amplification of the data errors in an inversion one can sacrifice the fit with the observations by neglecting small singular values.

Another way reducing error amplification in the inversion and to achieve a more realistic solution is to include an additional constraint in the cost function to minimize the magnitude **x** as well as fit to the observations. This is the approach used in the atmospheric inversion example presented in this chapter. One now finds the solution to new cost function as follows

$$\phi = \| \mathbf{Ax} - \mathbf{b} \| + \theta^2 \mathbf{x}^T \mathbf{x}$$

$$\frac{\partial \phi}{\partial \mathbf{x}} = 2\mathbf{A}^T\mathbf{Ax} - 2\mathbf{A}^T\mathbf{b} + 2\theta^2\mathbf{x} = 0$$

$$(\mathbf{A}^T\mathbf{A} + \theta^2 I)\mathbf{x} = \mathbf{A}^T\mathbf{b}$$

$$(\mathbf{V}\Lambda^2\mathbf{V}^T + \theta^2\mathbf{I})\mathbf{x} = \mathbf{V}\Lambda\mathbf{U}^T\mathbf{b}$$

$$\mathbf{V}^T\mathbf{x} = (\Lambda^2 + \theta^2\mathbf{I})^{-1}\Lambda\mathbf{U}^T\mathbf{b} \quad \text{(A19)}$$

By adding the minimum magnitude term to the cost function one modifies the singular values of the original problem by the following equation

$$\lambda'_i = \frac{\lambda_i^2 + \theta^2}{\lambda_i} = \lambda_i(1 + \frac{\theta^2}{\lambda_i^2}) \quad \text{(A20)}$$

The large singular values ($\lambda_I >> \theta$) are not significantly affected by the additional constrain but for the small singular values ($\lambda_I < \theta$) the value is changed to θ^2/λ_I. This reduces the error amplification of the inverse solution at the expense of reducing the fit to the data and independently resolving individual members of **x**.

A.2 WEIGHTING OF THE INVERSE PROBLEM

Through the use of weighting matrices one can include additional information in the inversion. Two different weighting matrices are used, the data weighting matrix **Q** and the parameter-weighting matrix **W**. The matrix **Q** provides a way to include errors in the data in the inverse problem. In solving the inverse problem it is desirable to normalize these equations according to the errors in each of the elements in **b**. By doing so one can force equations with small errors to be given more importance than data with large errors when inverting the problem. This can be accomplished by defining the **Q** as

$$\mathbf{Q} = \begin{bmatrix} 1/\delta b_1 & 0 & 0 & 0 \\ 0 & 1/\delta b_2 & 0 & 0 \\ 0 & 0 & \dots & 0 \\ 0 & 0 & 0 & 1/\delta b_N \end{bmatrix} . \tag{A21}$$

Similarly one can define a parameter-weighting matrix **W** which includes information on the expected magnitude of the different model parameters. Without including such information the inversion will tend to produce unrealistic estimates of solution. This can be accomplished by defining the **W** as

$$\mathbf{W} = \begin{bmatrix} \| x_1 \|^{1/2} & 0 & 0 & 0 \\ 0 & \| x_2 \|^{1/2} & 0 & 0 \\ 0 & 0 & \dots & 0 \\ 0 & 0 & 0 & \| x_M \|^{1/2} \end{bmatrix} . \tag{A22}$$

With the two weight matrices, the original inverse problem becomes the same as the previously presented problem,

$$\begin{aligned} \mathbf{QAW\,W^{-1}x} &= \mathbf{Qb} \\ \mathbf{B\quad m} &= \mathbf{e} \end{aligned} . \tag{A23}$$

Once **m** is determined **x** is determined by

$$\mathbf{x} = \mathbf{Wm} \quad . \tag{A24}$$

A.3 NON-LINEAR PROBLEM

Since many of the real applications of data assimilation require solving a non-linear inverse problem we will explore this issue. Following our earlier notation we can write the forward non-linear problem as mapping from the solution space to the data space as

$$\mathbf{b}_{\mathrm{i}} = \mathbf{A}_{\mathrm{i}}\mathbf{x} \tag{A25}$$

where $\mathbf{b}_I$ represents a data point created from the forward integration of the non-linear problem. One can expand the non-linear forward problem about an initial set of **x** values $\mathbf{x}_0$ as

$$\mathbf{b}_i = \mathbf{A}_i\mathbf{x}_o + \sum_{k=1}^{M}\frac{\partial A_i}{\partial x_k}\delta x_k + \frac{1}{2}\sum_{j=1}^{M}\sum_{k=1}^{M}\frac{\partial^2 A_i}{\partial x_j \partial x_k}\delta x_k \delta x_j \quad . \tag{A26}$$

If we apply the above equation to all the data equations, the second term on the rhs gives the Jacobian or sensitivity matrix and the third on the rhs gives the Hessian matrix. At a minimum the Jacobian equals zero and the Hessian contains information on the co-variance of the solution vector **x**. If one neglects the second and higher order terms, an updated solution for x can be determined by solving

$$\mathbf{b}_i = \mathbf{A}_i\mathbf{x}_o + \sum_{k=1}^{M}\frac{\partial A_i}{\partial x_k}\delta x_k$$

$$b_i - \mathbf{A}_i\mathbf{x}_o = c_i = \sum_{k=1}^{M} S_{i,k}\delta x_k = \sum_{k=1}^{M} S_{i,k} y_k \quad . \tag{A27}$$

To solve **x** requires iterative solving

$$\mathbf{c} = \mathbf{S}\mathbf{y}$$

$$\mathbf{x} = \mathbf{x}_o + \mathbf{y} \tag{A28}$$

to obtain new estimates of **x** with no guarantee that the inversion will be stable and converge. A more typical representation of non-linear inversion problem is the least squares fit to data. One then defines a cost function, *J*, which represents the fit to the observations,

$$J = \| \mathbf{b} - \mathbf{A}\mathbf{x} \|^2 . \tag{A29}$$

If one proceeds in the direction steepest decent, $-\partial J/\partial x_i$, one can iterative update the solution until optimal solution is obtained, $\partial J/\partial x_i$,=0

Several different methods exist that attempt to use the gradient information to carry out the iteration (gradient methods). The large computational demand in applying these methods is in the computation of the gradient of the cost function with respect to the model parameters. The adjoint method is an efficient means of calculating the gradient (Example 2). An alternative approach is to use the stochastic approach that randomly guess potential **x** to find the optimal solution (Example 3).

For a complete solution of the inverse problem one needs an estimates of the uncertainty in the optimal parameters, **x**. By expanding the cost function, *J*, in a power series about the optimal solution, $\mathbf{x}^*$ gives

$$J=J_{\min}-1/2\,(\mathbf{x}-\mathbf{x}^*)\,\mathrm{H}\,(\mathbf{x}-\mathbf{x}^*)$$

where

$$H_{i,j}=\frac{\partial^2 J}{\partial x_i \partial x_j} \quad \text{for i, j=1,..., m.} \tag{A30}$$

In the expansion of J the third and higher-order terms are neglect and the matrix **H** is called the Hessian. If the neglected terms are small, then the uncertainties in **x** are normally distributed with zero mean and with an error-covariance matrix, **C**, defined as the inverse of the Hessian, $\mathbf{C} = \mathbf{H}^{-1}$ [*Thacker*, 1987; *Tziperman and Thacker*, 1989a]. The error-covariance matrix provides information on the probability distribution of the optimal solution. The diagonal elements of the matrix provide a measure of the variance in the optimal solution. The off-diagonal elements of the covariance matrix indicate the degree to which pairs of model parameters are correlated. To show the relationship between the various elements of **x**, a correlation matrix, **R**, can be calculated from the error-covariance matrix. The elements of **R** are related to the element of the covariance matrix, **C**, as follows

$$R_{i,j}=\frac{C_{i,j}}{\sqrt{C_{i,i}C_{j,j}}} \quad . \tag{A31}$$

$R_{i,j}$ values near 1 or –1 indicate strong linear correlation between parameters and values near zero indicate little or no correlation.

APPENDIX B. RULES FOR VECTOR DIFFERENTIATION

First consider two column vectors,

$$\mathbf{x}=\begin{bmatrix}x_1 & x_2 & \dots & x_M\end{bmatrix}^T$$
$$\mathbf{y}=\begin{bmatrix}y_1 & y_2 & \dots & y_M\end{bmatrix}^T$$

The vector product of these two vectors can be differentiated with respect to vector **x** to obtain:

$$\frac{\mathrm{d}}{\mathrm{d}\mathbf{x}}\left(\mathbf{y}^{\mathrm{T}}\mathbf{x}\right)=\mathbf{y}$$
$$\frac{\mathrm{d}}{\mathrm{d}\mathbf{x}}\left(\mathbf{x}^{\mathrm{T}}\mathbf{y}\right)=\mathbf{y}$$

Expanding on the results above, if **x** is multiplied by matrix **W** the vector differentiation with respect to **x** of the resulting matrix equation is

$$\frac{d}{d\mathbf{x}}\left(\mathbf{x}^T\mathbf{W}^T\right) = \mathbf{W}^T$$

$$\frac{d}{d\mathbf{x}}\left(\mathbf{W}^T\mathbf{x}\right) = \mathbf{W}$$

$$\frac{d}{d\mathbf{x}}\left(\mathbf{x}^T\mathbf{W}^T\mathbf{x}\right) = \mathbf{W}\mathbf{x} + \mathbf{W}^T\mathbf{x}$$

The last equation above can be simplified if **W** is symmetric matrix to give

$$\frac{d}{d\mathbf{x}}\left(\mathbf{x}^T\mathbf{W}^T\mathbf{x}\right) = 2\mathbf{W}\mathbf{x}$$

Finally, if **W**= **I** (identity matrix) then the previous equation can be written as

$$\frac{d}{d\mathbf{x}}\left(\mathbf{x}^T\mathbf{W}^T\mathbf{x}\right) = 2\mathbf{x}$$

REFERENCES

Anderson, G.C., R.K. Lam, B.C. Booth, and J.M. Glass, A description and numerical analysis of the factors affecting the processes of production in the Gulf of Alaska: Final report. Research Unit 58, *Environmental assessment of Alaska continental shelf, Annual Report of Principal Investigators for the year ending March 1977*, pp. 477-798, NOAA 03-5-022067, 1977.

Athias, V., P. Mazzega, and C. Jeandel, Selecting a global optimization method to estimate the oceanic particle cycling rate constants, *Journal of Marine Research, 58* (5), 675-707, 2000.

Aumont, O., J.C. Orr, P. Monfray, G. Madec, and E. Maier-Reimer, Nutrient trapping in the equatorial Pacific: The ocean circulation solution, *Global Biogeochemical Cycles, 13* (2), 351-369, 1999.

Bacastow, R., and E. Maier-Reimer, Dissolved organic carbon in modeling oceanic new production, *Global Biogeochemical Cycles, 5*, 71-85, 1991.

Battle, M., M.L. Bender, P.P. Tans, J.W.C. White, J.T. Ellis, T. Conway, and R.J. Francey, Global carbon sinks and their variability inferred from atmospheric O-2 and delta C-13, *Science, 287* (5462), 2467-2470, 2000.

Bousquet, P., P. Peylin, P. Ciais, C. Le Quere, P. Friedlingstein, and P.P. Tans, Regional changes in carbon dioxide fluxes of land and oceans since 1980, *Science, 290* (5495), 1342-1346, 2000.

Cias, P., P.P. Tans, J.W.C. White, M. Trolier, R.J. Francey, J.A. Berry, D.R. Randall, P.J. Sellers, J.G. Collatz, and D.S. Schimel, Partitioning of Ocean and Land uptake of CO_2 as inferred by $d^{13}C$ measurements from the NOAA/CMDL global air sampling network, *JGR, 100*, 5051-5070, 1995.

Eknes, M., and G. Evensen, An Ensemble Kalman filter with a 1-D marine ecosystem model, *Journal of Marine Systems, 36* (1-2), 75-100, 2002.

Evans, G.T., and J.S. Parslow, A model of annual plankton cycles, *Biol. Oceanogr., 3*, 327-347, 1985.

Fasham, M.J.R., H.W. Ducklow, and S.M. McKelvie, A nitrogen-based model of plankton dynamics in the oceanic mixed layer, *Journal of Marine Research, 48*, 591-639, 1990.

Fiadeiro, M.E., and G. Veronis, Obtaining velocities from tracer distributions, *Journal of Physical Oceanography, 4*, 1734-1746, 1984.

Fowler, S.W., and G.A. Knauer, Role of large particles in the transport of elements and organic compounds through the oceanic water column, *Progress Oceanography, 16*, 147-194, 1986.

Frost, B., Grazing control of phytoplankton stock in the open subarctic Pacific Ocean: a model assessing the role of mesozooplankton, particularly the large calanoid copepods *Neocalanus spp., Marine Ecology Progress Series, 39*, 49-68, 1987.

Frost, B.W., A modelling study of processes regulating plankton standing stock and production in the upper ocean subarctic Pacific Ocean, *Progress Oceanography, 32*, 17-56, 1993.

Ghil, M., and P. Malanotte-Rizzoli, Data Assimilation in meterology and oceanography, *Advances in Geophysics, 33*, 141-266, 1991.

Gloor, M., N. Gruber, J. Sarmiento, C.L. Sabine, R.A. Feely, and C. Rodenbeck, A first estimate of present and preindustrial air-sea CO2 flux patterns based on ocean interior carbon measurements and models, *Geophysical Research Letters, 30* (1), art. no.-1010, 2003.

Gruber, N., M. Gloor, S.M. Fan, and J.L. Sarmiento, Air-sea flux of oxygen estimated from bulk data: Implications for the marine and atmospheric oxygen cycles, *Global Biogeochemical Cycles, 15* (4), 783-803, 2001.

Gurney, K.R., R.M. Law, A.S. Denning, P.J. Rayner, D. Baker, P. Bousquet, L. Bruhwiler, Y.H. Chen, P. Ciais, S. Fan, I.Y. Fung, M. Gloor, M. Heimann, K. Higuchi, J. John, T. Maki, S. Maksyutov, K. Masarie, P. Peylin, M. Prather, B.C. Pak, J. Randerson, J. Sarmiento, S. Taguchi, T. Takahashi, and C.W. Yuen, Towards robust regional estimates of CO2 sources and sinks using atmospheric transport models, *Nature, 415* (6872), 626-630, 2002.

Kasibhatla, P.S., Inverse Methods in global biogeochemical cycles, in *Geophysical Monograph*, pp. 324, American Geophysical Union, Washington DC, 2000.

Krüger, J., Simulated Annealing: A tool for data assimilation into an almost steady model state, *Journal of Physical Oceanography, 23*, 679-688, 1993.

Langenfelds, R.L., R.J. Francey, L.P. Steele, M. Battle, R.F. Keeling, and W.F. Budd, Partitioning of the global fossil CO2 sink using a 19-year trend in atmospheric O-2, *Geophysical Research Letters, 26* (13), 1897-1900, 1999.

Lawson, C.L., and R.J. Hanson, *Solving Least Squares Problem*, 1054 pp., Prentice-Hall, Englewood Cliff, N.J., 1974.

Le Dimet, F.X., and O. Talagrand, Variational algorithms for the analysis and assimilation of meteorological observations: Theoretical aspects, *Tellus, 38*, 97-110, 1986.

Matear, R.J., Parameter optimization and analysis of ecosystem models using simulated annealing: A case study at Station P, *Journal of Marine Research, 53*, 571-607, 1995.

Matear, R.J., Effects of Eddy Parameterizations and Numerical Advection Schemes on Ocean Ventilation and Anthropogenic CO_2 uptake of a ocean general circulation model, *Ocean Modelling*, *3*, 217-248, 2001.

Matear, R.J., and A.C. Hirst, Climate Change Feedback on the Future Oceanic CO_2 uptake, *Tellus*, *51B* (3), 722-733, 1999.

Matear, R.J., and G. Holloway, Modeling the inorganic phosphorus cycle of the North Pacific using an adjoint data assimilation model to assess the role of dissolved organic phosphorus, *Global Biogeochemical Cycles*, *9*, 101-119, 1995.

McAllister, C.D., Zooplankton studies at Ocean Weather Station P in the Northeast Pacific Ocean, *J. Fish. Res. Bd. Can.*, *18*, 1-29, 1961.

Menke, W., *Geophysical Data Analysis, Discrete Inverse Theory*, 260 pp., Academic Press, Orlando Florida, 1984.

Metzl, N., B. Tilbrook, and A. Poisson, The annual fCO(2) cycle and the air-sea CO2 flux in the sub- Antarctic Ocean, *Tellus*, *51B* (4), 849-861, 1999.

Morse, P.M., and H. Feshbach, *Methods of Theoretical Physics, 2 Vols.*, McGraw-Hill, Hew York, 1953.

Najjar, R., J. Sarmiento, and J. Toggweiler, Downward transport and fate of organic matter in the ocean: simulations with a general circulation model, *Global Biogeochemical Cycles*, *6*, 45-76, 1992.

Prunet, P., J.-F. Minster, D. Ruiz-Pino, and I. Dadou, Assimilation of surface data in a one-dimensional physical-biogeochemical model of the surface ocean, 1, Method and preliminary results, *Global Biogeochemical Cycles*, *10*, 111, 1996.

Randerson, J.T., C.B. Field, I.Y. Fung, and P.P. Tans, Increases in early season ecosystem uptake explain recent changes in the seasonal cycle of atmospheric CO2 at high northern latitudes, *Geophysical Research Letters*, *26* (17), 2765-2768, 1999.

Rayner, P.J., I.G. Enting, R.J. Francey, and R. Langenfelds, Reconstructing the recent carbon cycle from atmospheric CO2, delta C-13 and O-2/N-2 observations, *Tellus Series B-Chemical and Physical Meteorology*, *51* (2), 213-232, 1999.

Rayner, P.J., and R.M. Law, The interannual variability of the global carbon cycle, *Tellus Series B-Chemical and Physical Meteorology*, *51* (2), 210-212, 1999.

Roy, T., P. Rayner, R.J. Matear, and R. Francey, Southern hemisphere ocean CO2 uptake: reconciling atmospheric and oceanic estimates, *Tellus Series B-Chemical and Physical Meteorology*, *55* (2), 701-710, 2003.

Schlitzer, R., Carbon export fluxes in the Southern Ocean: results from inverse modeling and comparison with satellite based estimates, *Deep Sea Research II*, *Special Volume on the Southern Ocean*, 1623-1644, 2002.

Spitz, Y.H., J.R. Moisan, M.R. Abbott, and J.G. Richman, Data assimilation and a pelagic ecosystem model: parameterization using time series observations, *Journal of Marine Systems*, *16* (1-2), 51-68, 1998.

Stephens, K., Productivity measurements in the northeast Pacific with associated chemical and physical data, 1958-1964, pp. 168pp, 1964.

Stephens, K., Primary production data from the northeast Pacific Ocean, January 1964 to December 1965, pp. 14pp., 1966.

Stephens, K., Primary production data from the northeast Pacific Ocean, January 1966 to December 1967, pp. 58, 1968.

Stephens, K., Primary production data from the northeast Pacific Ocean, January 1967 to December 1969, pp. 16pp., 1970.

Takahashi, T., S.C. Sutherland, C. Sweeney, A. Poisson, N. Metzl, B. Tilbrook, N. Bates, R. Wanninkhof, R.A. Feely, C. Sabine, J. Olafsson, and Y. Nojiri, Global sea-air CO2 flux based on climatological surface ocean pCO(2), and seasonal biological and temperature effects, *Deep-Sea Research Part Ii-Topical Studies in Oceanography*, *49* (9-10), 1601-1622, 2002.

Tans, P., I. Fung, and T. Takahashi, Observational constraints on the global atmospheric CO_2 budget, *Science*, *247*, 1431-1438, 1990.

Thacker, W.C., Three lectures on fitting numerical models to observations, pp. 64, GKSS-Forschungszentrum Geesthacht, GmbHGeesthacht, Federal Republic of Germany, 1987.

Thacker, W.C., Fitting models to inadequate data by enforcing spatial and temporal smoothness, *Journal of Geophysical Research*, *93*, 10,655- 10,665, 1988.

Tziperman, E., and W.C. Thacker, An optimal control/adjoint- equation approach to studying the oceanic general circulation, *Journal of Physical Oceanography*, *19*, 1471-1485, 1989a.

Tziperman, E., and W.C. Thacker, An optimal-control/ adjoint-equations approach to studying the oceanic general circulation, *Journal of Physical Oceanography*, *19*, 1471-1485, 1989b.

Winguth, A.M.E., D. Archer, J.C. Duplessy, E. Maier-Reimer, and U. Mikolajewicz, Sensitivity of paleonutrient tracer distributions and deep-sea circulation to glacial boundary conditions, *Paleoceanography*, *14* (3), 304-323, 1999.

Printed in the United States
56439LVS00001B/160-162